Emerging Radiation Detection

Kris Iniewski • Harish Gadey
Editors

Emerging Radiation Detection

Technology and Applications

Springer

Editors
Kris Iniewski
Redlen Technologies
Port Moody, BC, Canada

Harish Gadey
Idaho National Laboratory
Kennewick, WA, USA

ISBN 978-3-031-63896-1 ISBN 978-3-031-63897-8 (eBook)
https://doi.org/10.1007/978-3-031-63897-8

This Springer imprint is published by the registered company Springer Nature Switzerland AG
The registered company address is: Gewerbestrasse 11, 6330 Cham, Switzerland

Contents

About the Editors

Kris Iniewski is a director of detector architectures and applications at Redlen Technologies Inc., a detector company based in British Columbia, Canada. During his 18 years at Redlen, he has managed the development of highly integrated CZT detector products in medical imaging and security applications. Prior to Redlen, Kris held various management and academic positions at PMC-Sierra, University of Alberta, SFU, UBC, and University of Toronto. Dr. Iniewski has published over 150+ research papers in international journals and conferences. He holds 25+ international patents granted in USA, Canada, France, Germany, and Japan. He wrote and edited 75+ books for Wiley, Cambridge University Press, Mc Graw Hill, CRC Press, and Springer. He is a frequent invited speaker and has consulted for multiple organizations internationally.

Harish Gadey is a Spent Fuel Analyst at the Idaho National Laboratory (INL) specializing in the areas of spent fuel management and radiation detection with over 8 years of hands-on experience in Monte Carlo–based simulations, spent fuel research, detector development, and systems engineering. Dr. Gadey is proficient in Geant4, MATLAB, Python, and CAD software packages. He has worked on a number of detector development efforts, including several projects in the areas of radioxenon noble gas detection. Currently, Dr. Gadey is focusing on reactor backend systems analysis, SNF analysis, and health physics code modernization efforts. He has led and contributed to over 30 publications and conference presentations.

Radiation Interactions with Matter

Lily Ranjbar

Radiation interactions with matter are the foundation of nuclear physics and its various applications. Understanding these interactions for fields like medical imaging and diagnostics, radiation therapy, radiation protection, and nuclear power generation is crucial. This chapter covers the fundamental principles of radiation interactions with matter, addressing both charged and uncharged particles.

The chapter begins by providing a comprehensive explanation of how photons interact with matter. After that, it discusses the interactions of charged particles and introduces the concept of stopping power for them. This concept helps to describe how particles lose energy as they move through a material. Additionally, the chapter goes over factors that influence these interactions, such as the type of particle, its energy, and the density of the material.

The chapter also explores neutron interactions, which are significant topics in nuclear physics and engineering. It discusses the processes of elastic and inelastic scattering and neutron capture, emphasizing the importance of neutron energies and target nuclei in these interactions.

1 Interactions of Photons with Matter

Photons are particles with no electric charge and no rest mass and travel at the speed of light. They interact with matter through multiple mechanisms, each governed by the photon's energy and the composition of the material encountered. There are four major interactions for photons:

L. Ranjbar (✉)
Oregon State University, Corvallis, OR, USA
e-mail: ranjbarl@oregonstate.edu

K. Iniewski, H. Gadey (eds.), *Emerging Radiation Detection*,
https://doi.org/10.1007/978-3-031-63897-8_1

- Photoelectric effect
- Compton scattering (incoherent scattering)
- Pair production
- The Rayleigh interaction (coherent scattering)

1.1 Photoelectric Effect

The photoelectric effect occurs when a photon interacts with an atom and transfers all of its energy to an inner shell electron, causing the electron to be ejected from the atom. This electron is called a photoelectron. The photon is completely absorbed by an atom during this process. The photoelectric effect is more likely to occur at lower photon energies (usually less than 100 KeV) and is the primary interaction for high-Z materials (Z: atomic number) (Fig. 1) [1].

The energy of the ejected *photoelectron* (KE_e) is equal to the incident photon energy (E_γ) minus the electron's binding energy (B_e) [2]. For photoelectric absorption to occur, the energy of the photon must be greater than or equal to the binding energy of the electron.

The energy of the photoelectron is:

$$KE_e = E_\gamma - B_e;\ E_\gamma \geq B_{e.}$$

When the electron is ejected from the atom, it ionizes the atom and creates a vacancy in its inner shell. This causes electrons from the outer shells to cascade into the inner shell, resulting in the emission of either a characteristic X-ray or the competing mechanism *Auger electron emission.*

The *Auger effect* is a physical phenomenon where the filling of an inner shell vacancy leads to the emission of an electron from the same atom (Fig. 2) [3].

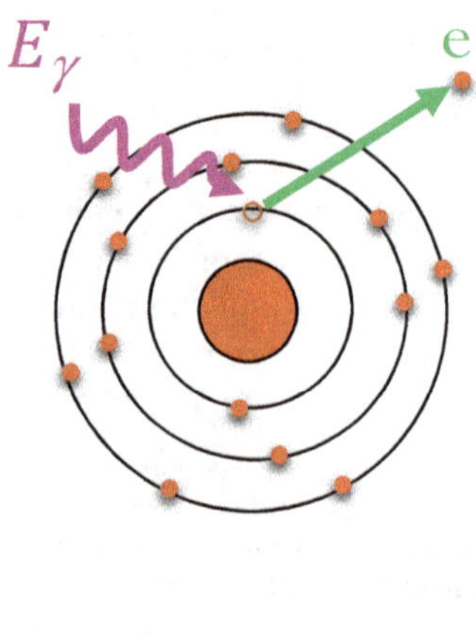

Fig. 1 Photoelectric effect (e, photoelectron; E_γ, photon energy)

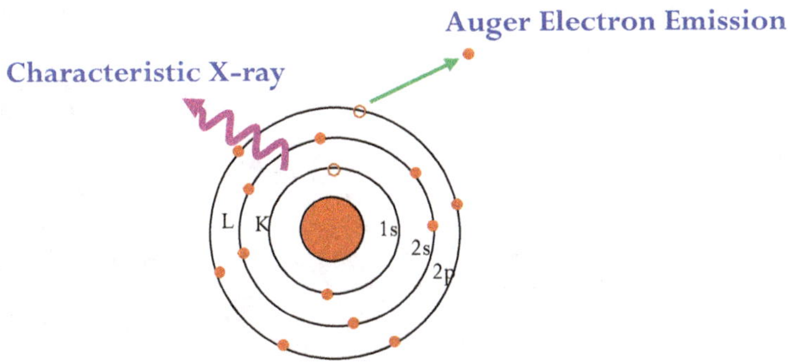

Fig. 2 Characteristic X-ray or Auger electron emission after a photoelectric event

1.2 Compton Scattering

Compton scattering is a phenomenon that occurs when a photon interacts with an outer shell electron. During this interaction, some of the energy of the photon is transferred to the electron, which causes the photon to change its direction. As a result, the scattered photon moves in a different direction and has less energy than the original photon. This phenomenon is more common at intermediate photon energies and is characterized by a reduction in photon energy and a change in scattering angle (Fig. 3) [2, 4].

In Compton scattering, the electron's binding energy is usually negligible, and energy and momentum must be conserved.

The energy of the scattered photon ($E_{\gamma}{}'$) is [5]

$$E_{\gamma}{}' = \frac{E_{\gamma}}{1 + \left(\frac{E_{\gamma}}{m_e c^2}\right)(1 - \cos\theta)}$$

E_{γ}: the energy of the incident photon
$E_{\gamma}{}'$: the energy of the scattered photon
θ: scattering angle
$m_e c^2$: electron's rest mass (0.511 MeV)

Then the kinetic energy of the recoiled electron (KE_e) is [6]

$$KE_e = E_{\gamma} - E_{\gamma}{}' = E_{\gamma}\left[\frac{\left({E_{\gamma}}/{m_e c^2}\right)(1 - \cos\theta)}{1 + \left({E_{\gamma}}/{m_e c^2}\right)(1 - \cos\theta)}\right]$$

The likelihood of a Compton interaction is directly proportional to the number of electrons that a photon can interact with; hence, it increases as the atomic number (Z) increases. The probability generally increases as the photon energy is lowered but then decreases as the energy rises [7].

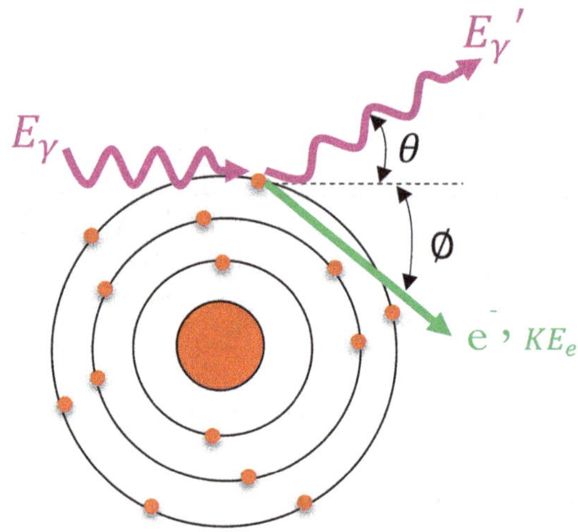

Fig. 3 Compton scattering interaction

1.3 Pair Production

At very high photon energies (usually exceeding 1.022 MeV), an interaction between photons and the electric field of an atomic nucleus can occur, leading to the creation of an electron-positron pair. This interaction results in the disappearance of the photon. For pair production to occur, the energy of the photon must be greater than or equal to the rest mass energy of the electron-positron pair ($E_\gamma \geq 2E_0$ ($E_0 = m_ec^2 = 0.511$ MeV)), which is the threshold energy required for pair production. The excess energy of the photon is transferred to the electron-positron pair as kinetic energy [8].

The sum of the kinetic energies of the electron and positron is equal to the difference between the energy of the incoming photon and the rest mass of two electrons [6].

The sum of the kinetic energies of electron and positron is as follows: $E_{e+} + E_{e-} = E_\gamma - 2E_0$.

Electrons and positrons have opposite electric charges, which results in their repulsion by the electromagnetic field of the atomic nucleus or electron. Consequently, they get ejected from the interaction location. The newly formed electron and positron can lose their energy by ionizing atoms in the material medium or by emitting bremsstrahlung photons as they slow down. The positron can also encounter a free electron in the material, leading to annihilation by reaction. This process produces two gamma photons, each with an energy of 0.511 MeV (Fig. 4) [9].

Pair production is a significant phenomenon in high-energy physics and medical imaging diagnostics, such as positron emission tomography (PET).

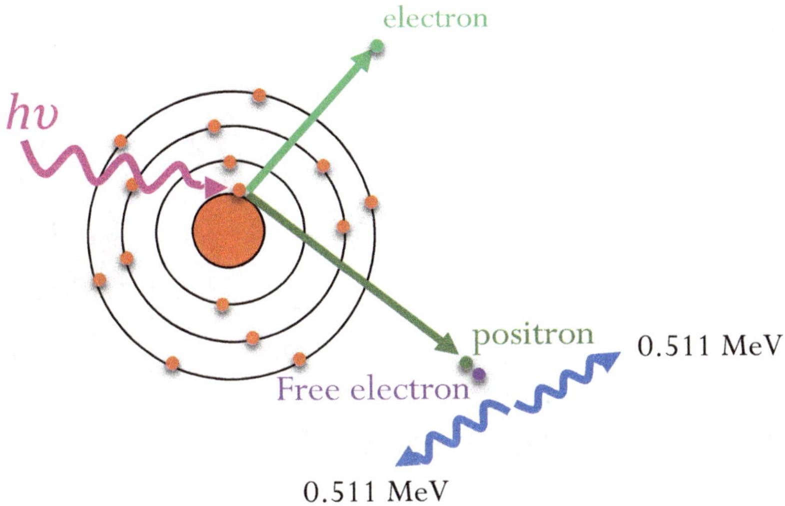

Fig. 4 Pair production and pair annihilation resulting in the creation of two 0.511 MeV photons

1.4 Coherent (Rayleigh) Scattering

Coherent scattering is a process in which a photon interacts with an entire atom, causing the atom's electrons to momentarily oscillate [2]. The photon is then re-emitted in a different direction but with the same amount of energy. This phenomenon is more commonly observed at low photon energies and is responsible for the scattering of X-rays in diagnostic imaging [10].

2 Charged Particle Interactions with Matter

Charged particles are particles that carry electrical charge. Examples include electrons, positrons, alpha particles, and recoils from other reactions, such as photons or neutron interactions. Charged particles interact with matter through two primary mechanisms: *excitation* and *ionization*. *Excitation* occurs when an electron in the inner shell of an atom obtains enough energy to move into a higher-energy state, leaving the atom in an excited state. In order to return to a stable state, the electron can either travel back to its original state and release the excess energy as an X-ray, or the excited atom can break apart into smaller atoms [2, 11].

Ionization occurs when an outer shell electron is ejected from the atom, creating a free electron and a positively charged ion. This free electron has enough energy to travel through the medium until it loses all of its energy and comes to a stop, while a positive ion can capture a free electron from the medium to regain stability [6].

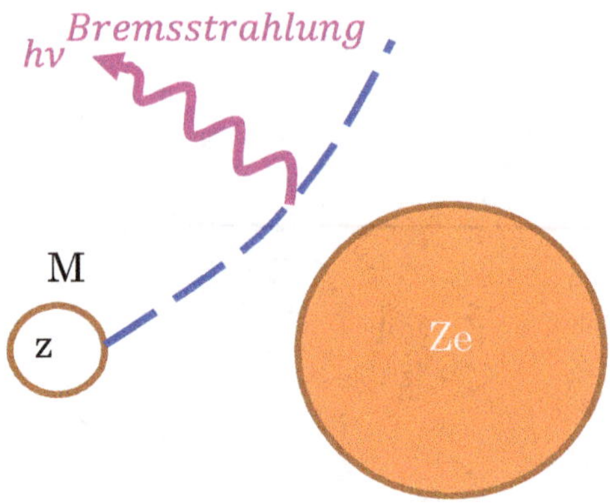

Fig. 5 Bremsstrahlung radiation production

2.1 Collisional and Radiative Losses

As light charged particles, especially electrons, travel through a medium, they lose energy through two mechanisms—*collisional losses* and *radiative losses*. In the collisional loss mechanism, electrons lose energy by interacting with the orbital electrons in the medium, leading to excitation or ionization of the atoms. This energy loss mechanism is commonly referred to as collisional loss. During this process, electrons collide with particles of the same mass, which can cause significant scattering angles. Hence, the path of the electron is not straight but very tortuous, unlike the path of a heavy charged particle [2].

In radiative loss, a light-charged particle, such as an electron, loses energy by interacting with the nucleus of an atom. During this process, the charged particle might undergo an acceleration change, leading to the emission of electromagnetic radiation, which is commonly known as *bremsstrahlung* or *braking radiation* [12]. In simple words, bremsstrahlung radiation is produced when the acceleration of charged particles is altered. Bremsstrahlung radiation has a range of energies. The maximum energy of these photons is equal to the kinetic energy of the electron. The amount of bremsstrahlung radiation produced increases with the kinetic energy of the electron and the atomic number of the material it interacts with (Fig. 5) [13].

2.2 Stopping Power, Linear Energy Transfer, and Range

There are two main terms used to describe the changes in a particle's energy: *stopping power (S)* and *linear energy transfer (LET)*. Stopping power refers to the amount of energy lost by a particle as it travels through a medium over a certain distance. On the other hand, LET describes the amount of energy that a particle transfers to the medium per unit length as it passes through it. These two terms are quite similar, but they differ in the way that LET takes into account the energy that is absorbed locally, while stopping power only measures the rate at which a particle loses its energy as it passes through a medium [14]. However, assuming that there are no radiative losses or delta rays (where electrons ejected by ionization can ionize other target atoms), LET and stopping power are the same quantity [9].

When a charged particle moves through a medium, it interacts with the medium and continuously loses energy. Eventually, the particle comes to a stop after traveling a certain distance, which is known as the *range*. The range value is determined by various factors, such as the type and energy of the particle, as well as the material through which the particle is moving [2].

The electron range is crucial for determining the thickness required for shielding or the depth at which the radiation dose is delivered. Two terms are used for electron range: *mean range* and *extrapolated range*. The mean range is the distance covered by half of the particles (above background), and the extrapolated range is the estimated maximum range distance obtained by extrapolating from the linear portion of the range curve [15].

There are two common empirical range equations for electrons for two different energy ranges. These two empirical equations overlap and represent approximately the same electron range.

The range of beta particles, when expressed in terms of milligram per centimeter squared, is independent of the absorbing material and only depends upon energy (E_e, electron energy in MeV; R, electron range in mg/cm^2) [4].

- For 0.01 MeV < E_e < 3 MeV: $R = 412 * (E_e)^{1.265-0.0954*\ln(E_e)}$(unit: mg/cm^2)
- For $E_e \geq 0.6$ MeV: $R = 542 * E_e - 133$

Electron energy and its range can also be related using the third equation:

- For $R < 1200\ {}^{\text{mg}}/_{\text{cm}^2}$: $\ln(E_e) = 6.63 - 3.2376 * (10.2146 - \ln(R))^{1/2}$

3 Neutron Interactions with Matter

Neutrons are electrically neutral particles, meaning they do not interact with electric fields; instead, they penetrate the electron cloud and directly interact with the nucleus. Understanding neutron interactions is fundamental for applications in nuclear reactors, particle physics experiments, and neutron shielding and dosimetry.

Neutrons interact with matter through several processes, each with distinct characteristics and implications. These interactions include scattering interactions and radiative capture/absorption. The probability of each reaction depends on the neutron kinetic energy and the target nuclide [16].

3.1 Elastic Scattering

Elastic scattering is a type of collision similar to that of billiard balls. In this process, a neutron collides with an atomic nucleus, transferring some of its kinetic energy

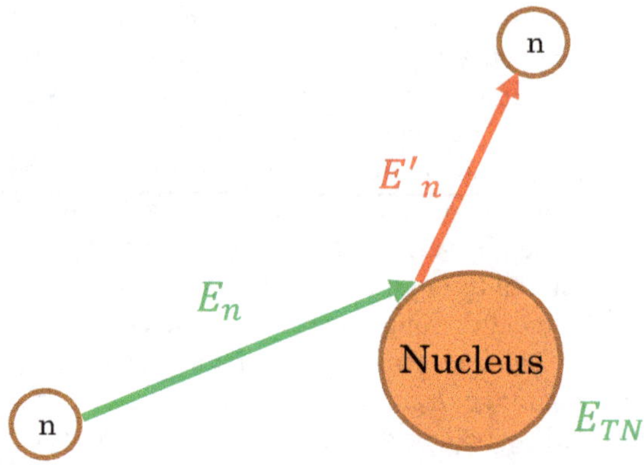

Fig. 6 Elastic Scattering of neutrons

to the target nucleus without changing its identity. The scattered neutron changes direction, following the laws of energy and momentum conservation, which means that the total kinetic energy and momentum remain the same before and after the collision. Elastic scattering is dominant at lower neutron energies and plays a crucial role in neutron moderation in nuclear reactors. When a neutron scatters off a target nucleus with the same mass, the most energy can be transferred. Therefore, a moderator is more effective if it has a lower atomic mass number such as water of paraffin (Fig. 6) [17].

$$\text{Conservation of energy} : E_n = E'_n + E_{TN}$$

E_n: neutron's energy before elastic scattering
E'_n: neutron's energy after elastic scattering
E_{TN}: the energy of the target nucleus

3.2 *Inelastic Scattering*

In inelastic scattering, a neutron collides with an atomic nucleus and transfers some of its energy to the nucleus. This energy excites the nucleus to a higher energy state. The remaining energy is transferred to the target nucleus as excitation energy. As the nucleus returns to its ground state, it emits gamma radiation. Inelastic scattering collision does not conserve kinetic energy and momentum. The total kinetic energy after the collision is lower than before the collision. This type of scattering is not common. It usually involves high Z nuclei and high-energy neutrons (Fig. 7) [9, 17].

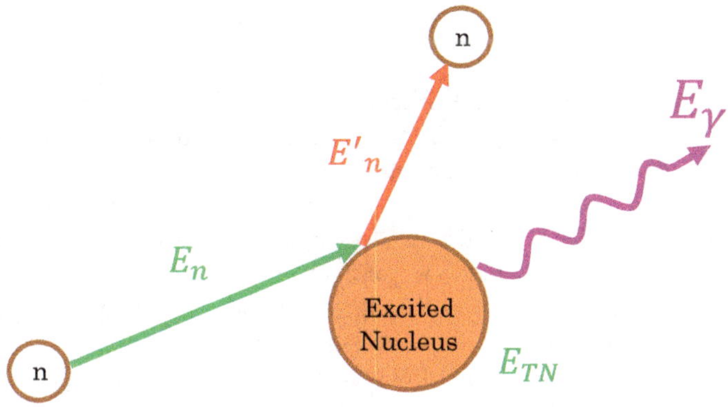

Fig. 7 Inelastic scattering of neutrons

3.3 Scattering Cross Section

The scattering cross section (σ_s) is a fundamental concept in nuclear physics that quantifies the likelihood of a neutron interacting with a target nucleus through scattering. It represents the effective area for interaction between the neutron and the target nucleus and is a crucial parameter in understanding neutron transport and reactor design.

The scattering cross section can vary significantly with neutron energy and the target nucleus. It depends on factors such as the mass and charge of the target nucleus, the velocity of the incident neutron, and the angle of scattering. Generally, the scattering cross section increases with increasing neutron energy and tends to be higher for lighter nuclei compared to heavier ones [9, 18].

3.4 Neutron Capture (Absorption)

Neutron capture, or neutron absorption, usually occurs with thermal neutrons. This process involves the absorption of neutrons by atomic nuclei, resulting in the formation of a compound nucleus. The compound nucleus is in an excited state and will eventually return to the ground state by emitting one or more gamma rays, which are often high in energy. Depending on the nucleus and neutron energy, various reaction channels can occur, such as (n,γ), (n,p), (n,α), or (n,fission). Neutron absorption is crucial for nuclear reactions and serves as the foundation for neutron activation analysis [19].

4 Conclusion

This chapter presented a detailed overview of radiation interactions with matter, which are fundamental principles in nuclear physics and have numerous applications in various fields. By understanding these interactions, professionals working in areas like medical imaging, radiation therapy, radiation protection, and nuclear power generation can make informed decisions and advancements. The chapter explored various photon interactions, such as the photoelectric effect, Compton scattering, pair production, and coherent scattering, as well as charged particle interactions and neutron interactions. It also explained concepts such as stopping power, linear energy transfer, range, and scattering cross-section, which are essential for further study and practical application.

References

1. K. Buchtela, *Encyclopedia of Analytical Science: Radiochemical Methods I Gamma-ray Spectrometry* (Elsevier, 2005)
2. H. Kolanoski, N. Wermes, *Particle Detectors, Fundamentals and Principles* (Oxford University Press, 2022)
3. J.E. Martin, *Physics for Radiation Protection: A Handbook*, 2nd edn. (Wiley, 2006)
4. M.G. Stabin, *Radiation Protection and Dosimetry: An Introduction to Health Physics*. Springer.
5. L.J. Verhey, P.L. Petti, *Principles of Radiation Physics* (Elsevier, 2010)
6. J.E. Turner, Interaction of ionizing radiation with matter. *Health Phys.* **86**(3) (2004)
7. D. Bourland, *Clinical Radiation Oncology: Radiation Oncology Physics*, 3rd edn. (ScienceDirect, 2012)
8. G. Choppin, J.-O. Liljenzin, J. Rydberg, C. Ekberg, *Radiochemistry and Nuclear Chemistry: Absorption of Nuclear Radiation* (ScienceDirect, 2013)
9. T. James, *Atoms, Radiation, and Radiation Protection* (Wiley-VCH, Weinheim, 2007)
10. M. Berger, Q. Yang, A. Maier, *Medical Imaging Systems: An Introductory Guide* (National Library of Medicine, 2018)
11. N. Tsoulfanidis, S. Landsberger, *Measurement and Detection of Radiation*, 4th edn. (Taylor & Francis Group)
12. F.H. Attix, *Introduction to Radiological Physics and Radiation Dosimetry* (Wiley, 2004)
13. N. Barthe, S. Maîtrejean, N. Carvou, A. Cardona, Chapter 9 – High-resolution beta imaging, in *Handbook of Radioactivity Analysis*, vol. 2, (2020)
14. S.H. Park, J.O. Kang, Basics of particle therapy I: physics. Radiat. Oncol. J. **29**, 135 (2011)
15. T. Tabata, V. Moskvin, P. Andreo, V. Lazurik, Y. Rogov, Extrapolated ranges of electrons determined from transmission and projected-range straggling curves. Radiat. Phys. Chem **64**(3), 161
16. E. Podgorsak, *Radiation Physics for Medical Physics* (Springer, 2009)
17. K. Shultis, R. Faw, *Fundamentals of Nuclear Science and Engineering* (Taylor & Francis Group)
18. W. Langel, Introduction to neutron scattering. *ChemTexts* **9**(4), 12 (2023). https://doi.org/10.1007/s40828-023-00184-7
19. W.H. Jin, C. Seldon, M. Butkus, W. Sauerwein, H.B. Giap, A review of boron neutron capture therapy: its history and current challenges. Int. J. Part. Ther. **9**(1), 71–82 (2022). https://doi.org/10.14338/IJPT-22-00002.1

Photon-Counting X-Ray Detectors (PCXDs)

Daehee Lee

1 Introduction

For decades, X-ray technology has played a vital role in various fields such as medical diagnostics, industrial inspection, physics, and space science [1]. However, this technology has encountered several limitations due to the use of traditional, energy integration detectors (EIDs). EIDs operate by converting X-rays into light through a scintillator, accumulating light in a manner that limits the precise measurement of individual X-ray photon energy and location, leading to original position loss [2].

The development and application of photoconductor materials represent a significant step in overcoming these limitations. Materials like cadmium telluride (CdTe) or cadmium zinc telluride (CZT) possess the capability to directly convert X-ray photons into electrical charge. This opens the possibility for precise measurement of individual photon energy [3]. Furthermore, detectors based on photoconductors not only produce images with higher-energy resolution but also assist in improving spatial resolution. By analyzing photon energy individually, it becomes possible to provide more detailed energy-specific X-ray images.

The advancement of photon-counting X-ray detectors (PCXDs) has been significantly accelerated by the miniaturization and innovative designs of readout signal processing circuits. These innovations have drastically reduced noise in the final images and enabled more precise signal processing, substantially enhancing the quality of X-ray imaging. The process of obtaining PCXD images by combining photoconductors with readout signal processing circuits demonstrates the potential

D. Lee (✉)
UC Davies, Davies, CA, USA
e-mail: mnmlee@ucdavis.edu

K. Iniewski, H. Gadey (eds.), *Emerging Radiation Detection*,
https://doi.org/10.1007/978-3-031-63897-8_2

to further improve image resolution through the adjustment of energy-specific signal information [4].

This chapter begins by examining the fundamental principles and advantages of PCXDs, followed by an in-depth analysis of the core elements enabling this technology: photoconductors and application-specific integrated circuits (ASICs). It explores the combined technology of photoconductors and ASICs, such as bump bonding, and examines various examples of PCXDs produced, exploring their practical application possibilities. Additionally, it analyzes the characteristics and limitations of PCXDs, seeks solutions to technical problems, and investigates the potential revolutionary applications of this technology in medical diagnostics, material science, security inspection, and more. The technological advancements are expected to significantly expand the application possibilities across diverse fields.

2 Basic Operations of PCXD and Advantages over EIDs in X-Ray Imaging

Since the discovery of X-rays by Wilhelm Röntgen in 1895, digital X-ray imaging technology has played a pivotal role especially in a medicine field. X-ray detection principles can be broadly divided into indirect and direct methods as shown in Fig. 1. The indirect method first converts X-rays into light, which is then transformed into an electrical signal, whereas the direct method converts X-rays directly into charge to generate images. These methods are further differentiated by their signal processing techniques: the indirect integration method accumulates signals over time to form an image, and the counting method processes individual electrical signals corresponding to each photon [5].

Traditionally, X-ray detectors have employed the indirect integration method such as EID, primarily consisting of scintillator materials combined with CMOS (complementary metal-oxide semiconductor) or CCD (charge-coupled device) sensors [6]. In this configuration, X-ray photons are absorbed by the scintillator, generating visible light that is converted into charge by the sensor to form images. However, due to the isotropic scintillation light, this method suffers from reduced spatial resolution (image blurring). Additionally, the accumulation process precludes tracking the energy of individual incident X-rays.

PCXDs have been developed to overcome these limitations and enhance image quality and resolution. These detectors integrate a photoconductor for direct conversion of incident X-ray into charge and a signal processing unit for handling the generated charge. This configuration allows for the optimization of each processor and independent system optimization, leading to performance enhancements. The photoconductor directly converts X-rays into charge without a scintillation conversion process, and the generated charge, which is proportional to the energy, is transformed into an electrical signal. The signal processing unit then processes these **energy-proportional** electrical signals, acquiring energy information.

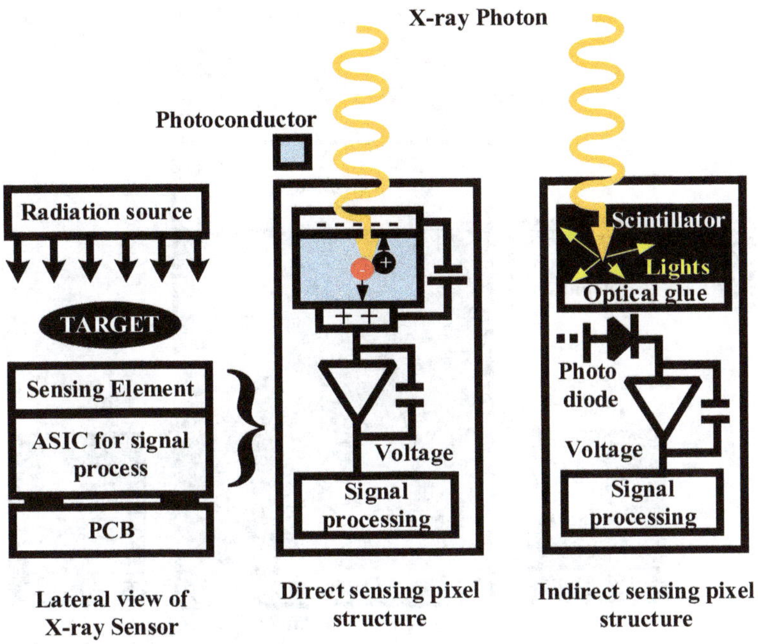

Fig. 1 Comparison of direct and indirect methods for X-ray imaging

Consequently, PCXDs not only count individual photons but also obtain energy information, providing capabilities not possible with EIDs [7].

Based on the efficient counting and energy measurement of incident X-ray photons, PCXDs enable the acquisition of energy-specific images from a single X-ray imaging, as illustrated in Fig. 2. Based on efficient physical data acquisition in detection, PCXDs offer several benefits over EIDs, including the following:

– Absence of Electronic Noise

One of the fundamental advantages of PCXDs is their capability to minimize electronic noise, a common issue with EIDs. PCXDs are designed with a threshold energy level (~20 keV) for X-ray photon detection, effectively filtering out lower energy levels and background electronic noise [8]. This feature is particularly advantageous in scenarios with low X-ray flux, where EIDs suffer from noise dominance, leading to degraded image quality. In contrast, PCXDs maintain signal integrity even at low doses, significantly enhancing image clarity and stability. This facilitates low-dose scans and improves diagnostics in obese patients through reduced noise and higher-quality images.

– Enhanced Spatial Resolution

Thanks to the structural of PCXDs, the utilization of smaller pixels is feasible without the need for opaque interlayers required in EIDs to prevent optical cross

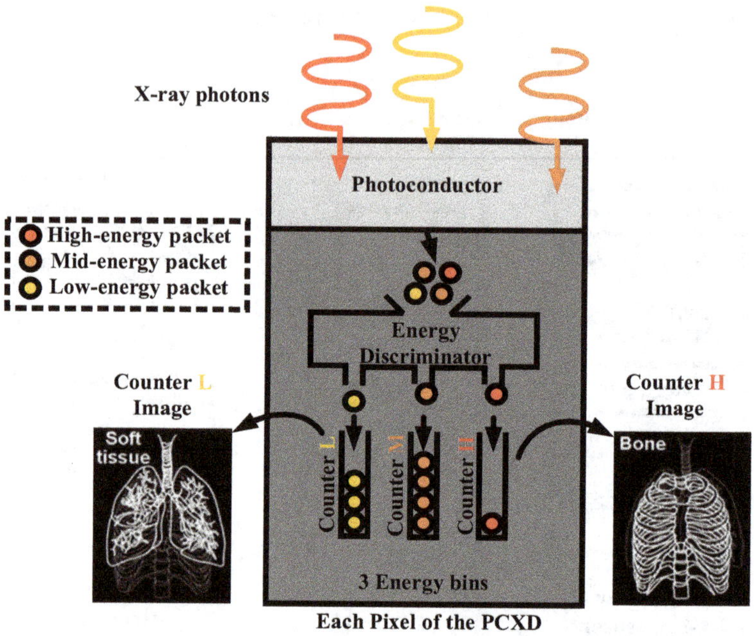

Fig. 2 Energy-specific acquired images based on the energy discrimination capability of PCXD

talk. The elimination of these interlayers not only improves dose efficiency but also significantly enhances spatial resolution. The capability for finer pixelation with PCXDs results in improved resolution in images, crucial for the accurate evaluation of small anatomical structures. This heightened resolution is particularly evident in applications requiring precise imaging, such as evaluations of the inner ear, lung imaging, and computed tomography (CT) angiography of small vessels [9].

– Improved Contrast-to-Noise Ratio (CNR)

PCXDs offer an enhanced CNR, a critical aspect in contrast-enhanced scans. Unlike EIDs, PCXDs count all X-ray quanta exceeding a specific energy threshold equally, enabling more distinct low-contrast differentiation. This feature is particularly beneficial in scans using iodinated contrast agents, leading to improved image quality or the possibility of reducing radiation or contrast agent dosage, thereby providing significant clinical advantages [10].

– Spectral Imaging Capabilities

PCXDs excel in spectral imaging, capable of collecting multiple signals at different energy thresholds. This multi-threshold capability enables material decomposition and more precise diagnostics. The ability to perform high-quality spectral imaging opens advanced diagnostic procedures, such as the evaluation of pulmonary embolism or coronary artery disease, while potentially reducing the volume of contrast agents used [11].

– Radiation Dose Reduction

One of the most significant advantages of PCXDs is their ability to substantially reduce radiation doses without compromising image quality. Studies indicate that PCXDs can achieve comparable or even superior image quality at significantly lower radiation doses compared to EID-based CT scans. This reduction is made possible by the detectors' efficiency and the absence of electronic noise, which allows for clearer images at lower doses. The dose reduction capability of PCXDs is particularly valuable in lung imaging, oncological evaluations, and examinations of pediatric and obese patients, where minimizing radiation exposure is crucial [12].

PCXDs represent a pivotal advancement in the field of X-ray imaging, enhancing noise reduction, spatial resolution, contrast sensitivity, and dose efficiency. These advancements hold the promise of transforming clinical imaging practices, enabling more precise diagnostics, safer imaging procedures, and a broader range of applications in medical diagnostics.

3 Photoconductors for PCXD

Photoconductors play a crucial role in high-precision radiation imaging systems, especially in PCXDs, serving as the initial stage in efficiently processing incident X-ray. This technology utilizes various materials, such as silicon (Si), cadmium telluride (CdTe), and cadmium zinc telluride (CZT), along with exploring additional materials like gallium arsenide (GaAs) and mercury iodide (HgI_2), to delve into their characteristics, radiation response efficiency, applications, electrical properties, and the main challenges they face [13]. This section aims to provide a comprehensive understanding of photoconductors' operation in PCXDs, highlighting their pivotal role in advancing radiation imaging by exploring the operational dynamics and potential of various photoconductive materials and analyzing the technological advancements of each candidate for PCXDs.

3.1 Principles of Photoconductor Operation

When radiation impinges on a photoconductor, electron-hole pairs are generated in proportion to the X-ray photon energy. The amount of charge produced is also affected by the photoconductor. The movement of the charge is governed by the voltage applied to the photoconductor. As illustrated in Fig. 3a, a continuous metal plate (cathode) is applied in the direction of the X-ray photon incidence, typically with a positive (+) bias, and segmented metal (anode) connected below forming an individual pixel, with a negative (−) bias applied. In this process, the internally

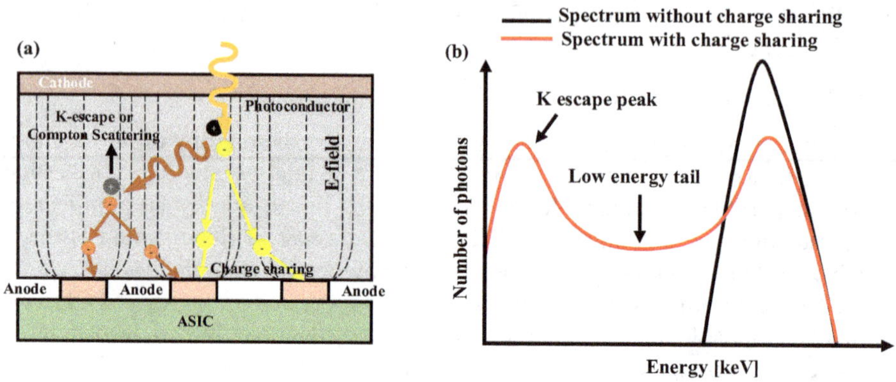

Fig. 3 Operation of photoconductors when X-ray photon is incident (**a**) and changes in the energy spectrum due to Compton scattering, K-escape, and charge sharing (**b**)

generated electrons and holes move toward anode and cathode, respectively [14]. For simplicity, in Fig. 3a, only a single electron and hole were indicated. This movement of electrons or holes generates a current at the anode, and summed current signal by a movement of both charges is processed by the signal processing unit below. The signals acquired at the anode are utilized as pixel signals for X-ray imaging.

The interaction between X-rays and photoconductors involves phenomena such as Compton scattering and the photoelectric effect. The photoelectric effect is crucial for generating electron-hole pairs, whereas Compton scattering describes the process by which an X-ray loses energy as it progresses, dispersing signals across neighbor pixels. The effects of Compton scattering, which lead to a degradation in image spatial resolution, are depicted in Fig. 3b. This shows how scattering across various pixels occurs, highlighting the impact on spatial resolution. Additionally, within the photoconductor, the loss of electrons due to the photoelectric effect and their subsequent replacement generate characteristic X-rays. These produced characteristic X-rays (k-escape) create electron-hole pairs at adjacent pixel locations, distorting the energy spectrum in a manner similar to Compton scattering and causing comparable resolution degradation. Materials such as CdTe are especially prone to generating characteristic X-rays upon photoelectron production.

Following the generation of electron-hole pairs, they drift under the influence of a bias voltage, leading to electron cloud diffusion. This diffusion, driven by the repulsion among same electrical charges, results in an increase in cloud size, a phenomenon known as the "charge sharing." Charge sharing, where signal and energy information are disseminated between adjacent pixels, significantly distorts energy information. The extent of charge diffusion is influenced by the photoconductor's physical properties and operational conditions, with factors like

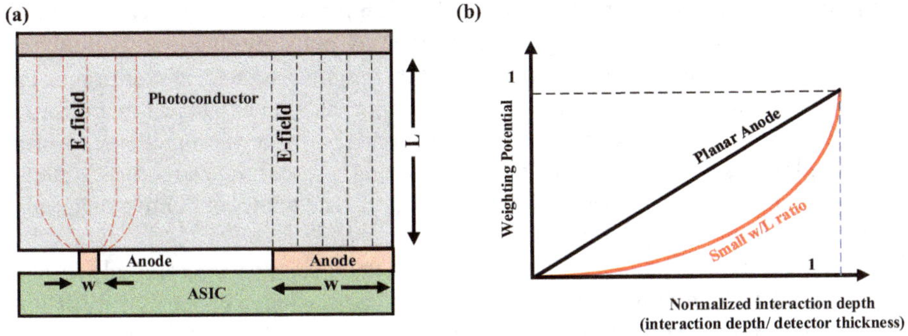

Fig. 4 Changes in the internal electric field of a photoconductor with varying pixel size (anode) (**a**) and the corresponding signal weighting near the anode (**b**)

the thickness of the photoconductor and pixel size playing crucial roles. The degree of charge spread (σ) over time can be represented by Eq. 1:

$$\sigma = d\sqrt{2\frac{k_B}{q}\frac{T}{V_{Bias}}} \quad (1)$$

which is dependent on d the drift length, T the temperature, k_B the Boltzmann constant, q the elementary charge, and V_{Bias} the applied bias voltage [15]. This equation indicates the estimation of cloud size under specific conditions, for instance, at a temperature of 300 K and with a 1 mm thickness of a CdTe under a 300 V bias voltage, where the diffusion's full width at half maximum (FWHM) measures 2.35 $\sigma = 31$ μm. As pixel sizes decrease, the impact of energy distortion due to charge sharing becomes increasingly significant. Figure 3b visualizes the changes in the energy spectrum caused by charge sharing.

To enhance spatial resolution, PCXDs frequently employ small pixel sizes, which introduces the "small pixel effect." However, charge sharing more pronounced as pixel size decreases relative to the detector's thickness and significantly impacts the electric field within the photoconductor, as schematically shown in Fig. 4a. The reduction in pixel size leads to an increased contribution of electrons to the final signal, as depicted in Fig. 4b [16]. This shift results in a diminished impact from holes and a substantial increase in the electron signal contribution. Known as the small pixel effect, this change amplifies charge sharing in adjacent pixel areas, where intersecting electric fields cause the division of incident electrons among neighboring pixels. Consequently, this leads to the obliteration of peaks corresponding to specific X-ray energies and a pronounced increase in continuous spectrum values in the lower-energy region.

Technical advancements are critical in enhancing the performance of photoconductor-based X-ray imaging systems. For instance, improving the charge transport properties of photoconductive materials can shorten charge transit times and reduce charge diffusion. Addressing the mobility disparity between electrons and holes can also mitigate charge sharing, thereby boosting image resolution and accuracy. Additionally, contemporary research has investigated alterations in the structural design of photoconductors, incorporating inter-pixel barriers that obstruct charge diffusion to neighboring pixels. These barriers, by confining charges within designated pixels, effectively reduce charge sharing, presenting a structural strategy to enhance imaging performance without modifying the photoconductive materials' electrical properties [17].

Simultaneously, the advancement of sophisticated signal processing algorithms plays a vital role in countering the challenges associated with charge diffusion. By employing these signal processing techniques, image distortions due to charge sharing can be corrected, producing clearer images. Notably, these algorithms are especially effective in mitigating spectral distortions in the low-energy region, thus significantly enhancing the diagnostic capabilities of the imaging system [18, 19].

3.2 *Various Photoconductor Materials*

Silicon, known for generating electrical signals in response to light or radiation, are integral to a variety of photoelectronic devices and detectors. Si, an indirect bandgap semiconductor with a bandgap of approximately 1.12 eV, is highly efficient in absorbing photons across visible light to the near-infrared region. Despite its relatively low absorption rate for high-energy radiation due to its low atomic number ($Z = 14$), Si's superior charge carrier mobility and the feasibility of producing high-purity, crystalline wafers on a large scale render it particularly suitable for detecting low-energy X-rays in high-resolution imaging systems. However, efficiency diminishes with high-energy radiation, making it ideal for applications requiring detection within the range of 100–500 μm thickness. Typically, Si detectors operate at a bias voltage of 10–100 V, optimizing charge mobility and collection efficiency while maintaining low sensitivity to high-energy X-rays, prompting the preference for higher atomic number materials in such scenarios [20, 21].

CdTe, a direct bandgap semiconductor boasting a bandgap of about 1.44 eV, excels in the absorption and detection of high-energy radiation, including X-rays and gamma rays. Its high atomic number (Cd = 48, Te = 52) enables efficient high-energy radiation absorption, making it a choice for medical imaging, radiation detection, and nuclear physics experiments. Employed typically at thicknesses of 1–2 mm and operating at bias voltages ranging from 100 to 600 V, CdTe combines absorption efficiency with charge collection proficiency, though it confronts challenges such as performance degradation at elevated temperatures and handling precautions due to its toxic cadmium content and relative expense [20–22].

CZT, renowned for its high-energy resolution and charge collection efficiency, is utilized in thicknesses of 2–5 mm to effectively detect high-energy radiation. With a bias voltage range similar to CdTe (100–600 V), optimized based on material thickness and energy range, CZT is highly valued in high-performance applications

Table 1 Characteristics of photoconductors used in PCXDs

Material	Si	CdTe	CZT	HgI_2	GaAs
Atomic number	14	48, 52	48, 30, 52	80,53	31,33
Density (g/cm^3)	2.33	5.85	5.78	6.4	5.32
Bandgap (eV)	1.12	1.44	1.57	2.15	1.43
Pair creation energy (eV)	3.62	4.43	4.6	4.2	4.2
$\mu_e\tau_e$ (cm^2/V)	>1	3.3×10^{-3}	1×10^{-3}	$>10^{-3}$	8×10^{-5}
$\mu_h\tau_h$ (cm^2/V)	~1	2×10^{-4}	3×10^{-4}	1×10^{-5}	4×10^{-6}

such as medical diagnostics, nuclear safety, and space radiation detection. Its selection over other materials often hinges on balancing resolution enhancement, material and thickness choices, and bias voltage optimization [20, 21, 23].

HgI_2 is distinguished by its exceptional absorption rates for high-energy radiation, attributed to its very high atomic number. However, its application has been constrained by concerns over mechanical stability and temperature sensitivity. Advancements are aimed at overcoming these limitations to harness its full potential in high-energy radiation detection [23].

GaAs, featuring excellent charge mobility and energy resolution, stands out in high-resolution radiation imaging applications. Despite its higher cost and toxicity issues necessitating careful handling, GaAs is pursued for its unmatched capabilities in sophisticated imaging systems, with research focusing on enhancing its stability and cost-effectiveness for broader applications [24].

Table 1 summarizes important characteristics of five photoconductors. The necessary information includes the atomic number, density, bandgap, pair creation energy, etc.

4 ASICs for PCXD

4.1 Basic Circuits for Photon-Counting Operations

In PCXDs, the circuits process electrical signals proportional to the incident X-ray photon energies, facilitated by charges transferred within photoconductors. Figure 5 illustrates the basic circuit components of a PCXD, including the output waveforms of each element. Electrons reaching the pixel pad electrode accumulate on the first stage charge-sensitive amplifier's (CSA) feedback capacitor (C_F), producing an electric signal pulse at the CSA output that is proportional to the electron count. A feedback resistor (R_F), connected in parallel with the capacitor, serves as a reset mechanism, ensuring the voltage returns to the baseline for subsequent signal processing. The transfer function $Z(s)$ from an input current (I_{Det}) to the output voltage (V_{CSA}), representing each charge packet by input time (t_{PD}) and magnitude (I_{Det}), is given by Eq. 2. Here, g_m denotes the transconductance of the

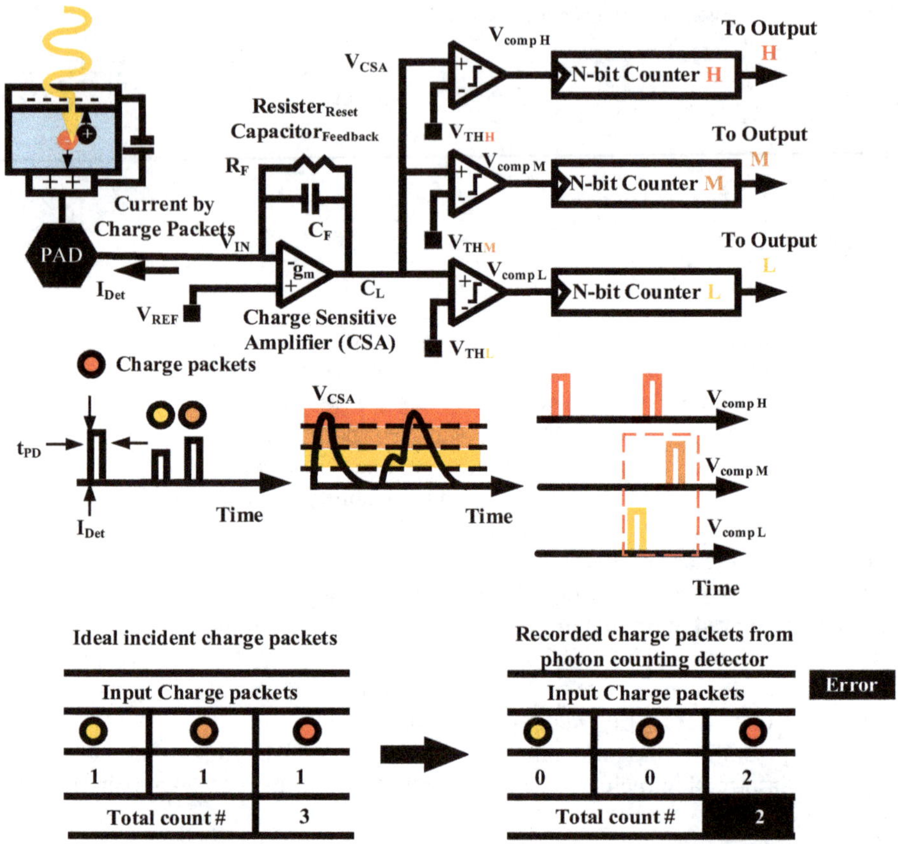

Fig. 5 Basic circuit for PCXD and the electrical signals at each node. Pulse pileup due to signal overlap and the resulting loss and distortion of signal counting

amplifier constituting the CSA. Transfer function, $Z(s)$, includes a dominant pole and a nondominant pole at $\frac{1}{R_F C_F}$ and $\frac{g_m}{\left(C_L + C_{In} + \frac{C_L C_{In}}{C_F}\right)}$, respectively, with the high-frequency zero $\frac{g_m}{C_F}$ being negligible [25].

$$\frac{V_{out}}{I_{Det}}(s) = Z(s) = \frac{R_F\left(1 - \frac{sC_F}{g_m}\right)}{1 + s\left(R_F C_F + \frac{C_L + C_{In}}{g_m}\right) + s^2\left(C_L C_{In} + C_L C_F + C_{In} C_F\right)\frac{R_F}{g_m}} \tag{2}$$

Figure 5 displays signals from individual photon interactions caused by three distinct X-ray photons, represented as V_{CSA}. The initial signal quickly rises before being reset by R_F. This pulse train signal is then potentially shaped by additional circuits to improve noise performance and processing speed, depending on the requirement. However, incorporating shaping circuits within the limited pixel size for resolution enhancement poses a challenge due to spatial constraints. Thus, the inclusion of shaping circuits must be judiciously considered, considering various factors.

Furthermore, pulse trains undergo energy measurement through a subsequent comparator. A photon-counting detector typically includes at least one comparator, each set to a specific threshold, enabling the acquisition of energy-specific information with multiple comparators. Applying various threshold levels to the electronic signal (pulse train) generates data with different spectral origins, allowing the PCXDs to measure not just the presence of X-rays but their "color." Each time the signal exceeds a threshold, the comparator's rising edge increments the counter value, with thresholds set high enough above the noise floor to limit noise-induced counts. Figure 5 also shows a comparator using three different comparison voltages, enabling selective counting and energy measurement of pulses. The pulse train includes a noise floor completely blocked by threshold level comparison, ensuring that counts do not display common noise, thus showing actual zero when X-rays do not strike the sensor. However, noise inherent in the actual circuit operation remains unavoidable, necessitating circuit designs optimized for noise characteristics.

Additionally, as the energy of X-ray photons is measured, each energy bin number can be weighted differently according to its energy [26]. This capability allows the detector to distinguish between the effects of a single high-energy photon and multiple low-energy photons. Since low-energy photons provide more contrast to the X-ray image, it is possible to reduce the dose applied to the patient without diminishing the image contrast. This feature offers a dose reduction effect depending on the patient's condition and age, as well as the applied dose of X-rays or contrast agent (since all photons from low to high energy are equally counted, ensuring good contrast). In contrast, EID contributes less from low-energy photons, resulting in lower contrast. As the energy information of human organs and soft tissues is distributed across the lower-energy X-ray spectrum, this can enhance the contrast of soft tissues.

As previously discussed in the photoconductor section, PCXDs do not guarantee perfect energy measurement of X-ray photon. Charge sharing between electrodes significantly distorts the influx of energy. This can be addressed by employing charge sharing compensation logic or by selecting larger pixel sizes to an appropriate level for the application.

Errors in acquired energy and counts are exacerbated by pulse pileup issues, as illustrated in Fig. 5. Single pulses processed by the CSA require a certain amount of time for signal processing in the circuit. However, in high X-ray flux environments, the successive input of X-ray photons inevitably leads to signal overlap. In the field of computerized tomography, the typical rate of X-ray photons entering photon-counting detectors can reach up to 10^9 photons/second/mm^2; hence smaller pixels support enhanced measurement capabilities by individually and parallel processing every photon signal [27, 28]. Superpositioned signals, as shown in Fig. 5, result in not only the loss of counts but also distortion of energy. An intuitive method to resolve these issues is to reduce the size of pixels. However, as observed earlier, smaller pixels introduce another constraint due to charge sharing, affecting the final image's spatial resolution. Therefore, optimizing pixel size for an application must consider charge sharing, pulse processing, and signal speed comprehensively.

4.2 Bump Bonding and Wire Bonding

The fabrication of PCXDs requires a high degree of technical sophistication, starting with the intricate linkage between photoconductors and ASICs. Central to this intricate connection process is bump bonding technology, which involves forming minute metallic or conductive bumps on the semiconductor chip's surface [29]. These bumps facilitate a physical and electrical connection to printed circuit boards (PCBs) or other semiconductor devices, utilizing various materials and methods such as lead soldering, gold-gold thermocompression, or copper thermocompression. A significant benefit of this approach is its high connection density and exceptional electrical and mechanical properties, allowing for tight integration between the detector's volume and its readout electronics. Consequently, this integration achieves optimal low-noise and high-bandwidth performance relative to power consumption. However, one of the drawbacks is the high cost of connection processes for low volumes.

Following the bump bonding process, PCXDs should be connected to PCBs for operation and final signal acquisition. While wire bonding has traditionally been used for such connections, requiring additional physical space, and restricting the placement of detectors in proximity to increase active area, a new method utilizing through-silicon Via (TSV) technology has been developed to allow for the expansion of all four sides of the photon-counting detectors [30]. TSV technology is a revolutionary approach in the manufacturing of high-performance electronic devices like PCXDs. It creates vertical connection pathways through the silicon wafer, enabling direct electrical connections between different layers of the chip, as illustrated in Fig. 6b. The TSV fabrication process involves drilling fine holes into the silicon wafer and filling these holes with metal (typically copper) to form electrical pathways. Formed TSVs also offer a route for effective thermal dissipation within the chip, mitigating potential overheating issues during high-performance operations. Additionally, TSV technology allows for the minimization of chip area while maintaining functionality, thereby reducing the overall size of the device. In PCXDs, a distinct advantage is the ability to arrange chips on all four sides, significantly enhancing scalability compared to traditional wire bonding to PCBs.

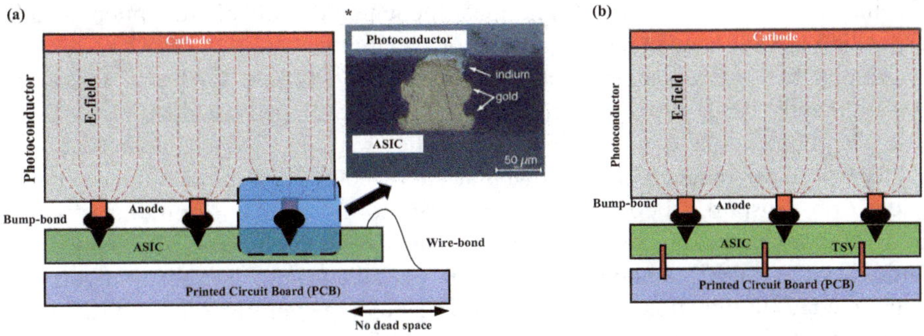

Fig. 6 Photoconductor and ASIC connected via bump bonding. Methods of connecting the assembly to the PCB for final signal reading: (**a**) coupling by wire bonding (a microscopic photo from [29]) and (**b**) coupling by the through-silicon via (TSV) method

4.3 Various PCXD Implementations

In this section, a variety of PCXDs that have been reported were introduced. The PILATUS3 ASIC and IBEX ASIC represent cutting-edge integrated circuits engineered for applications in high-resolution spectral computed tomography (CT). Manufactured through a 0.25 μm CMOS technological process, the PILATUS3 features a pixel size of 172 μm × 172 μm [31], capable of handling up to 10 Mphotons/second/pixel. By reassessing the output of the pulse shaper following a delay that is slightly longer than the duration of a pulse from a single photon, this architecture ensures that the detection rate continues uninterrupted, even in the presence of multiple photon arrivals. Such advancements in the PILATUS3 significantly reduce the electronic noise within the analog front end. The combined electronic noise from the analog front end, which includes the CSA, shaper, and discriminator input stages, amounts to roughly 2.51 mV rms. This level of noise correlates to an energy resolution of about 4.0 keV full width at half maximum (FWHM).

IBEX ASIC represents a more recent development, crafted using a 0.11 μm CMOS process, and supports a pixel dimension of 75 μm × 75 μm, offering similar capabilities of up to 10 Mphotons/second/pixel as the PILATUS3 [32]. Thanks to its reduced pixel size, the IBEX facilitates higher-resolution imaging and exhibits decreased sensitivity to photon clustering. In comparison to the PILATUS3, the IBEX introduces several additional features, including dual independent energy thresholds per pixel, pixel binning capabilities, and adjustable gain for a broad energy spectrum. Notably, each pixel contains two independent comparators, enabling the use of polychromatic X-ray sources. The dynamic range is expanded by adjusting the gain via different feedback capacitor values in the CSA, allowing for the measurement of X-rays up to 140 keV. This suite of enhancements positions the IBEX ASIC as a sophisticated tool for advancing high-resolution spectral

CT imaging, minimizing the limitations encountered with photon pileup while expanding the dynamic range and resolution capabilities.

Among the various sensor materials utilized in PCXDs, CZT and CdTe are predominantly used. However, one of the ASICs developed specifically for Si as a target photoconductor is the KTH/Prismatic ASIC [33, 34]. Unlike CZT and CdTe, where the induced charge signal at the detector front is predominantly due to electron transport, silicon relies on hole transport. A primary limitation of using Si as a photoconductor material is its low absorption coefficient. To counter this limitation, the KTH/Prismatic detector employs a silicon strip detector configuration with an edge-illumination setup. The detector module comprises five ASICs wire-bonded to the Si sensor. Si is organized into 16 strip contacts running from top to bottom, each containing 50 strips extending from left to right. The average ENC/channel is measured at 192 electrons in an open input condition and 214 electrons with a 3-pF capacitive load at the front end. This ASIC boasts a count rate capability of 17 Mcps/channel, with an estimated average nonparalyzable dead time of 20.2 ± 5.2 ns.

The Medipix3 ASIC validated the use of TSV technology for connecting the readout ASIC to the PCB, marking a pioneering step toward designing readout chips that can be seamlessly tiled on all four sides [30]. TSV technology reduces the need for wire bonds and minimizes the required dead spaces between chips on large surfaces. Building on this validation, the Medipix4 and Timepix4 chips were constructed to allow for a four-side expandability [35]. As the readout logic is not confined to one edge of the chip, this affords greater flexibility in selecting readout architectures, marking a significant milestone in the design and implementation of PCXDs. Both bump bonding and TSV technologies have facilitated a paradigm shift in detector design, enabling finer pixel pitch for improved resolution and closer integration with readout electronics to maximize performance. Specifically, TSV technology substantially enhances the scalability of detector arrays, which is critical for meeting the demands of large-scale experiments and applications in various fields. This innovation in PCXD technology underpins the advancement toward achieving higher precision and flexibility in scientific and medical imaging applications.

5 Applications and Future of PCXDs

5.1 Application Areas for PCXDs

Unlike traditional EIDs that measure the cumulative energy of X-ray photons, PCXDs count and measure the energy of individual photons. This capability facilitates numerous applications across various domains, including medical imaging, security, industrial inspection, and scientific research, which are detailed in this section.

- Medical Imaging

– Diagnostic Radiology

PCXDs enhance diagnostic radiology by enabling dual-energy or multi-energy imaging techniques, offering superior tissue characterization compared to traditional modalities. This technology allows for the extraction of material-specific images by exploiting the unique X-ray attenuation profiles of different tissues and materials at various photon energy levels. For example, in mammography, PCXDs employ spectral imaging to distinguish calcium deposits from glandular breast tissue more precisely, thus facilitating early breast cancer detection through improved visualization of microcalcifications [36].

– Computed Tomography (CT)

In CT imaging, the PCXD has led to the development of spectral CT technology, allowing for not only the high spatial resolution but also the acquisition of images at multiple energy information simultaneously. This capability enhances the differentiation of materials with close atomic numbers by utilizing K-edge subtraction imaging, where the sudden increase in attenuation at the K-edge of a contrast agent can be used to isolate its signal. Figure 7 illustrates the principle of K-edge imaging. It involves comparing images at different energy levels with a multi-energy discriminable PCXD, capitalizing on the abrupt changes in image information at a material's K-edge for enhanced material differentiation [10, 12, 38].

Consequently, CT images with PCXD significantly enhance the contrast-to-noise ratio in contrast-enhanced studies, facilitating effective visualization of vascular structures and delineation of tumors from surrounding tissues [39].

- Security Screening

PCXDs are employed in advanced X-ray backscatter systems, providing high-resolution, material-specific images that can differentiate between organic, inorganic, and metallic compounds [40]. The integration of machine learning algorithms with PCD data enhances the automatic detection and classification of contraband items, including explosives and narcotics, based on their spectral signatures. This technical synergy facilitates faster and more accurate threat assessments in real time, crucial for maintaining high throughput in security checkpoints.

- Industrial Inspection

In industrial inspection, PCXDs are instrumental in achieving micron-level resolution in X-ray imaging. The utilization of phase-contrast imaging techniques, enhanced by the spectral sensitivity of PCXDs, allows for the visualization of minute defects within complex components. This capability is essential for verifying the integrity of solder joints in electronic assemblies and identifying hairline fractures in aerospace components, surpassing traditional NDT methods.

- Scientific Research

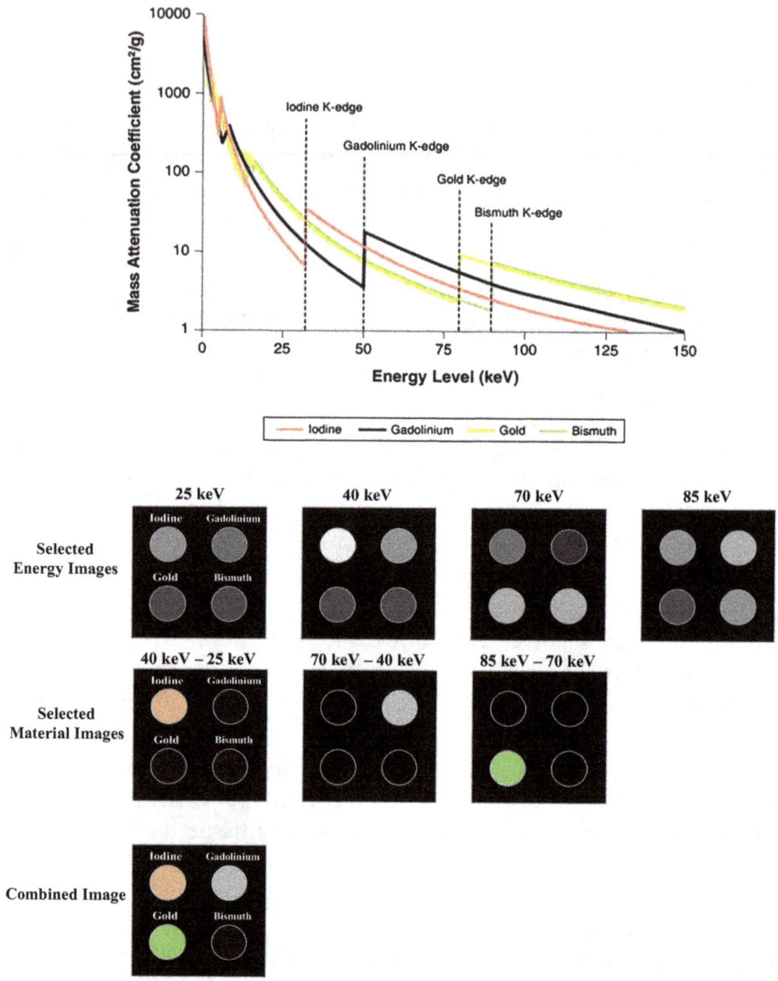

Fig. 7 Mass attenuation profiles for different materials (image from [41]). PCXD images acquired in limited energy ranges based on discontinuous K-edge information. K-edge imaging utilizing the variation in mass attenuation coefficients across different energies

– Material Science

PCXDs are pivotal in synchrotron radiation facilities, enabling the detailed study of materials under extreme conditions of pressure, temperature, and magnetic fields. The high photon flux and spectral purity of synchrotron sources, combined with the energy-resolving capabilities of PCXDs, allow for unprecedented insights into the electronic and magnetic phenomena at the atomic scale. This integration is key to understanding high-temperature superconductivity, magnetoresistance, and other quantum material properties.

– Astrophysics

In astrophysics, PCXDs equipped with high spectral resolution are essential for the next generation of space telescopes, aiming to observe the high-energy universe with unprecedented clarity [42–44]. These detectors facilitate the study of the emission lines from hot gas in galaxy clusters, the iron line from accretion disks around black holes, and the spectral features of X-ray binaries. The technical advancements in PCXD technology, including enhanced cooling systems for noise reduction and increased detector area for greater photon capture, are instrumental in pushing the boundaries of cosmic discovery.

This overview underscores PCXDs' transformative impact across various fields, heralding new exploration and innovation avenues.

5.2 Technical Limitations and Future Directions for PCXDs

Despite the various advantages of PCXDs, there are areas where performance enhancement is necessary. This section examines technical challenges that require overcoming and explores areas that must be addressed to facilitate the broader application of PCXD in medical imaging.

Firstly, charge sharing, a phenomenon where charges disperse among adjacent pixels, diminishes PCXD's spectral measurement capabilities and dose efficiency. Mitigating this effect through technical approaches such as improved digital and charge summing methods, as well as applying AI for charge sharing correction, can enhance spectral imaging accuracy and performance in material decomposition imaging [45–48]. Engineering efforts, alongside advanced electronic circuit design and software algorithm development, are crucial. Successfully addressing this could maximize the potential in sophisticated applications like K-edge contrast agent imaging.

Secondly, the spectral resolution of PCXD is partly limited by Compton scattering and K-escape events within the photoconductor. To improve spectral response, these phenomena must be corrected through precise calibration techniques. High spectral resolution plays a crucial role in spectral imaging with iodine or other K-edge contrast agents, allowing for improved image quality at lower radiation doses. Recent studies have explored AI-based correction algorithms for spectral data, aiming to address these limitations [49, 50].

Thirdly, addressing high count rates is essential, especially in CT fields where rapid image acquisition during rotation is unavoidable, necessitating swift signal processing by PCXD. Although Siemens reported a commercial photon-counting CT in 2022, other major companies are yet to release their commercial detectors, with active research suggesting an imminent commercialization phase. Innovations like cross-detection signal processing methods have been introduced to overcome high count rates [37, 51]. Figure 8a illustrates a key operation of the cross-detection method. By comparing the original V_{CSA} signal with its copied signal, V_{CCSA},

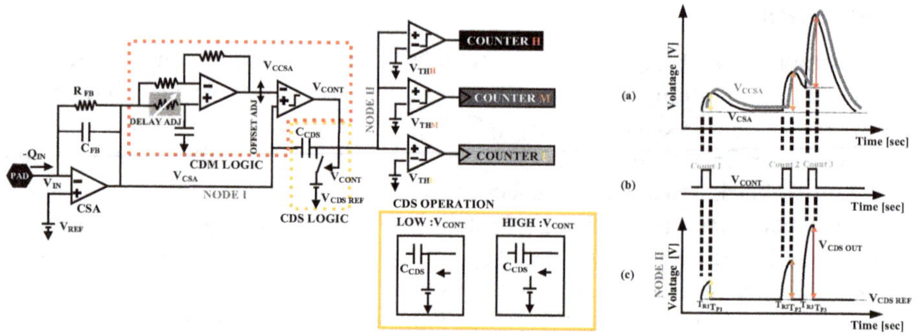

Fig. 8 Method for eliminating pulse pileup distortion using the cross-detection technique and correlated double sampling method. Energy distortion is significantly reduced, and the count rate is concurrently enhanced. (Image from [37])

this method enables not only the measurement of incoming photons but also the extraction of accurate energy information, as shown in Figs. 8b, c.

Research shows that comparing the original and duplicate signals can significantly reduce energy distortion and count loss while dramatically increasing signal processing speed. Such studies indicate a promising future for commercial photon-counting CT.

Moreover, optimizing data processing and storage is imperative for PCCT due to its generation of vast amounts of high-resolution and multispectral channel data. High-performance computing resources and advanced data compression techniques are needed for efficient data handling, analysis, and storage. AI- and machine learning-based image reconstruction and noise reduction technologies can accelerate data processing and enhance image quality [52–54]. Additionally, the development of standardized spectral data formats is crucial for ensuring compatibility across different software and systems, enhancing scalability in research and clinical applications.

Addressing these technical challenges is crucial, but preparing for the clinical application and evaluation of new technologies in medical imaging is equally important. Broad clinical studies and evaluations across large patient cohorts are necessary to demonstrate the clinical value of PCXD technology. Such research will clarify the distinct advantages provided by PCXD's spectral separation and high-resolution imaging in real clinical settings, expecting significant performance improvements over dual-energy approach with the EIDs [55, 56]. However, in-depth studies are essential to verify whether these technological advancements make a meaningful difference clinically.

These research and development efforts will significantly broaden the clinical applicability of PCXDs, facilitating the integration of new technologies for optimized patient care. The successful clinical application of PCXDs promises to bring innovations in patient treatment through high image quality, reduced radiation doses, and enhanced diagnostic capabilities.

6 Summary

Photon-counting X-ray detector (PCXD) technology marks a significant advancement in X-ray imaging, offering substantial benefits over traditional energy-integrating detectors (EIDs). By enabling the counting of individual X-ray photons and the measurement of their specific energies, PCXDs unlock new possibilities in material differentiation, enhanced image quality, and reduced radiation exposure. For example, their application in mammography has led to improved contrast between calcifications and surrounding tissues, facilitating early breast cancer detection. This highlights the technology's potential to transform diagnostic radiology.

In this chapter, the core operation and advancements in PCXD technology focus on three key components: photoconductors, application-specific integrated circuits (ASICs), and integration techniques like bump bonding and through-silicon vias (TSVs). These components are crucial to PCXD's rapidly becoming the standard in X-ray imaging applications. Transitioning from EID to PCXD technology overcomes several limitations of traditional methods, including the need for multiple measurements and the challenge of distinguishing between materials with similar attenuation profiles. PCXDs enhance material identification capabilities and medical imaging technical prowess through K-edge subtraction imaging, which discriminates materials by atomic number.

Despite significant progress with various PCXDs, opportunities for further advancements remain, including higher count rates, improved noise reduction, and better management of pulse pileup effects. Addressing these challenges necessitates ongoing research and development, focusing on refining the technology to realize its full potential. Innovations in photoconductive materials, ASIC design enhancements, and optimization of bonding and integration techniques are pushing the boundaries of PCXD technology. These developments not only enhance existing systems' performance but also pave the way for new applications beyond medical fields, such as security screening, industrial inspection, and scientific research.

In conclusion, PCXD technology's evolution from concept to widespread application represents a significant leap in the X-ray imaging field. As we approach further discoveries and enhancements, the continued evolution of PCXD technology promises to redefine the landscape of medical diagnostics and more. With ongoing research aimed at overcoming technical limitations and exploring new applications, the future of PCXD technology is bright, heralding a new era of precision, efficiency, and innovation in X-ray imaging.

References

1. X. Ou et al., Recent development in X-ray imaging technology: future and challenges. Research **2021**, 9892152 (2021). https://doi.org/10.34133/2021/9892152

2. C.H. McCollough, P. Rajiah, J.P. Bois, T.N. Winfree, R.E. Carter, K. Rajendran, E.E. Williamson, J.E. Thorne, S. Leng, Comparison of photon-counting detector and energy-integrating detector CT for visual estimation of coronary percent luminal stenosis. Radiology **309**(3), e230853 (2023).
3. S. del Sordo, L. Abbene, E. Caroli, A.M. Mancini, A. Zappettini, P. Ubertini, Progress in the development of CdTe and CdZnTe semiconductor radiation detectors for astrophysical and medical applications. Sensors **9**(5), 3491–3526 (2009). https://doi.org/10.3390/s90503491
4. T.G. Schmidt, Optimal 'image-based' weighting for energy-resolved CT. Med. Phys. **36**(7), 3018–3027 (2009). https://doi.org/10.1118/1.3148535
5. Institute of Electrical and Electronics Engineers. and S. Medical Imaging Conference (2011) Valencia, Nuclear Science Symposium and Medical Imaging Conference (NSS/MIC), 2011 IEEE : date, 23–29 Oct. 2011. IEEE, 2011
6. M.S. Kim, K.T. Lim, G. Kim, G. Cho, A comparative study of scintillator combining methods for flat-panel X-ray image sensors. Nucl. Instrum. Methods Phys. Res. A **882**, 84–90 (2018). https://doi.org/10.1016/j.nima.2017.11.023
7. D. Lee, K. Lim, K. Park, C. Lee, S. Alexander, G. Cho, An innovative method to reduce count loss from pulse pile-up in a photon-counting pixel for high flux X-ray applications. J. Instrum. **12**(3), 1–15. Institute of Physics Publishing (2017). https://doi.org/10.1088/1748-0221/12/03/P03006
8. Y. Nakamura, T. Higaki, S. Kondo, I. Kawashita, I. Takahashi, K. Awai, An introduction to photon-counting detector CT (PCD CT) for radiologists. Jpn. J. Radiol. **41**(3), 266–282 (2023). https://doi.org/10.1007/s11604-022-01350-6. Springer
9. J. van der Bie et al., Photon-counting CT: review of initial clinical results. Eur. J. Radiol. **163**, 110829 (2023). https://doi.org/10.1016/j.ejrad.2023.110829. Elsevier Ireland Ltd
10. T. Flohr, B. Schmidt, Technical basics and clinical benefits of photon-counting CT. Investig. Radiol. **58**(7), 441–450 (2023). https://doi.org/10.1097/RLI.0000000000000980. Lippincott Williams and Wilkins
11. V. Mergen et al., Virtual noncontrast abdominal imaging with photon-counting detector CT. Radiology **305**(1), 107–115 (2022). https://doi.org/10.1148/radiol.213260
12. A. Donuru et al., Photon-counting detector CT allows significant reduction in radiation dose while maintaining image quality and noise on non-contrast chest CT. Eur. J. Radiol. Open **11**, 100538 (2023). https://doi.org/10.1016/j.ejro.2023.100538
13. T. Schulman and Yliopistopaino, Si, CdTe and CdZnTe radiation detectors for imaging applications. [Tom Schulman], 2006
14. H. Hayashi, N. Kimoto, T. Asahara, T. Asakawa, C. Lee, A. Katsumata, *Photon Counting Detectors for X-Ray Imaging: Physics and Applications* (Springer, 2021). https://doi.org/10.1007/978-3-030-62680-8
15. C. Brönnimann, P. Trüb, Hybrid pixel photon counting X-ray detectors for synchrotron radiation, in *Synchrotron Light Sources and Free-Electron Lasers*, (Springer, Cham, 2015), pp. 1–29. https://doi.org/10.1007/978-3-319-04507-8_36-1
16. Q. Zhang, C. Zhang, Y. Lu, K. Yang, Q. Ren, Progress in the development of CdZnTe unipolar detectors for different anode geometries and data corrections. Sensors (Switzerland) **13**(2), 2447–2474 (2013). https://doi.org/10.3390/s130202447
17. B. Yu and Institute of Electrical and Electronics Engineers, 2012 IEEE Nuclear Science Symposium and Medical Imaging Conference record (NSS/MIC) : October 29–November 3, 2012, Disneyland Hotel, Anaheim, California, USA
18. M. Campbell et al., Towards a new generation of pixel detector readout chips. J. Instrum. **11**, C01007 (2016). https://doi.org/10.1088/1748-0221/11/01/C01007. Institute of Physics Publishing
19. D. Pennicard, R. Ballabriga, X. Llopart, M. Campbell, H. Graafsma, Simulations of charge summing and threshold dispersion effects in Medipix3. Nucl. Instrum. Methods Phys. Res. A. **636**(1), 74–81 (2011). https://doi.org/10.1016/j.nima.2011.01.124
20. M. Chmeissani, B. Mikulec, Performance limits of a single photon counting pixel system. Nucl. Instrum. Methods A **460**, 81–90 (2001)

21. J. Fink, H. Krüger, P. Lodomez, N. Wermes, Characterization of charge collection in CdTe and CZT using the transient current technique. Nucl. Instrum. Methods Phys. Res. A **560**(2), 435–443 (2006). https://doi.org/10.1016/j.nima.2006.01.072
22. Institute of Electrical and Electronics Engineers, 2011 IEEE international electron devices meeting: IEDM 2011: Washington, DC, USA, 5–7 December 2011. IEEE, 2011
23. N.E. Hartsough, J.S. Iwanczyk, E. Nygard, N. Malakhov, W.C. Barber, T. Gandhi, Polycrystalline mercuric iodide films on CMOS readout arrays. IEEE Trans. Nucl. Sci. **56**, 1810–1816 (2009). https://doi.org/10.1109/TNS.2009.2023478
24. B. Ghammraoui, S. Gkoumas, S.J. Glick, Characterization of a GaAs photon-counting detector for mammography. J. Med. Imaging **8**(03), 033504 (2021). https://doi.org/10.1117/1.jmi.8.3.033504
25. H.S. Kim et al., An asynchronous sampling-based 128 × 128 direct photon-counting X-ray image detector with multi-energy discrimination and high spatial resolution. IEEE J. Solid-State Circuits **48**(2), 541–558 (2013). https://doi.org/10.1109/JSSC.2012.2221196
26. E. Samei, N.J. Pelc, *Computed Tomography: Approaches, Applications, and Operations* (Springer, Cham, 2019). https://doi.org/10.1007/978-3-030-26957-9
27. M. Lee et al., Pulse pileup correction method for gamma-ray spectroscopy in high radiation fields. Nucl. Eng. Technol. **52**(5), 1029–1035 (2020). https://doi.org/10.1016/j.net.2019.12.003
28. T. Flohr, S. Ulzheimer, M. Petersilka, B. Schmidt, Basic principles and clinical potential of photon-counting detector CT. Chin. J. Acad. Radiol. **3**(1), 19–34 (2020). https://doi.org/10.1007/s42058-020-00029-z. Springer
29. T. Takahashi et al., High-resolution cdte detector and applications to imaging devices. IEEE Trans. Nucl. Sci. **48**, 287–291 (2001)
30. T. Tick, M. Campbell, TSV processing of Medipix3 wafers by CEA-LETI: a progress report. J. Instrum. **6**(11) (2011). https://doi.org/10.1088/1748-0221/6/11/C11018
31. T. Loeliger, C. Bronnimann, T. Donath, M. Schneebeli, R. Schnyder, P. Triib, *The New PILATUS3 ASIC with Instant Retrigger Capability*. 2012 IEEE Nuclear Science Symposiwn and Medical Imaging Conference Record (NSS/MIC).
32. M. Bochenek et al., IBEX: versatile readout ASIC with spectral imaging capability and high count rate capability. IEEE Trans. Nucl. Sci. **65**(6), 1285–1291 (2018). https://doi.org/10.1109/TNS.2018.2832464
33. X. Liu, F. Grönberg, M. Sjölin, S. Karlsson, M. Danielsson, Count rate performance of a silicon-strip detector for photon-counting spectral CT. Nucl. Instrum. Methods Phys. Res. A. **827**, 102–106 (2016). https://doi.org/10.1016/j.nima.2016.04.087
34. M. Gustavsson, F.U. Amin, A. Björklid, A. Ehliar, C. Xu, C. Svensson, A high-rate energy-resolving photon-counting ASIC for spectral computed tomography. IEEE Trans. Nucl. Sci **59**(1), 30–39 (2012). https://doi.org/10.1109/TNS.2011.2169811
35. X. Llopart et al., Timepix4, a large area pixel detector readout chip which can be tiled on 4 sides providing sub-200 ps timestamp binning. J. Instrum. **17**, C01044 (2022). https://doi.org/10.1088/1748-0221/17/01/C01044. IOP Publishing Ltd
36. H. Kim, M. Lee, D. Kim, D. Lee, H.J. Kim, Evaluation of photon-counting spectral mammography for classification of breast microcalcifications. Radiat. Phys. Chem. **162**, 39–47 (2019). https://doi.org/10.1016/j.radphyschem.2019.04.028
37. D. Lee, K.T. Lim, K. Park, C. Lee, G. Cho, A new cross-detection method for improved energy-resolving photon counting under pulse pile-up. Nucl. Instrum. Methods Phys. Res. A. **867**, 154–162 (2017). https://doi.org/10.1016/j.nima.2017.05.042
38. D. Richtsmeier, C.A.S. Dunning, K. Iniewski, M. Bazalova-Carter, Multi-contrast K-edge imaging on a bench-top photon-counting CT system: acquisition parameter study. J. Instrum. **15**(10), P10029 (2020). https://doi.org/10.1088/1748-0221/15/10/P10029
39. G. Jost, M. McDermott, R. Gutjahr, T. Nowak, B. Schmidt, H. Pietsch, New contrast media for K-edge imaging with photon-counting detector CT. Investig. Radiol. **58**(7), 515–522 (2023). https://doi.org/10.1097/RLI.0000000000000978. Lippincott Williams and Wilkins

40. G. Beldjoudi, V. Rebuffel, L. Verger, V. Kaftandjian, J. Rinkel, An optimised method for material identification using a photon counting detector. Nucl. Instrum. Methods Phys. Res. A. **663**(1), 26–36 (2012). https://doi.org/10.1016/j.nima.2011.09.002
41. I. Danad, Z.A. Fayad, M.J. Willemink, J.K. Min, New applications of cardiac computed tomography dual-energy, spectral, and molecular CT imaging. JACC Cardiovasc. Imaging **8**(6), 710–723 (2015)
42. B. C. P. et al., Development of Raspberry Pi-based Processing Unit for UV Photon-Counting Detectors, Jan. 2024, [Online]. Available: http://arxiv.org/abs/2401.01443
43. J. Ingjald, *Photon Counting in Astronomy: Evaluation of Avalanche Photodiode Detectors* (Lund Observatory, Lund University, Lund, 2010).
44. F. Zappa et al., Photon counting arrays for astrophysics. J. Mod. Opt. **54**(3–4), 163–189 (2007). https://doi.org/10.1080/09500340600742320
45. D. Pennicard, R. Ballabriga, X. Llopart, M. Campbell, H. Graafsma, Simulations of charge summing and threshold dispersion effects in Medipix3. Nucl. Instrum. Methods Phys. Res. A **636**(1), 74–81 (2011). https://doi.org/10.1016/j.nima.2011.01.124
46. J. Cammin, J. Xu, W.C. Barber, J.S. Iwanczyk, N.E. Hartsough, K. Taguchi, A cascaded model of spectral distortions due to spectral response effects and pulse pileup effects in a photon-counting x-ray detector for CT. Med. Phys. **41**(4), 041905 (2014). https://doi.org/10.1118/1.4866890
47. A. Smith, J. Atlas, A. Atharifard, Correcting charge sharing distortions in photon counting detectors utilising a spatial-temporal CNN, in *Lecture Notes in Computer Science (including subseries Lecture Notes in Artificial Intelligence and Lecture Notes in Bioinformatics)*, Springer Science and Business Media Deutschland GmbH, (2023), pp. 75–90. https://doi.org/10.1007/978-3-031-25825-1_6
48. P.T. Talla, P. Bartl, T. Michel, J. Durst, G. Anton, A modified spectrum reconstruction method for the charge summing mode of Medipix3, in *Nuclear Instruments and Methods in Physics Research, Section A: Accelerators, Spectrometers, Detectors and Associated Equipment*, May 2011. https://doi.org/10.1016/j.nima.2010.06.144
49. M. Li, D.S. Rundle, G. Wang, *X-ray Photon-Counting Data Correction Through Deep Learning*, Jul. 2020, [Online]. Available: http://arxiv.org/abs/2007.03119
50. Z. Shi, Z. Wang, F. Kong, Q. Cao, N. Wang, J. Qi, An x-ray crosstalk correction method using FCNN for a novel energy resolving scheme in spectral CT. Phys. Med. Biol. **66**(11), 115022 (2021). https://doi.org/10.1088/1361-6560/abfbf1
51. D. Lee, K. Park, K.T. Lim, G. Cho, Energy-correction photon counting pixel for photon energy extraction under pulse pile-up. Nucl. Instrum. Methods Phys. Res. A. **856**, 36–46 (2017). https://doi.org/10.1016/j.nima.2017.02.084
52. A. Eguizabal, O. Öktem, M.U. Persson, *Deep Learning for Material Decomposition in Photon-Counting CT*, Aug. 2022, [Online]. Available: http://arxiv.org/abs/2208.03360
53. A.S. Wang, N.J. Pelc, Spectral photon counting CT: imaging algorithms and performance assessment. IEEE Trans. Radiat. Plasma. Med. Sci. **5**(4), 453–464 (2021). https://doi.org/10.1109/TRPMS.2020.3007380. Institute of Electrical and Electronics Engineers Inc
54. F.I. Baffour et al., Photon-counting detector CT with deep learning noise reduction to detect multiple myeloma. Radiology **306**(1), 229–236 (2023). https://doi.org/10.1148/radiol.220311
55. M.D. Holbrook, D.P. Clark, C.T. Badea, Dual source hybrid spectral micro-CT using an energy-integrating and a photon-counting detector. Phys. Med. Biol. **65**(20) (2020). https://doi.org/10.1088/1361-6560/aba8b2
56. K. Rajendran et al., First clinical photon-counting detector CT system: technical evaluation. Radiology **303**(1), 130–138 (2022). https://doi.org/10.1148/RADIOL.212579

LYSO-GAPD Photon Counting Detectors for DEXA Applications

Jihoon Kang and Donggeun Roh

1 Introduction

Osteoporosis is the most common bone disease in humans, representing a major public health threat. Osteoporosis is defined as "a skeletal disease characterized by compromised bone strength predisposing a person to an increased risk of fracture" [1]. According to the National Osteoporosis Foundation (NOF), approximately 54 million American adults, representing 50% of the US adult population, are affected by osteoporosis (10.2 million) and osteopenia (43.4 million) [2]. One-tenth of women aged 60, one-fifth of women aged 70, two-fifths of women aged 80, and two-thirds of women aged 90 worldwide are estimated to have osteoporosis [3]. Approximately one in three women over the age of 50 and one in 50 men over the age of 50 worldwide experience osteoporotic fractures [4–6]. The prevalence of osteoporosis, osteopenia, and fractures is expected to increase in all countries because the number of older people is increasing.

Bone densitometry is a noninvasive and inexpensive examination for the quantitative measurement of bone mass or density. Numerous methods have been proposed and utilized to measure bone mineral density (BMD) and strength [7–14]. DEXA is the most widely used densitometry method for BMD assessment because it gives very precise measurements at clinically relevant skeletal sites. Even though DEXA has inevitable disadvantages including a nonportable bulky system and radiation exposure (albeit in a very low dose), it is the standard test used to diagnose osteoporosis, according to the classification of the World Health Organization (WHO) [15, 16]. Since the first DXA system was introduced by Hologic in 1987 [17, 18], advances in technology have increased the need to overcome various technical

J. Kang (✉) · D. Roh
Chonnam National University, Yeosu, South Korea

K. Iniewski, H. Gadey (eds.), *Emerging Radiation Detection*,
https://doi.org/10.1007/978-3-031-63897-8_3

challenges and replace older-generation DEXA systems based on NaI scintillation crystals coupled with photomultiplier tube (PMT).

More than 30 years after its implementation, semiconductor detectors based on cadmium-zinc-telluride (CZT) or cadmium telluride (CdTe) are regarded as one of the most useful DEXA detectors. Currently, CZT or CdTe detectors with direct conversion mechanisms have been widely utilized on a commercial basis [19, 20]. This is due to their outstanding features of excellent energy resolution, high detection efficiency, room temperature operation, and compact four-side tillable scalability. These compound semiconductor detectors have a number of features that may render their prolonged implementation difficult. They are susceptible to irradiation damage, incomplete carrier collection, and a large dark current. Their high price and fragile structure also limit their applicability. As an alternative, an X-ray diagnostic detector combining high-performance scintillation crystals with a compact photosensor has attracted interest for use in DEXA applications. The first commercial DEXA scanner based on this technology was recently introduced [21].

This chapter focuses on DEXA instrumentation, the most common densitometry method currently, to review its principles and use techniques. The main technical features in creating the dual-photon beams and employing the photon counting detectors are described and compared. We also present recent results obtained from indirect photon counting detectors consisting of a Geiger mode avalanche photodiode (GAPD) coupled with widely utilized lutetium yttrium oxyorthosilicate (LYSO) for potential application in DEXA systems.

2 Basic Principles and Techniques for DEXA Instrumentation

Dual-energy X-ray absorptiometry exploits the difference in the transmission and attenuation of the tissue and bone at low and high energies. These X-rays transfer their energy to the body as they pass through it. The energy level should be sufficient to transmit the matter within the imaging area and still be efficiently detected by the DEXA detector (Fig. 1).

In the basic physics of the interaction of radiation with matter, the transmission of a monoenergetic photon beam having the same energy through matter is described using the following exponential equation:

$$I = I_0 \times e^{-(\mu_l)(x)} = I_0 \times e^{-(\mu_m \times \rho)(x)} = I_0 \times e^{-(\mu_m)(d_m)} \tag{1}$$

Where:

I is the beam intensity transmitted through the matter.

I_0 is the initial beam intensity recorded with no matter present.

μ_l [cm^{-1}] is the linear attenuation coefficient of the matter at the photon energy of interest and increases linearly with matter density ρ.

x [cm] is the thickness of the matter.

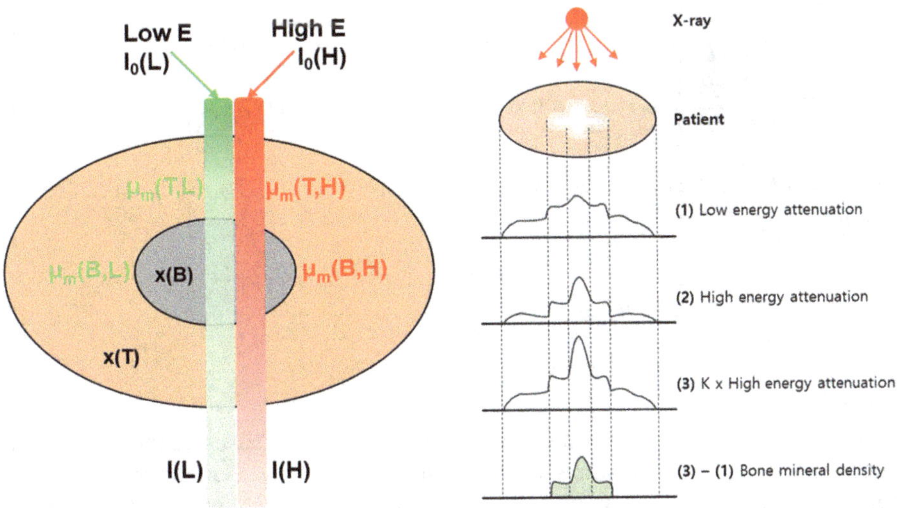

Fig. 1 The fundamental principle of DEXA. Bone density values can be calculated by the monoexponential attenuation process that transmits or attenuates X-rays through the body using low and high energy

μ_m[cm^2/g] is the mass attenuation coefficient of the matter and depends upon the effective atomic number Z_{eff} of the matter and incident photon energy E.
ρ[g/cm^3] is the mass density.
and d_m[g/cm^2] is the area density.

It should be noted that the negative sign signifies that the beam intensity decreases with increasing matter thickness. A fundamental assumption in calculating bone density values is that the body is composed of a two-compartment model of bone mineral and soft tissue. Bone mineral includes phosphorus and calcium molecules having relatively higher effective atomic numbers and physical density, whereas soft tissue is a mixture of muscle, fat, skin, and water. For a monoenergetic photon beam, the mass attenuation coefficient for a mixture of elements can be obtained from the values for its component elements with bone (*B*) and soft tissue (*T*). For polyenergetic photon beams, realistic X-ray photon beams with the different X-ray energies, the mass attenuation coefficient (u_m) for a mixture of elements can be obtained from the values for its component elements with the bone (*B*) and soft tissue (*T*) for each high energy (*H*) and low energy (*L*) as in the following equations:

$$\mu_m\,(mix, L)\,d_m(mix) = \mu_m\,(B, L)\,d_m(B) + \mu_m\,(T, L)\,d_m(T) \tag{2.1}$$

$$\mu_m\,(mix, H)\,d_m(mix) = \mu_m\,(B, H)\,d_m(B) + \mu_m\,(T, H)\,d_m(T) \tag{2.2}$$

Thus, from Eqs. 1, 2.1, and 2.2, the mathematical equations become the following:

$$\ln\left(\frac{I(L)}{I_0(L)}\right) = -\left(\mu_m\left(B, L\right) d_m(B) + \mu_m\left(T, L\right) d_m(T)\right) \tag{3.1}$$

$$\ln\left(\frac{I(H)}{I_0(H)}\right) = -\left(\mu_m\left(B, H\right) d_m(B) + \mu_m\left(T, H\right) d_m(T)\right) \tag{3.2}$$

Considering k factor, these equations are solved for d_m (B) for the area density of bone.

$$d_m(B) = -\frac{\ln\left(\frac{I(L)}{I_0(L)}\right) - \ln\left(\frac{I(H)}{I_0(H)}\right)(k)}{\left(\mu_m\left(B, L\right) - (k)\mu_m\left(B, H\right)\right)} \tag{4}$$

where k is $\left(\frac{\mu_m(T,L)}{\mu_m(T,H)}\right)$.

The k factor can be acquired from the patient scan by measuring the transmitted intensity of the beam at points at which there is no bone. Once the k factor is determined, the equation can be solved to calculate the area bone density, d_m (B).

3 Different Features for Dual-Photon Beams: Voltage Switching Versus K-Edge Filtering

In most DEXA systems, the X-ray tubes used are standard tungsten anode tubes with focal spot sizes on the order of 0.5–1 mm^2. However, significant differences in the technologies are present in all DEXA systems on the market. Differences are seen in the irradiating X-ray energy levels, the filtering X-ray energy ranges, the creation of dual-photon beams, the types of photon detectors employed, the calibration methods used, the use of bone edge detection algorithms, how the region of interest ranges are calculated, and the reference databases used. For all of these reasons, it is difficult to make quantitative comparisons of BMD values from different DEXA systems unless a cross-calibration study has been done.

Considerable feature should be how to create dual-photon beams to acquire DEXA images. One method is the voltage switching system utilized by Hologic (Marlborough, MA, USA). The X-ray tube potential is switched rapidly from low (~70 kVp) to high (~140 kVp), alternating at 60 Hz with an internal rotating disc of additional beam-filtering materials. Low- and high-energy beams are alternated with very short pulses (~8.3 ms) during DEXA image acquisition. An aluminum filter is used to harden the low- and high-energy beams. Additional copper or brass is used to remove the low-energy part of the spectrum in high-energy beams and minimize the overlap between low and high X-ray spectra in dual-energy beams. The tube voltage switches, beam-filtering materials, and photon detectors are all electronically and mechanically synchronized to sequentially acquire low- and high-energy information during DEXA scans.

The other method is the K-edge filtering system employed by Norland (Fort Atkinson, WI, USA) and GE Lunar (Madison, WI, USA). The X-ray tube unit is

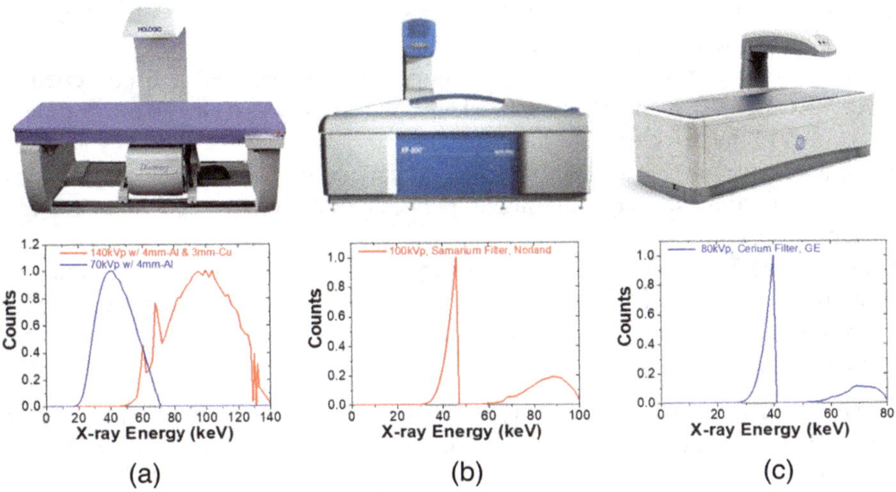

Fig. 2 Commercial DEXA systems with different approaches to creating dual-energy beams (top) and their representative energy spectra (bottom). (**a**) Discovery, Hologic, (**b**) XR-800, Norland, and (**c**) Prodigy, GE Lunar

operated in a steady direct current mode and the potential X-ray source is constant. One of several rare-earth filters is placed between the X-ray tube and the patient bed for the generation of low- and high-energy beams. These filters selectively absorb X-ray photons around the K-shell energy of the filter material due to the K-edge of the atomic structure of the element. With an X-ray voltage of 100 kVp, the samarium ($Z = 62$) filter module was used to separate high-energy from low-energy photons by the Norland Corporation. An X-ray tube with 80 kVp combined with a cerium ($Z = 58$) filter module by GE Lunar was used to split a single split X-ray spectrum into two components of low- and high-energy photons.

Figure 2 shows the representative DEXA systems and their simulated X-ray energy spectra to create dual-photon beams for each method. These representative energy spectra were generated using the tungsten anode spectral model using interpolating polynomial (TASMIP) algorithm [22]. In summary, each method generates different dual-energy beams and has inherent strengths and weaknesses. An additional discussion of these methods is presented in [23].

4 Different Features for Photon Counting Detectors: Direct Versus Indirect Mechanisms

Photon counting DEXA detectors must meet several property standards as follows:

1. Good energy resolution: The detector has to distinguish low- and high-energy from incident dual-energy X-ray.
2. High stopping power: All X-ray photons must be detected and converted to electric signals. It mainly is dependent upon density and the effective atomic number of the detector material.
3. High count rate: Detectors must be sufficient to handle incident photons. A shorter time constant of output pulse is preferred in high counting rate applications.
4. Short- and long-term stability: It is very important that the response must be uniform during DEXA scans. Although detector performance remains very stable over time, the gain of the detector and electronics invariably changes as the system ages.

There are currently two main types of DEXA detectors, direct and indirect. The main difference between the two types is the conversion process for incident photons.

For indirect detectors, a scintillation crystal converts the incident radiation to visible light photons that are subsequently detected by the photosensors in the form of electronic charges. These detectors are being used to detect the X-rays indirectly, and the total charge produced in the photosensors is proportional to the visible light emitted from a scintillator and the energy deposited by the incident radiation in the scintillator. The original DEXA scanner was the DPX system (Bravo and Duo) made by the GE Lunar Corporation. It employed an indirect detector, which was composed of a single NaI(Tl) crystal coupled with a PMT. This indirect detector enabled energy-selective radiation counting, and the output signal was proportional to the detected energy, providing affordable DEXA image acquisition. Light yield of NaI is approximately 38 photons/keV, and 2280 scintillation light is generated by interacting with incident radiation of 60 keV. Considering a 70% geometrical light collection efficiency at the scintillator and a 25% quantum efficiency at PMT, the entered light photons and converted initial photoelectrons in the photosensors are ~1600 photons and 400 photoelectrons, respectively. The Poisson statistical variation of this number is 20 (square root of 400), or the percentage uncertainty is 5% for 60 keV energy. This system continues to work well and achieve high precision in patient scanning. Despite its superior stopping power and detection efficiency of the scintillation detector, a number of technical problems may render its eternal implementation difficult. These include bulky size, multirow array scalability, poor energy resolution, and long-term stability. Also, there are clinical reasons to pursue an improved DEXA detector design, including reduced scan time and improved image resolution.

The easiest way to solve the technical hurdles and achieve the clinical requirements was to develop and employ a direct detector for a next-generation DEXA scanner. Currently, semiconductor detectors based on CZT or CdTe have been widely utilized on a commercial basis. Compared to the NaI(Tl)-PMT detector, the compactness of CZT allows a scalable design in two-dimensional arrays. Thus, it is possible to make multirow detectors for fan-beam scanners and reduce scan time. For direct detectors, a semiconductor can effectively absorb the incoming radiation and directly convert the X-ray energy into electrical signals large enough to permit the integration and counting of individual photons, which are read out by preamplifiers and backend electronics. Thus, it produces a signal proportional to the amount of absorbed radiation energy. Recent advances in flip-chip bonding and electronic packaging technology have facilitated direct X-ray detector assembly. Several direct detectors have been fabricated by bump-bonding CZT or CdTe detectors to a dedicated custom application-specific integrated circuit (ASIC) readout for various medical applications. Good energy resolution is one of the attractive features of improved image resolution. The conversion efficiency of X-ray photons to an electrical signal in CdTe and CZT can be much higher than that in scintillation detectors. Also, the created charges do not spread and do not escape to neighboring pixels in direct detectors, but the generated scintillation light is dispersed within the pixel volume and disappears by escape/absorption/extinction in the scintillation detectors. Thus, the overall SNR (signal-to-noise ratio) in the direct detection approach can be significantly better than that achievable by indirect detectors. The bandgap of these detectors is approximately 1.55 eV, and one ionization, creating an electron-hole pair, is produced per ~4.4 eV of radiation energy absorbed. Therefore, a 60 keV photon results in 13,600 electron-hole pairs. The Poisson statistical variation of this number is 117 (square root of 13,600), or percentage uncertainty is 0.8% for 60 keV energy. Considering a Fano factor of ~0.15, the statistical variation on 13,600 is actually closer to $0.15 \times 117 = 17$, and the theoretical energy resolution is 0.1%. In the current state of technology, it would be difficult to achieve this energy resolution in realistic systems where electron/charge trapping phenomena, current leakage, electronic noise, and temperature effects may influence the results. Experimental measurements revealed that CZT detectors could provide better energy resolution in a typical range of 2–5%, compared to ~20% by NaI-PMT detectors [24].

5 LYSO-GAPD Photon Counting Detector for DEXA Application

Semiconductor detectors such as CZT or CdTe, with excellent energy resolution, have been utilized widely on a commercial basis, as mentioned previously. As an alternative, X-ray diagnostic detectors, combining high-performance scintillation crystals with compact photosensors, have attracted interest for use in DEXA applications. The first commercial DEXA scanner was recently introduced [25–27].

A research team from the Chonnam National University (CNU), composed of the Research Center for Healthcare and Biomedical Engineering and the Medical Imaging System and Equipment Lab, has actively studied GAPDs as radiation detector photosensors since 2007. The characterization of DEXA scanners based on LYSO-GAPD photon counting detectors was conducted. The intrinsic performance of the DEXA detector module was characterized under a variety of conditions. Initial imaging studies were performed to demonstrate the feasibility of X-ray imaging with these prototypes [28–31].

5.1 LYSO-GAPD Photon Counting Detector Configurations

A wide range of inorganic scintillator materials is used to convert radiation (X-ray and gamma rays) to scintillation light for radiation detectors and medical applications. At present, lutetium-based oxide scintillation crystals are most commonly used in conventional radiation detection systems. Lutetium-yttrium oxyorthosilicate (LYSO) had a good stopping power for DEXA energy in the range of 20–100 keV. The estimated detection efficiency of >99.9% for 80 keV was adequate even with 2-mm-thick LYSO, whereas CZT should increase the thickness to 5 mm. Due to good stopping power of LYSO, a rotatable photon counting detector concept providing normal-resolution (NR) mode to the high-resolution (HR) mode could allow both clinical and research applications with one DEXA system (as described in Sect. 5.5).

The role of the photosensor is to convert the scintillation light generated in the scintillation crystals into a measurable electrical current. Various types of photosensors including photo multiplier tube (PMT), Avalanche photo-diode (APDs), and Geiger-mode avalanche photodiodes (GAPDs), have been utilized for the medical imaging systems. Recently, GAPDs also called solid-state photomultiplier (SSPM), silicon photomultiplier (SiPM), multipixel photon counters (MPPC), and micropixel avalanche photodiodes (MAPDs) have been utilized as next-generation photosensor in radiation detection systems, and many research groups have also demonstrated the usefulness of GAPDs in different medical applications. GAPDs consist of a densely packed matrix with many microcells (~1000–~10,000) ranging from 5×5 to 100×100 um^2 in size, and each microcell operates independently in a Geiger mode as an on/off switch for the photons. The amplitude of the output pulse is proportional to the total number of fired microcells, reflecting the number of absorbed photons. GAPDs provide several key properties [32, 33].

A LYSO-GAPD photon counting detector was consisted of 1×8 LYSO array with an individual crystal size of $3 \times 3 \times 2$ mm^3, arranged with a 3.2 mm pitch. The crystal pixels were polished on all faces and separated by white epoxy except for the surface facing GAPDs. The 1×8 GAPD arrays based on SiPM technology were used for the scintillation light readout. It had an active area of 3.07×3.07 mm^2 and 5676 microcells of 35×35 um^2. An individual LYSO pixel was coupled one to one to a separate pixel of the GAPD with optical grease. The black tape was applied as a shield from light.

5.2 *Readout Electronic Configurations for Photon Counting Detector*

The output signals from the DEXA detector were fed into custom-made current feedback charge-sensitive preamplifiers using high-gain operational amplifiers and then into shaping amplifiers for signal shaping and inversion. The amplified signals were digitized and recorded by the custom-made data acquisition (DAQ) system based on a field-programmable gate array (FPGA). The digitized output was postprocessed and converted into energy and counts. The GAPD was operated at a bias voltage of 30.0 V, exceeding the breakdown voltage by 2.5 V and corresponding to a gain of approximately 10^6. The detector module and readout electronics were operated at room temperature (25 °C) without additional cooling equipment.

The data acquisition system should be responsible for all digital signal processing, in both the energy resolving mode (for evaluating the intrinsic performance) and event counting mode (for characterizing the image performance). Normally, the energy of the radiation event could be defined by the voltage amplitude. The analog signals were fed into the analog-to-digital converter (ADC) board having a sampling rate of 125 MHz and an input range of −1.0–1.0 V. Then, these digitized signals were further processed in the FPGA block. Whenever a digital signal appears with a value larger than the threshold level, the FPGA block decides the next ~60 ns (longer than the rising portion of the event pulse) which belongs to the event pulses and then finds the maximum value in that period and writes it to the on-chip block RAM-based first-in-first-out (FIFO). Then, the next falling edge period is skipped.

In the energy resolving mode, the output data containing the pulse energy and arrival time information were recorded in the list-mode format (LMF) on a memory block. After the memory was filled, then the LMF data were transferred on to a hard disk. However, it would be difficult to completely process the tremendous number of incident photons because of the limited memory resources and dead time. There was a time gap between each cycle for processing, storing, and transferring output data. This issue frequently occurs with increasing the X-ray tube current.

To address the previously mentioned issues, the event counting mode could be employed for high photon flux in imaging acquisition. The analog signals were fed into four comparators blocks, including LL (low window at low energy), LH (high window at low energy), HL (low window at high energy), and HH (high window at high energy), to generate effective digital count signals if the analog signal voltage was above the specific threshold levels of each discriminator. Instead of storing each event with its energy value and arrival time, the event was counted before storing it in memory as the number of events in the predefined period. In this way, a counter block classifies incident photons into several energy bins based on their energy by comparing all pulses with several thresholds. The energy bins range from 2 to 6 in this readout electronics for photon counting detector. Each comparator offered a short propagation delay of ~3.5 ns. It had different threshold voltages adjusted by an external digital circuit (Fig. 3).

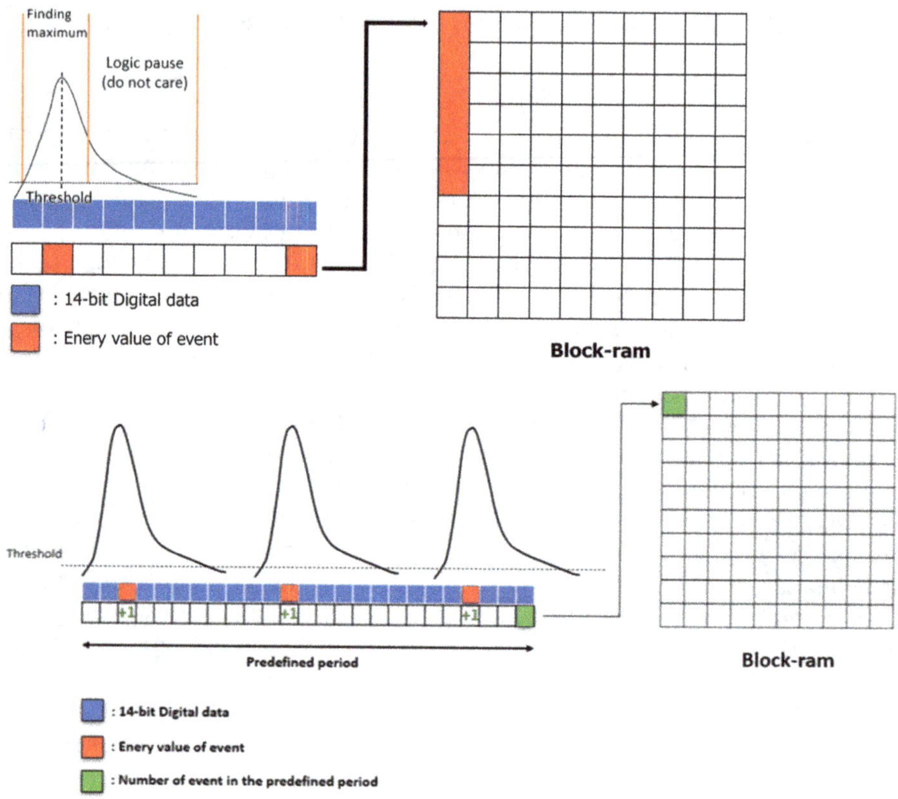

Fig. 3 Schematic diagram of energy-resolving mode (**a**) and event counting mode (**b**) for photon counting detector

5.3 *Experimental Setup*

The detector and X-ray tube were positioned at opposite sides. The collimator formed a fan-shaped beam of irradiation followed by a Ce-filter. The source-object distance (SOD) was 186 mm and the source-detector distance (SDD) was 584 mm.

The X-ray source consisted of an integrated monoblock system (XRB80; Spellman, Germany) with a high-voltage power supply (HVPS), filament supply, X-ray tube, and oil encapsulant. The X-ray tube, in a sealed tank, had a fixed tungsten anode and a focal spot size of 0.5 mm. The emitted X-ray beam angle was about 75 ° with a fan-shaped geometry. The maximum tube voltage and tube current were 80 kVp and 1.25 mA, respectively. The X-ray source was operated in a continuous mode with an intrinsic beam filter, including an amorphous of 3.3 mm, oil of 8 mm, beryllium of 0.8 mm, and glass 1 mm thick [34].

A lead collimator with a 3 mm thickness was installed on the X-ray tube. In addition, a K-edge cerium filter module was used to separate high-energy from low-energy photons by selectively absorbing X-ray photons around the K-shell energy of the filter material [35, 36]. It could generate effective energy peaks of 30 keV and 60 keV. The K-edge filter module consisted of six cerium sheet components 0.1 mm thick and a 50 mm × 50 mm surface area (Alfa Aesar, Ward Hill, MA, USA). It had different filter thicknesses ranging from 0.1 mm to 0.6 mm and allowed for X-ray beam intensity adjustment according to the patient thickness.

5.4 *Characterization of the Intrinsic Performance of the DEXA Detectors*

Figure 4a shows the measured gamma-ray energy spectra, which were acquired using Am-241 (60 keV) and Co-57 (122 keV) for 10 min for LYSO-GAPD detector module. The measured photopeak positions of Am-241 and Co-57 were 1100 mV and 2200 mV, respectively. The energy resolutions of Am-241 and Co-57 were 26.8% and 17.2%, respectively. It should be noted that the energy resolution value (17.2%) at 122 keV was considerably improved compared to that seen in previous studies (27%) [37, 38]. Figure 4b shows the dual-energy X-ray spectra for six different thicknesses of K-edge cerium filter. The X-ray tube was operated at 80 kVp and 0.1 mA, and filtered dual-energy spectra were acquired for 1 min. Each spectrum was normalized with respect to its maximum count value. The dual-energy peaks were located at around 650 mV and 1250 mV. They were clearly isolated, and the peak-to-valley ratio as a function of K-edge cerium filter thickness increased from 2.8 to 14.2 at low energy and from 1.7 to 3.8 at a high energy in the LYSO-GAPD detector.

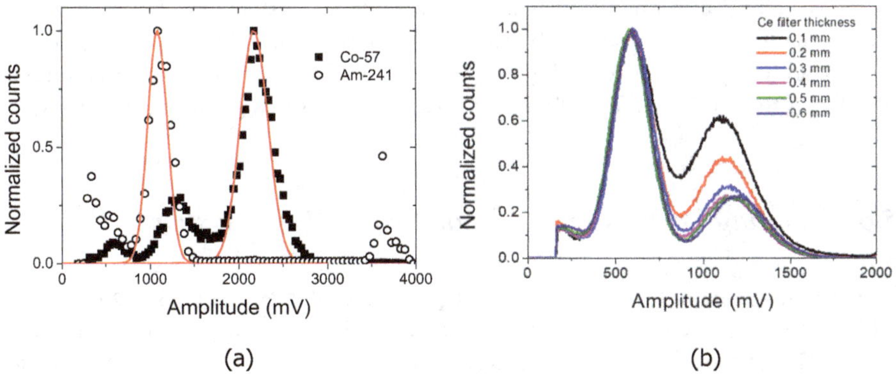

Fig. 4 Spectral response for gamma-ray and X-ray. (**a**) Acquired energy spectra with gamma-ray sources of Am-241(○) and Co-57(■) and (**b**) dual-energy X-ray spectra with different Ce-filter thicknesses

5.5 *Characterization of the Rotatable LYSO-GAPD DEXA Detector*

Rotatable LYSO-GAPD DEXA detector providing improved spatial resolution could be the innovative benefits of an X-ray diagnostic detector combining high-performance scintillation crystals with a compact photosensor in DEXA applications. The geometric configurations could be changed from the normal-resolution (NR) mode to the high-resolution (HR) mode without sacrificing intrinsic detection efficiency. The versatile transformation in the HR mode was enabled by rotating the module (90 degrees) on its own axis, as depicted in Fig. 5a. After this rotation, the detection area and attenuation thickness changed from 3 mm × 3 mm to 2 mm × 3 mm and from 2 mm to 3 mm, respectively. Therefore, the intrinsic resolution in this mode was improved by 33.3% compared to the NR mode. In contrast, conventional CZT or the CdTe DEXA detector should have a relatively thicker attenuation length of greater than 3 mm, which makes it difficult to obtain an HR image by the rotating DEXA detector [30]. No considerable changes of peak positions and peak-to-valley ratio were observed in the characterization of the rotatable LYSO-GAPD DEXA detector by measuring the dual-energy X-ray spectra for the NR mode and the HR mode. In contrast, considerable improvements in phantom images such as overall contrast and fine-spot detectability were observed in the HR mode irradiating X-ray on side surface of DEXA detector (Fig. 5b). It should be noted that spatial resolution was improved by reducing the detection area from 3 mm × 3 mm to 2 mm × 3 mm in the HR mode, but the count rate was also decreased by ~33%, as expected.

These versatile transformations of the geometric configurations would allow LYSO-GAPD photon counting detector to utilize for both clinical and research applications with one DEXA system. The NR mode could achieve spatial resolution and sensitivity equivalent to a dedicated clinical DEXA system. The HR mode could provide better spatial resolution by sacrificing more scanning time (~ 1.3 times) in the research application requiring high-resolution images. These were additional benefits of the LYSO-GAPD DEXA detector, because of its good stopping power.

5.6 *Imaging Performance of Fan-Beam DEXA System*

The imaging capability and dynamic range were evaluated using BMD phantoms with different thicknesses of aluminum and acryl to simulate the bone and soft tissue. The encapsulated spine phantom (Fig. 6a) consisted of an aluminum bar block with different thicknesses ranging from 4 mm to 12 mm with a step of 2 mm in a block of solid acrylic with a thickness of 120 mm. The aluminum step-wedge phantom (Fig. 6b) with 15 steps in increments of 0.4 mm had a surface area of 10 mm × 50 mm for each step and different thicknesses ranging from 1.2 mm to 6.8 mm. The lumbar spine phantom (Fig. 6c), which simulated the human spine

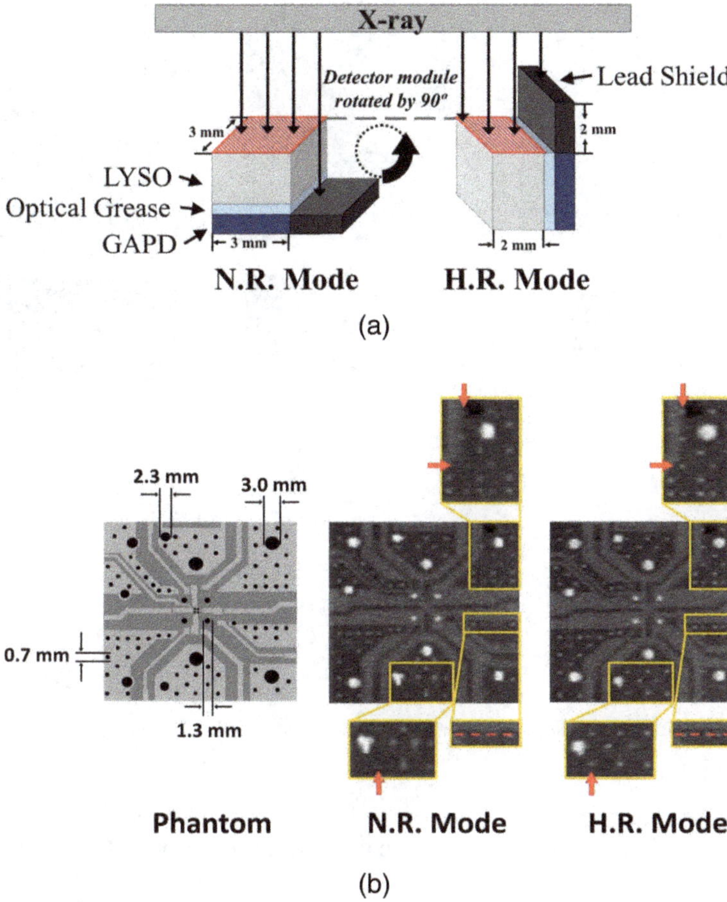

Fig. 5 Characterization of the rotatable LYSO-GAPD DEXA detector. (**a**) The rotatable LYSO-GAPD DEXA detector could be transformed from the normal-resolution (NR) mode to the high-resolution (HR) mode by rotating the detector on its own axis. (**b**) DEXA images were obtained in the NR mode (middle) and the HR mode (right) for the phantom with different hole sizes and line patterns (left)

from T12 to L5, consisted of aluminum of different thicknesses in the range of 4.0–9.5 mm. A fan-beam oriented parallel to the short axis of the phantom was irradiated, and a transverse pattern across the phantom was scanned. The X-ray source and detector were moved toward the short axis in 1 mm steps. The data were acquired for 0.05 seconds for each interval. The counted photons were separated into low- and high-energy bands. The scan time and pixel resolution were 70 seconds and 1.05 mm × 1.05 mm, respectively. These dual-energy images were converted into density images using a conventional dual-energy subtraction algorithm, as described in Fig. 1.

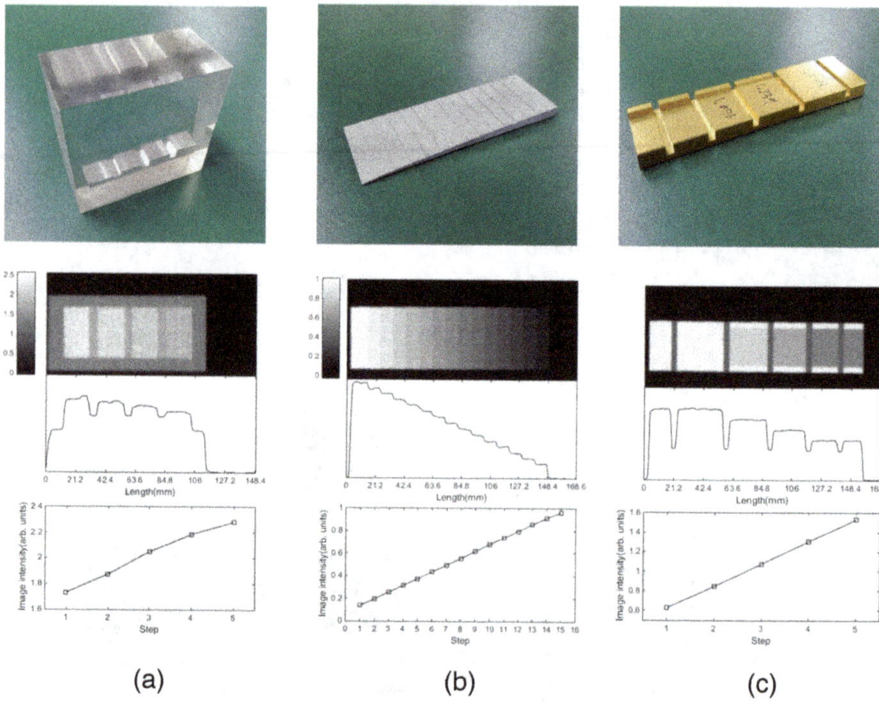

Fig. 6 Picture of DEXA phantoms and image performance by using LYSO-GAPD detector

Various phantoms were well imaged, and the shape, length, edge, and steps were well characterized. The image intensity increased linearly from 0.62 to 1.58 for the lumbar spine, from 0.14 to 0.96 for the step wedge phantom, and from 1.7 to 2.3 for the encapsulated spine. The pixel resolution achieved 1.05 mm × 1.05 mm, which was limited by the geometric magnification factor of the DEXA scanner and the pixel dimension of the LYSO-GAPD detector used in this study.

6 Summary

The chapter introduced DEXA instrumentation that has seen an exponential increase in use in clinical practice for bone densitometry as the number of older people has increased. DEXA technology has numerous strengths compared to other diagnostic methods, including its availability, short scan times, minimal radiation exposure, and excellent precision. In elderly adults, DEXA BMD is also a sufficiently robust predictor of osteoporotic fractures and can be used to define the disease.

DEXA detector converts incident X-ray photons to electrical signals, and then each pulse can be compared with the four different energy thresholds. The events falling within the selected energy window are binned into each low- and high-energy bands. In this manner of operation, DEXA detector works as a photon counting detector providing information on the number of counted photons separately for each energy bin.

The fundamental principle of DEXA is the measurement of differences in the transmission and attenuation of tissue and bone for low and high energies. There are different approaches to project the three-dimensional human body onto the two-dimensional image: creating low- and high-energy beams using voltage switching and K-edge filtering systems and converting incident X-ray photons to electrical signals using indirect and direct detectors.

CZT has been considered a representative detector for DEXA implementation due to its good stopping power, excellent energy resolution, and well-established techniques. The recent investigations of alternative LYSO-GAPD photon counting detector in DEXA application were described. The detail characterization of LYSO-GAPD detector was discussed. Moreover, the innovative benefits of an X-ray diagnostic detector combining high-performance scintillation crystals with a compact photosensor were characterized by the rotatable detector design. Various phantoms were well imaged in the fan-beam DEXA system based on LYSO-GAPD detector. These photon counting detectors could provide several potential merits including a relatively high-count rate owing to its fast decay time and small crystal volume due to its good stopping power. In addition, they could be potentially affordable and cost-effective because of the availability of suppliers for the components, enabling mass production.

References

1. National Institutes of Health, NIH consensus development panel on osteoporosis prevention, diagnosis, and therapy. Highlights of the conference. South. Med. J. **94**, 569–573 (2001)
2. https://www.nof.org/about-us/nof-background/
3. J.A. Kanis, *Assessment of Osteoporosis at the Primary Health Care Level* (WHO Collaborating Centre for Metabolic Bone Diseases, 2007)
4. L.J. Melton III, E.J. Atkinson, M.K. O'Connor, W.M. O'Fallon, B.L. Riggs, Bone density and fracture risk in men. J. Bone Miner. Res. **13**, 1915–1923 (1998)
5. L.J. Melton III, E.A. Chrischilles, C. Cooper, A.W. Lane, B.L. Riggs, How many women have osteoporosis? J. Bone Miner. Res. **20**, 886–892 (2005)
6. J.A. Kanis, O. Johnell, A. Oden, I. Sernbo, I. Redlund-Johnell, A. Dawson, D. Laet, B. Jonsson, Long-term risk of osteoporotic fracture in Malmö. Osteoporos **11**, 669–674 (2000)
7. I. Fogelman, G.M. Blake, Different approaches to bone densitometry. J. Nucl. Med. **41**, 2015–2025 (2000)
8. R.B. Mazess, Advances in single-and dual-photon absorptiometry. Osteoporosis **1**, 57–63 (1984)
9. R.B. Mazess, H.S. Barden, Single-and dual-photon absorptiometry for bone measurement in osteoporosis. Osteoporosis, 73–80 (1987)
10. R.B. Mazess, H. Barden, J. Vetter, M. Ettinger, Advances in noninvasive bone measurement. Ann. Biomed. Eng. **17**, 177–181 (1989)
11. R. Mazess, B. Collick, J. Trempe, H. Barden, J. Hanson, Performance evaluation of a dual-energy X-ray bone densitometer. Calcif. Tissue Int. **44**, 228–232 (1989)
12. D.J. Sartoris, D. Resnick, Dual-energy radiographic absorptiometry for bone densitometry: Current status and perspective. AJR Am. J. Roentgenol. **152**, 241–246 (1989)

13. P.C. Johnson, S.G. Jhingran, Precision of dual photon absorptiometry measurements. J. Nucl. Med. **27**, 1362–1365 (1986)
14. H.W. Wahner, W.L. Dunn, M.L. Brown, R.L. Morin, B.L. Riggs, Comparison of dual-energy x-ray absorptiometry and dual photon absorptiometry for bone mineral measurements of the lumbar spine. Mayo Clin. Proc. **63**, 1075–1084 (1988)
15. F. Cosman, S.J. de Beur, M.S. LeBoff, E.M. Lewiecki, B. Tanner, S. Randall, R. Lindsay, Clinician's guide to prevention and treatment of osteoporosis. Osteoporos. Int. **25**, 2359–2381 (2014)
16. World Health Organization, *Assessment of Fracture Risk and its Application to Screening for Postmenopausal Osteoporosis: Report of a WHO Study Group [Meeting Held in Rome from 22 to 25 June 1992]* (WHO, 1994)
17. FDA 510k K871553, October 10, 1987
18. T.L. Kelly, D.M. Slovik, R.M. Neer, Calibration and standardization of bone mineral densitometers. J. Bone Miner. Res. **4**, 663–669 (1989)
19. J. Wear, M. Buchholz, R.K. Payne, D. Gorsuch, J. Bisek, D.L. Ergun, J. Grosholz, R. Falk, CZT detector for dual-energy x-ray absorptiometry (DEXA). SPIE **4142**, 175–188 (2000)
20. T. Takahashi, S. Watanabe, Recent progress in CdTe and CdZnTe detectors. IEEE Trans. Nucl. **48**, 950–959 (2001)
21. https://www.gehealthcare.com/products/bone-and-metabolic-health/prodigy
22. G.M. Blake, I. Fogelman, Technical principles of dual energy x-ray absorptiometry. J. Nucl. Med. **27**, 210–228 (1997)
23. H. Wahner, P. Steiger, E. Von Stetten, Instruments and measurement techniques, in *The Evaluation of Osteoporosis: Dual Energy X-Ray Absorptiometry in Clinical Practice*, ed. by H.W. Wahner, I. Fogelman, (Springer, 1994)
24. G.F. Knoll, *Radiation Detection and Measurement* (Wiley, 2000)
25. https://www.hologic.com/
26. https://www.gehealthcare.com/
27. http://www.swissray.com/SRI/
28. J. Yang, B.J. Min, J. Kang, A feasibility study of LYSO-GAPD detector for DEXA applications. J. Instrum. **15**, P05017 (2020)
29. J. Yang, B.J. Min, J. Kang, A feasibility study of GAGG-GAPD detector for development of DEXA. J. Magn. **24**, 739–743 (2019)
30. H. Heo, J. Yang, J. Kang, Rotatable LYSO-GAPD DEXA detector for providing improved spatial resolution. J. Instrum. **16**, P12012 (2021)
31. M.-D. Hoang, J. Kang, Multichannel FPGA-based data acquisition system for fan-beam dual energy X-ray absorptiometry. J. Magn. **27**, 1–8 (2022)
32. J. Kang, Y. Choi, K.J. Hong, J.H. Jung, W. Hu, Y.S. Huh, H. Lim, B.T. Kim, A feasibility study of photosensor charge signal transmission to preamplifier using long cable for development of hybrid PET-MRI. Med. Phys. **37**, 5655–5664 (2010)
33. J. Kang, Y. Choi, K.J. Hong, W. Hu, J.H. Jung, Y. Huh, B.T. Kim, A small animal PET based on GAPDs and charge signal transmission approach for hybrid PET-MR imaging. J. Instrum. **6**, P08012 (2011)
34. https://www.spellmanhv.com/ko/high-voltage-power-supplies/XRB80N100
35. N.J. Crabtree, M.B. Leonard, B.S. Zemel, Dual-energy X-ray absorptiometry, in *Bone Densitometry in Growing Patients*, (Springer, 2007), pp. 41–57
36. J.M. Boone, J.A. Seibert, An accurate method for computer-generating tungsten anode x-ray spectra from 30 to 140 kV. Med. Phys. **24**, 1661–1670 (1997)
37. F. Habte, P.D. Olcott, C.S. Levin, A.M. Foudray, Prototype parallel readout system for position sensitive PMT based gamma ray imaging systems. IEEE Trans. Nucl. Sci. **54**, 60–65 (2007)
38. P.D. Olcott, J.A. Talcott, C.S. Levin, F. Habte, A.M.K. Foudray, Compact readout electronics for position sensitive photomultiplier tubes. IEEE Trans. Nucl. Sci. **52**, 21–27 (2005)

Duo-Spectral Imaging with Multilayered Energy-Integrating Detectors

Ho Kyung Kim

1 Introduction

Consider an X-ray *projection* of an object composed of linear-attenuation coefficients, $\mu(\mathbf{x})$, onto a flat-panel detector (FPD). The detector pixel signal (i.e., image signal) for X-ray beam attenuation through a ray, l, connecting the X-ray source and a pixel positioned at $\mathbf{s}$ in the FPD, can be expressed as

$$d(\mathbf{s}) \approx cb_x b_y \int_{\mathrm{kV}} q_0(\mathbf{s};\, E) e^{-\int_{l(\mathbf{x})} \mu(E,\mathbf{x})\, \mathrm{d}l} D(E)\, \mathrm{d}E, \tag{1}$$

where $\mathbf{x}$ and $\mathbf{s}$ denote the object and detector pixel coordinates, respectively, $q_0(E)$ indicates the X-ray spectrum (quanta mm^{-2} keV^{-1}), $D(E)$ represents an energy-dependent detector pixel response, assumed to be uniform across $\mathbf{s}$, $b_x \times b_y$ symbolizes the pixel size, and scalar c relates the energy measurement to the digital output. We mapped $\mu(\mathbf{x})$ onto the $\mathbf{s}$ domain using $q_0(E)$, as expected, aiming to obtain $\mu(\mathbf{s})$; however, $d(\mathbf{s})$ and $\mu(\mathbf{s})$ are considerably divergent. In particular, extracting $\mu(\mathbf{s})$ from $d(\mathbf{s})$ is challenging.

Equation (1) indicates that $\mu(\mathbf{x})$ is a prominent factor in the projection in *space-* and *energy-averaged* forms. The projection property of μ-averaging in space can conceal lesions with up and/or down normal structures or falsely depict superimposed normal structures as abnormalities. The property of μ-averaging in terms of energy can yield the same μ for different materials. Computed tomography (CT) and tomosynthesis "de-average" μ in space, while dual-energy imaging (DEI)

H. K. Kim (✉)
Radiation Imaging Laboratory, School of Mechanical Engineering, Pusan National University, Busan, Republic of Korea
e-mail: hokyung@pusan.ac.kr

K. Iniewski, H. Gadey (eds.), *Emerging Radiation Detection*,
https://doi.org/10.1007/978-3-031-63897-8_4

or multi-energy imaging de-averages μ in energy. A summary of the development history of these techniques is reported in the literature [1].

The image quality of DE images, particularly in terms of noise performance, has remained inferior to that of conventional radiographic images, as the DE-reconstruction procedure amplifies noise [2, 3]. In contrast, DE-based approaches isolate the *signal of interest* by suppressing the background signal, thereby enhancing its *conspicuity*. Therefore, the high signal-to-noise performance of a detector is fundamental for a satisfactory DEI, considering unavoidable noise amplification. As the detective quantum efficiency (DQE) represents the signal-to-noise ratio (SNR) power in images regarding a dose used for imaging, this metric is generally employed as an objective function for optimizing imaging techniques, including DEI [4–6].

Consider imaging target (t) material embedded in a uniform background (b). Assuming Poisson statistics, the corresponding contrast-to-noise ratio (CNR) in terms of attenuated fluence is expressed as $\text{CNR}^2 = \bar{q}_b M_t$, where $M_t = (1 - \bar{q}_t/\bar{q}_b)^2/(1 + \bar{q}_t/\bar{q}_b)$; $\bar{q}_j$ denotes the measured signal (or variance). For the interaction of the X-ray quantum with energy E in the detector, $D(E)$ in Eq. (1) can be factorized as the product of the quantum efficiency and energy absorption: $D(E) = \alpha(E)\varpi_1(E)$, where $\varpi_m(E) = \int_0^\infty \varpi^m(\varepsilon)p(\varepsilon, E)\,d\varepsilon$ and $p(\varepsilon, E)$ represents the probability density of depositing energy ε, given an interacting quantum with energy E. Accordingly, Tapiovaara and Wagner [7] demonstrated that

$$\text{DQE} \leq \left\langle \alpha(E)\frac{\varpi_1^2(E)}{\varpi_2(E)} \right\rangle_{\bar{q}_b(E)M_t(E)}, \tag{2}$$

where $\langle \cdots \rangle_{S(E)}$ implies the spectrum-weighted average. Here, $\varpi_1^2(E)/\varpi_2(E)$ is the Swank factor [8]. In the case of *photon-counting* detectors (PCDs), $\varpi(\varepsilon)$ may be constant, and the upper limit of DQE is α [9]. The *energy-integrating* detectors are less efficient than PCDs in terms of DQE [7, 9, 10]. However, PCDs are also not free from the Swank noise [11].

While the switching-kV double-shot or double-exposure technique dominates in DE radiography, the single-shot technique using a PCD renders superior DE image quality owing to its inherently better DQE performance. However, because the PCD suffers from lower quantum statistics in each energy bin, the choice of binning and weighting schemes can significantly influence the performance of the DEI [12]. Furthermore, the current research on PCDs is primarily restricted to CT applications rather than radiography [13–16].

As a *niche* approach between conventional kV-switching and PCD techniques for DE radiography, we demonstrate the potential of the "single-shot" DEI using a double-layered FPD structure. This sandwich-style detector was designed by configuring two detectors in tandem. The top or front detector acquires an image in a conventional manner that acts as a low-energy (LE) image. The image acquired from the bottom or rear detector utilizes the beam-hardening effect through the front-detector layer. Thus, relatively high-energy (HE) X-ray photons can be measured, and an HE image can be constructed. Sandwich detectors are immune to motion artifacts [17].

The concept of a sandwich detector is not new. For example, in 1986, Ishigaki et al. [18] reported a technique for single-shot DE subtraction radiography with two photostimulable phosphor plates separated by a filter. In 2005, this concept was employed for CT, called dual-layer CT [19]. In our laboratory, a small-scale sandwich detector was developed using two commercially available phosphor-coupled photodiode arrays that demonstrated DE mouse images [20, 21]. A prototype large-area (43 × 43 cm) sandwich detector using CsI:Tl-coupled amorphous silicon (*a*-Si) flat panels was introduced and investigated for its DE capabilities in radiography, fluoroscopy, and cone-beam CT (CBCT) [22].

The scope of this chapter is limited to our FPD-based sandwich detectors and findings pertaining to these devices. Section 2 presents two versions of sandwich detectors using phosphor-coupled photodiode arrays in terms of their fields of view. Section 3 elucidates the analytic models used to describe the signal and noise characteristics of the sandwich detector. Cascaded-systems analysis (CSA) is applied to the sandwich detector to determine the modulation-transfer function (MTF), noise-power spectrum (NPS), and DQE of each detector layer (Sect. 3.1) [23]. Furthermore, a simple linear model is presented to assess the experimental DQE, including the MTF and NPS, in the DE images (Sect. 3.2) [24, 25]. Section 4 presents the DE radiographic and tomographic images of a postmortem mouse [21, 26] obtained using sandwich detectors. Finally, this chapter concludes by discussing recent research on the development of sandwich detectors in Sect. 5.

2 Laboratory-Scale Designs of Sandwich-Style Detectors

2.1 Small-Area Design

A novel sandwich-style single-shot detector has been developed by stacking two FPDs [20, 21]. Two detector layers based on phosphor-coupled photodiode pixel arrays were configured as sandwich structures. The same photodiode array, manufactured using complementary metal–oxide–semiconductor (CMOS) technology (RadEye1™, Teledyne Rad-icon Imaging Corp., Sunnyvale, CA, US), was used for designing both detector layers. The array featured a 0.048-mm pixel arranged in a 512 × 1024 format to yield an imaging area of approximately 25 × 50 mm. In particular, Gd_2O_2S:Tb phosphor-based screens [27] have been used to convert X-rays into light. A relatively thin phosphor screen (~34 mg cm^{-2}) was employed for the construction of the front (or top) detector, considering the transmission of HE X-ray photons, whereas a thick screen (~67 mg cm^{-2}) was leveraged for the rear (or bottom) detector, considering the efficient detection of HE X-ray photons. To increase the spectral energy separation (ΔE) between the two-detector-layer measurements, a thin Cu sheet was placed between them, as illustrated in Fig. 1(a-1).

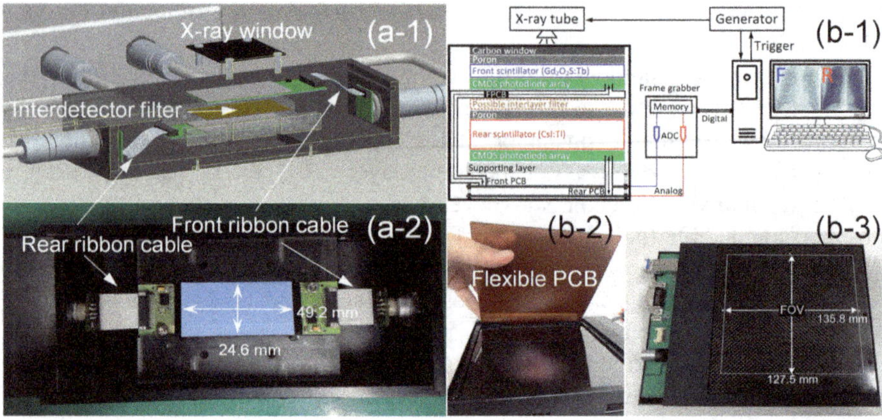

Fig. 1 Illustration of two versions of sandwich-detector designs. (**a**) Small-area design: (**a-1**) CAD drawing and (**a-2**) photograph indicating the inside of the detector. (**b**) Large-area design: (**b-1**) schematic diagram showing the structural architecture of the detector and (**b-2**) photograph showing the flexible PCB and (**b-3**) appearance of the detector

The two detector layers were placed in a lightproof Al box with a polycarbonate window. Figure 1(a-2) displays an internal view of the detector housing.

The maximum frame rate of the sandwich detector was 4.5 frames per second. For the X-ray exposure, the sandwich detector simultaneously generated a single pair of projection images. We utilized an electronic module (Shad-O-Snap™, Teledyne Rad-icon Imaging Corp., Sunnyvale, USA), in which two-channel analog signals from the sandwich detector were processed, digitized, and prepared for transmission to a personal computer (PC). This electronic module offers an Ethernet interface for the PC. Refer to references [20, 21] for more details.

2.2 *Large-Area Design*

We rebuilt a sandwich detector with a larger imaging area of ~136 × 128 mm using the same indirect-conversion detector scheme. The front phosphor was a commercial Gd_2O_2S:Tb screen with a thickness of 0.1 mm (DRZ™, Mitsubishi Chemical, Japan), while the rear phosphor was a CsI:Tl with a thickness of 0.28 mm (Rayence Co., South Korea). The same CMOS photodiode-array panels (Xmaru1314CF™, Rayence Co., Korea) were utilized to detect light quanta emitted from the phosphor layers. The photodiode array was based on 0.0495-mm active-pixel sensor technology in a 2744 × 2567-pixel format. The photodiode array could be operated in 2 × 2 binning mode with an effective pixel pitch of 0.099 mm. The frame readout time was ~30 ms, and a faster 4 × 4 pixel-binning operation with a pixel pitch of 0.198 mm was feasible, resulting in a readout time of ~10 ms.

The layer configuration of the sandwich detector is illustrated in Fig. 1(b-1). Considering a Si wafer is fragile, a photodiode-array wafer is typically realized on a rigid substrate, under which a printed circuit board (PCB) is placed for analog signal processing and interfacing with an operating computer. The key feature of the large-area design involves using a bare CMOS wafer-scale chip in the front layer without a substrate. Electrical connections to the PCB for pixel control and signal readout are made through a flexible PCB layer, as shown in Figs. 1(b-1) and (b-2). The thickness of the flexible PCB was equivalent to ~0.1 mm of Cu sheet. The interface between the two-detector PCBs and the host PC was realized using a frame grabber, in which two-channel analog signals from the sandwich detector were simultaneously processed, digitized, and prepared for transmission to a PC. Figure 1(b-3) shows the appearance of the developed detector.

The imaging performance of the large-area sandwich detector outperformed that of the small-area ones described in Sect. 2.1. For example, the DQE(0) values of the front- and rear-detector layers were measured as ~0.26 and 0.29, respectively, at the tube voltage of 60 kV and the exposure level of ~1 mR. Refer to the reference [28] for more information regarding the developed large-area sandwich detector.

3 Modeling Sandwich-Detector Responses

3.1 Cascaded-Systems Analysis

The transmission of X-ray photons through the front layer is a prerequisite for the functioning of a sandwich detector in DEI, resulting in additional noise in the detector. In a phosphor-coupled photodiode detector, X-ray photons that remain "unattenuated" by a phosphor can interact *directly* with the photodiode layer [29], dependent on the finite thickness of the phosphor, which is generally selected based on the tradeoff between X-ray absorption (quantum efficiency) and optical quantum scattering (spatial resolution). The probability of this direct interaction between X-ray photons and the photodiode is low, given that the photodiode active region is usually extremely thin (~ μm). However, the number of secondary quanta (i.e., charge carriers) per directly interacting X-ray photon is larger than that produced by conventional *indirect* interactions (i.e., charge carriers in the photodiode induced by the optical quanta generated in the phosphor) [30]. The average energy required to generate a single optical quantum in a phosphor is ~18 eV, whereas that required to create an electron–hole pair in Si is ~3.6 eV. Kim [31] demonstrated that the noise-power densities resulting from these direct interactions are independent of the spatial frequencies, which is detrimental, particularly in the high-spatial-frequency region where the number of secondary quanta decreases.

As the sandwich detector requires X-ray transmission through the front-detector layer, the direct interaction of X-ray photons in the active regions of the photodiode is inevitable, with corresponding potential increments in the image signal and noise.

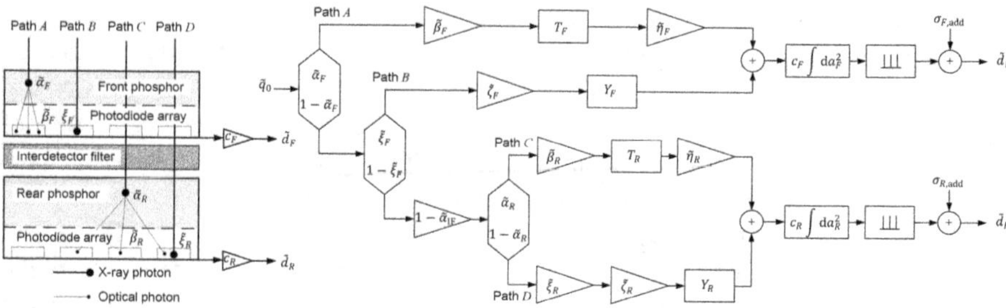

Fig. 2 (Left) Schematic illustration of signal detection in the flat-panel sandwich detector. (Right) Cascaded-model block diagram describing signal and noise propagation in the flat-panel sandwich detector. Path *A* conveys the conversion of X-ray quanta into optical quanta in the front phosphor and their detection in the front photodiode array, where α_F is the probability of X-ray interaction in the front phosphor, β_F is the mean conversion gain to optical quanta, T_F denotes the optical quantum spreading in the front phosphor, and η_F is the probability of an optical quantum liberating an electron–hole pair (charge carriers) in the Si photodiode. The subsequent integral ($c_F \int da_F^2$) represents the integration of all liberated charge carriers in the detector element (a_F^2), followed by sampling (III) and addition of readout noise ($\sigma_{F,\text{add}}$). Path *B* denotes the detection of charge carriers released by direct interactions in the photodiode array. Considering the fraction, $1 - \alpha_F$, of the incident X-ray quanta that do not interact in the front phosphor, a fraction, ξ_F, interacts directly in the photodiode, liberating ζ_F charge carriers with each interaction. The spatial spreading of charge carriers is characterized by Y_F. Among the X-ray photons passing through the front detector, a fraction, $1 - \alpha_{\text{IF}}$, passes through the interdetector metal filter. Path *C* represents the charge carriers liberated in the rear detector from X-ray interactions in the rear phosphor, and Path *D* describes the direct X-ray interactions in the rear photodiode. The overhead tilde (e.g., $\tilde{x}_j$) designates a random variable. Subscripts *F* and *R* represent the front- and rear-detector layers, respectively

Han et al. [21] developed a cascaded-systems model that describes the conversion of $\tilde{q}_0$ (mean $\overline{q}_0$) incident X-ray photons per mm^2 to signal $\tilde{d}$ (mean $\bar{d}$) per pixel, as illustrated in Fig. 2. The authors presented the signal and noise in each detector layer in terms of mean pixel values and their variances. Kim et al. [23] extended the CSA to represent the spatial-frequency-dependent MTF, NPS, and DQE of each detector layer, validated with the measured MTF(u), NPS(u), and DQE(u), where u represents the spatial frequency corresponding to the space variable x.

Figure 3 shows the CSA examples for the performance of each detector layer configuring the small-area sandwich detector as functions of kV and t_{IF} (thickness of the interdetector Cu filter), i.e., the main parameters of sandwich-detector operation and design, respectively. As illustrated in Figs. 3(a-1) and (a-2), the large-area gain of front detector G_F was nearly independent for a given range of kV and t_{IF}. In contrast, the rear-detector gain G_R increased with kV because of the corresponding increments in the transmittance τ_R through the front-detector layer, including t_{IF}, increasing the interactions in the rear detector, whereas it decreased with t_{IF} because of decreasing α_R and τ_R.

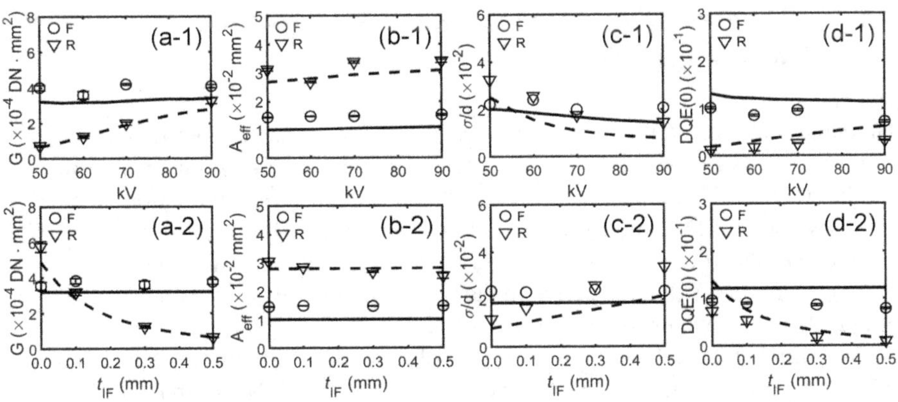

Fig. 3 Summary of the performance of each detector layer configuring the sandwich detector (the small-area design). Columns (**a**)–(**d**) represent the gain, effective aperture, relative noise (i.e., the inverse of SNR), and zero-frequency DQE or DQE(0), respectively. Row (1) demonstrates the performance as a function of kV and row (2) as a function of interdetector-filter (Cu) thickness t_{IF}. The symbols, including error bars, represent the measurements and the line plots represent the CSA calculation results. The upper row plots correspond to the sandwich detector with $t_{\mathrm{IF}} = 0.3$ mm, and the lower row plots correspond to the measurements for 60 kV. All the measurements were performed for similar exposure levels of ∼20 mR at the entrance surface of the sandwich detector

The influences of kV and t_{IF} on $A_{F,\mathrm{eff}}$ and $A_{R,\mathrm{eff}}$ were nearly negligible, as shown in Figs. 3(b-1) and (b-2). Here, $A_{j,\mathrm{eff}}$ represents the *effective aperture*, defined as the reciprocal of the volume under the squared 2D MTF, which describes the spatial resolving power [32]. The CSA estimated a slight increase in $A_{F,\mathrm{eff}}$ with increasing kV, possibly owing to the relative increase in the contribution of charge diffusion in the photodiode array $Y_F(u)$ to $\mathrm{MTF}_F(u)$.

The effects of kV and t_{IF} on detector noise are summarized in Figs. 3(c-1) and (c-2), respectively. The effects on the front-detector layer were almost negligible. In contrast, the relative noise of the rear-detector layer decreased with increasing kV and increased with t_{IF}. According to a detailed analysis [23], direct interaction-induced noise is the dominant noise source of the front detector, whereas in the rear detector, the magnitude of the direct noise is slightly larger than or comparable to that of the indirect noise. Therefore, the noise relative to the signal of the rear detector (i.e., $\sigma_R/\overline{d}_R$) is sensitive to the behavior of $\overline{d}_R$ for variations in kV and t_{IF}. The dynamic trends of the relative noise corresponding to the rear detector with increasing kV and t_{IF} are opposite to those of the gains, as shown in Fig. 3(a-1) and (a-2).

The DQE(0) performances of both detector layers are shown in Fig. 3(d-1) and (d-2), and the dynamic trends with increasing kV and t_{IF} are similar to those of the gain performance.

3.2 An Experimental Framework for Assessing the Dual-Energy DQE

The sandwich detector may be regarded as a single detector outputting multiple images, such as LE, HE, and DE images, for an incident X-ray fluence $\overline{q}_0$ (mm^{-2}), similar to that of multichannel PCDs. The weighted log-subtraction method is typically used for DE reconstruction, where the average DE signal from a sandwich detector is related to the front (F)- and rear (R)-detector signals such that $\overline{d}_{DE} = \overline{d}_F^w/\overline{d}_R$, where w is the weight parameter determining the contribution of the front- or rear-detector images to the reconstruction.

Assuming that the detector responses, including the DE-reconstruction operation, are linear and shift-invariant, we can define the MTF, NPS, and DQE for the images obtained from a sandwich detector. The resulting formulas are represented as follows, omitting the detailed derivation procedures:

$$\mathrm{MTF}(u) = \left| \frac{w\mathrm{MTF}_F(u) - \mathrm{MTF}_R(u)}{w-1} \right|, \tag{3}$$

$$\mathrm{NPS}(u) = \overline{d}_{DE}^2 \left[w^2 \frac{\mathrm{NPS}_F(u)}{\overline{d}_F^2} + \frac{\mathrm{NPS}_R(u)}{\overline{d}_R^2} - 2w \frac{\mathrm{NPS}_{FR}(u)}{\overline{d}_F \overline{d}_R} \right], \tag{4}$$

and

$$\mathrm{DQE}(u) = \frac{w^2 \mathrm{MTF}_F^2(u) + \mathrm{MTF}_R^2(u)}{\overline{q}_0 \mathrm{NNPS}(u)}, \tag{5}$$

where the normalized NPS (NNPS) is given by $\mathrm{NPS}(u)/\overline{d}_{DE}^2$. $\mathrm{NPS}_{FR}(u)$ represents the cross-NPS between the front- and rear-detector images, which can be measured by subtracting $\mathrm{NPS}_F(u)$ and $\mathrm{NPS}_R(u)$ from the NPS evaluated on an image formed by summing the two images [33]. Refer to references [24, 25] for a detailed derivation.

Figure 4a displays the DE MTF curves for the sandwich detector (small-area design with $t_{\mathrm{IF}} = 0$ mm), including the MTFs of each detector layer, along with two different DE reconstructions ($w = 0.3$ and $w = 0.7$ for soft- and bone-tissue imaging, respectively). Figure 4a also includes the results calculated using Eq. (3) for comparison. Analyzing DE edge-knife images acquired from the rear detector presents challenges owing to poor relative noise characteristics and their amplification during subtraction. Nevertheless, considering the measurement uncertainty, the single-shot DE MTF model described in Eq. (3) describes these measurements.

Both the measurements and the theoretical model of the single-shot DE MTF clearly demonstrate that tissue imaging diminishes the contrast-transfer capability of higher-spatial-frequency contents. Conversely, bone imaging exhibits restrained

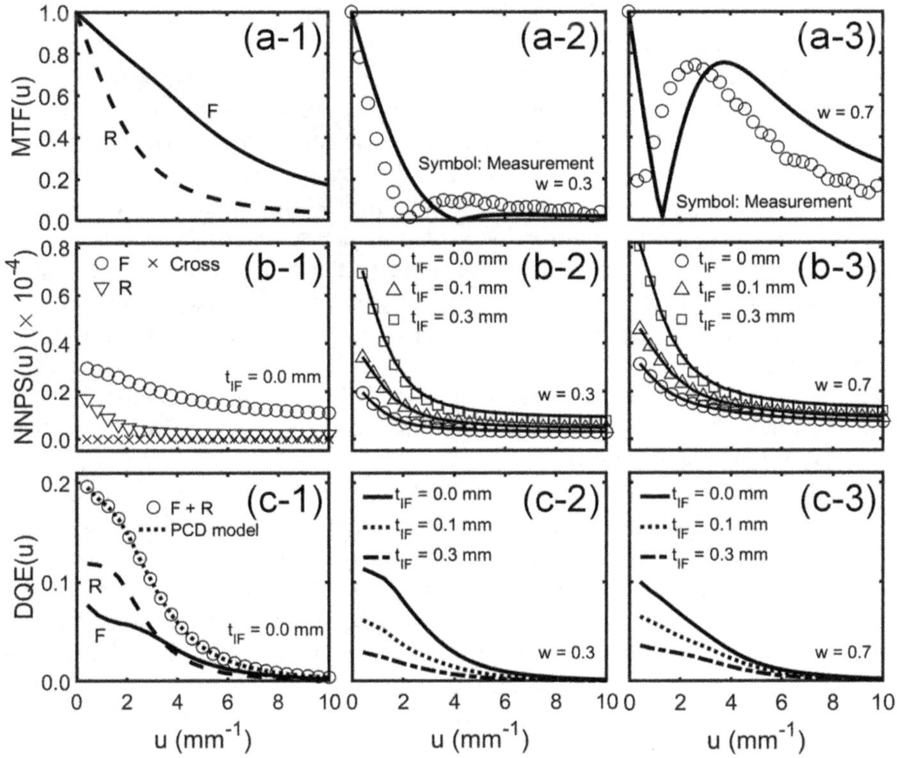

Fig. 4 Comparison of the MTFs, NNPSs, and DQEs of the small-area sandwich detector measured for the RQA-3 beam quality. (**a**) MTF: (**a-1**) compares the MTFs of the front- and rear-detector layers. (**a-2**) and (**a-3**) compare the DE MTFs obtained using Eq. (3) (line plots) and directly from the reconstructed DE images with $w = 0.3$ and $w = 0.7$ (symbols), respectively. (**b**) NNPS: (**b-1**) compares the NNPSs of the front- and rear-detector layers, including the cross NNPS. (**b-2**) and (**b-3**) compare the DE NNPSs obtained using Eq. (4) (line plots) and directly from the reconstructed DE images with $w = 0.3$ and $w = 0.7$ (symbols), respectively, for various t_{IF} values. (**c**) DQE: (**c-1**) compares the DQEs of the front- and rear-detector layers. Symbol plot represents a simple sum of front- and rear-detector DQEs, and the dotted line represents the PCD DQE model. (**c-2**) and (**c-3**) compare the DE DQEs obtained using Eq. (5) with $w = 0.3$ and $w = 0.7$, respectively, for various t_{IF} values. DQE measurements were performed for the air kerma of ~18 μGy at the entrance surface of the sandwich detector

contrast in low-frequency ones. Although not shown here, the influence of t_{IF} on DE MTF appears to be marginal. An in-depth analysis of the single-shot DE MTF model is presented in the reference [24].

Figure 4b illustrates a comparison of the measured DE NNPS results (symbol plots) with the results calculated using Eq. (4) (line plots) for the sandwich detector with various t_{IF} values and the two DE reconstructions. The measurements and theoretical predictions are in excellent agreement. We observed that an increase in t_{IF} led to an escalation in the noise-power density of the rear detector, consequently elevating the resulting DE NNPS. Notably, the spectral shape of the DE NNPS for the bone imaging primarily mirrored that of the rear detector. Additionally, the cross-NNPS remained negligibly small irrespective of t_{IF}. Two images from the sandwich-detector configuration have been theoretically proven to be independent [34]. Sones and Barnes [35] introduced a noise correlation model for sandwich detectors and validated its independence based on experimental results.

The higher weighting of the rear-detector images, known for relatively higher noise levels, during DE reconstruction leads to increased noise in the DE images. Furthermore, a thick t_{IF} exacerbates the noise in the rear-detector images owing to reduced photon incidence on the rear-detector layer. Consequently, the noise in the DE bone images acquired from the detector with a thick t_{IF} is significantly pronounced.

As shown in Fig. 4c, we calculated the DE DQE using Eq. (5). Notably, the DE DQE was more significantly influenced by the detector design parameter t_{IF} than the reconstruction parameter w.

We present the DQE of the sandwich detector calculated using the framework developed for PCDs [36] in Fig. 4(c-1) (dotted line). Given that the cross-NNPS of the sandwich detector was nearly zero, as indicated in Fig. 4(b-1), the DQE was almost equivalent to the sum of the DQEs of each detector layer (symbols). Intriguingly, even when DE reconstruction was performed using $w = 1$, implying equal weighting between the LE and HE images, the framework for the PCD DQE in Fig. 4(c-1) surpassed that of the DE DQE in Fig. 4(c-2) and (c-3).

The effect of the interdetector filter is not positive, as shown above, on the DE DQE performance of the small-area sandwich detector owing to the increase in noise. The RadEye1™ photosensor is based on a silicon photodiode array fixed on a ceramic board. A previous study estimated its spectral transmission factor to be equivalent to an Al sheet with a thickness of ~3 mm for x-ray spectra ranging from 50 to 120 kV [37]. For this reason, additional Cu filtration might not be beneficial for the DE DQE of the RQA-3 beam quality, which was applied to obtain the results shown in Fig. 4, and rather may be harmful by increasing noise. However, the design of the interdetector filter is crucial for a customized sandwich detector with bare silicon photodiode arrays, such as the large-area design shown in Sect. 2.2. Thin K-edge materials (e.g., Mo, Rh, Ag, Sn, and Ce) or their combinations are another option for designing an interdetector filter.

4 Applications

4.1 Radiography

Han et al. [21] demonstrated preclinical DE images showing bone- and soft-tissue components in a postmortem mouse using both double- and single-shot methods. Although the double-shot images (40/70 kV) exhibited superior SNR, the single-

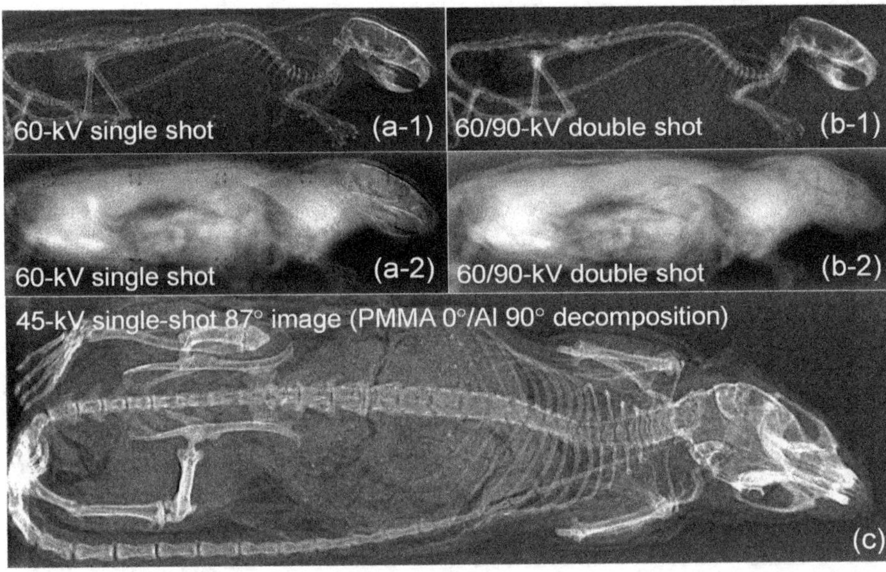

Fig. 5 DE images obtained for a postmortem mouse phantom using a large-area sandwich detector. (**a-1**) and (**b-1**) compare bone images obtained using the single- and double-shot methods. (**a-2**) and (**b-2**) compare soft-tissue images. In the case of double-shot imaging, only the front-detector layer was used for image acquisition. (**c**) Example of the basis-material (PMMA and Al) decomposition image

shot images (both 50 and 70 kV) were acceptable and advantageous in terms of reduced susceptibility to motion artifacts. The single-shot method exhibited superior SNR at higher spatial frequencies (e.g., edge regions) than the double-shot method. The rear detector, which utilized a thicker phosphor layer, yielded a more blurred image than the front detector. Subtraction of the two images resulted in a form of "unsharp masking," which enhanced the edges in the resultant image [refer to Eq. (3)]. Moreover, the suppression power of the bone tissue at 50 kV was not as desirable as that at 70 kV.

The edge-enhancement properties of the sandwich detector were examined, with a doubtful advantage identified. Although these properties help improve the visibility of details with high-spatial-frequency content, they can concurrently act as halo artifacts [26, 38, 39], generally along boundaries. This artifact becomes prominent near the edges if a low-pass filter is applied to the rear-detector image for noise reduction [3], where the rear-detector image suffers from quantum noise owing to fewer detected quanta. Noise-reduction approaches enhance the SNR in DE images; however, they increase edge artifacts [3, 40].

Figure 5 shows the single-shot DE images obtained for a mouse using a large-area sandwich detector in comparison with the double-shot method using only the front detector. The contrast and noise performances of the two methods were similar. A closer inspection of the mouse images obtained using the sandwich detector revealed mesh-like patterns across the entire image and repeated dotted patterns, similar to the arrangement on small square vertices. These patterns stem from the flexible PCB layers between the two detector layers. A flexible PCB is a thin plastic sheet cladded with thin Cu-mesh sheets on both sides, substantiating the former ghost patterns. The latter ghost patterns result from via-holes, positioned at specified locations in commercial flexible PCBs. Although these patterns were hidden in the rear-detector projection images with a typical gain-offset correction method, they were manifested in the DE images owing to the subtraction operation. This adverse artifact should be suppressed for the practical application of the sandwich detector. Figure 5c shows an example of a DE reconstruction using the basis-material decomposition approach [41]. Polymethyl methacrylate (PMMA) and Al were selected as the basis materials.

Recently, the feasibility of single-shot DEI was investigated in the pursuit of motion artifact-free subtraction angiography [37]. For the feasibility test, two plastic phantoms with linear arrays of cylindrical holes containing iodinated water were prepared. One consisted of cylinders with the same diameter and different iodine concentrations, whereas the other consisted of cylinders with different diameters and the same iodine concentration. The concentration and size discrimination capabilities of single-shot DEI were evaluated by investigating the phantom images. While the image noise relative to the signal was almost independent of the mass thickness of the iodine, the iodine detectability improved with the mass thickness. The detectability performance at lower tube voltages (e.g., 60 kV) was superior to that at higher voltages (e.g., 90 kV). The results obtained in that study demonstrated the possibility of performing subtraction angiography using a sandwich detector immune to object or patient motion.

4.2 Tomography

Figure 6 presents the CBCT images obtained from the small-area sandwich detector for a postmortem mouse. Image reconstruction was performed using the FDK algorithm [42]—the most feasible and widely used cone-beam-filtered backprojection (FBP) method. The DE CBCT was to apply the FDK to DE-reconstructed projection views obtained over the 2π angular range. Columns (a) and (b) show the reconstructed images from the front and rear detectors, respectively, and column (c) shows the DE bone-reconstruction images. The cross-sectional images of the mouse head are compared in row (1), reflecting the characteristics observed in the projection data of the sandwich detector. The visibility of the bone structures reconstructed using the DE method was higher than that obtained using the high-spatial-resolution front detector. Row (2) compares the 3D-rendered images of a mouse head, and row (3) presents an enlarged view of the forefoot. The details of the acropodia in the DE image, as shown in Fig. 6(c-3), as well as the joints, are clearly shown in the image.

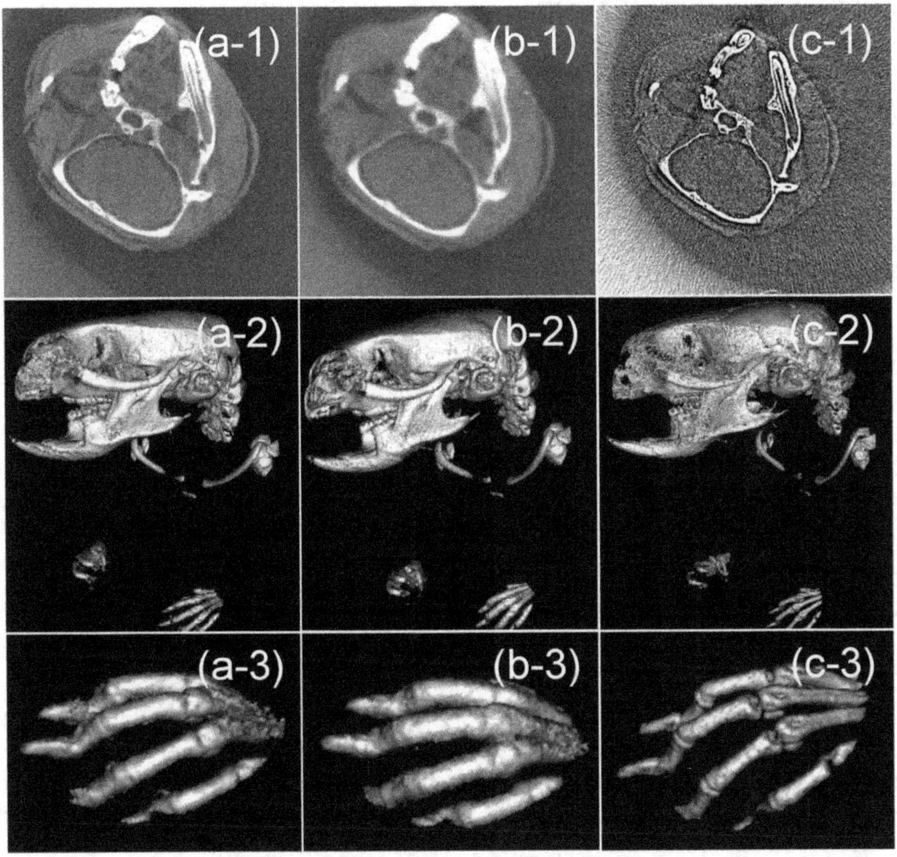

Fig. 6 Row (1) compares the cross-sectional images reconstructed using the projection datasets obtained from (**a**) front- and (**b**) rear-detector layers of the small-area sandwich detector and (**c**) their resultant DE projection dataset (L/W = $\mu / \pm 2\sigma$). Row (2) compares the volumetric images reconstructed using the projection datasets obtained from (**a**) front- and (**b**) rear-detector layers of the sandwich detector and (**c**) their resultant DE projection dataset. Row (3) compares the enlarged views of left forefoot images shown in row (2)

The DE images of a sandwich detector possess band-pass filter characteristics similar to those of conventional FBP-reconstructed images, further related to the low-contrast performance of DE images for large-area objects and the relatively high contrast of bone details. Ha et al. [43] showed that the DE MTF of CBCT can be similarly characterized by Eq. (3). Combining DE images with images obtained from the front detector may restore the low-contrast performance of DE images for large-area objects. In contrast, the edge-enhancement effects of the sandwich detector can render artifacts near the edges in the resultant DE images [for example, black sinks around bone structures, as displayed in Fig. 6(c-1)], similar to the halo artifacts in edge-enhanced images obtained using the unsharp masking algorithm [38].

For a single X-ray exposure, back-to-back detector layers with different phosphor thicknesses acquired X-ray projections with different X-ray energy spectra and spatial-resolution characteristics. Bone-selective imaging allows for different energy measurements, and different spatial-resolution characteristics enhance bone details in DE reconstruction. Figure 6 shows the potential of microtomography with a sandwich detector for high-resolution bone-selective imaging without post-processing segmentation procedures.

However, because shading artifacts (i.e., black sinks around the feature edges) result from the signal intensity drop, they cause distortions in the reconstructed voxel values. Therefore, the single-shot method might not be suitable for quantitative CT. However, in certain industrial tomographic applications, the exact densities of parts of objects are not necessarily required. Alternatively, the shapes or boundaries of features in the test objects, such as cracks or voids, are frequently sufficient for defect inspection.

5 Concluding Remarks

The most significant advantage of the sandwich detector is its ability to acquire two different spectral images in a single exposure, thus making it resistant to motion artifacts. Specifically, as the two detector layers forming the sandwich structure generally employed phosphors of different thicknesses, corresponding to different spatial-resolution characteristics, the weighted combination of the two images yielded a resulting DE image spectrally weighted in energy and spatial frequency. The former energy weighting was the same as that of conventional double-shot DEI. However, the latter frequency weighting is a distinguishing characteristic, similar to unsharp masking in digital image processing. For the widespread use of sandwich detectors, a correct understanding of this characteristic is required to determine whether it could be a treatment (image enhancement) or threat (artifact) to images.

Using two detectors may cost twice as much as using only one detector. Therefore, the cost–benefit should be prioritized for commercializing sandwich detectors. The production cost of the current *a*-Si thin-film transistor (TFT)-based FPD is much lower than that in the late 1990s to the early 2000s, which was the early stage of FPD development. This decline in cost is driven by several factors. In particular, active-matrix liquid-crystal display (AMLCD) technology, which is the underlying technology of the *a*-Si FPD, has rapidly evolved into advanced display technologies, and existing LCD manufacturing facilities have competitively explored new markets other than displays, such as amorphous Si and oxide TFT panels for X-ray imaging. Manufacturing sandwich detectors is expected to cost less than that required for the two detectors. These aspects warrant a thorough characterization of the benefits of simultaneously using two detectors compared to the use of separate detectors.

This chapter concludes by introducing recent research on the development of sandwich detectors and their applications. In 2014, the author's research group stacked two conventional indirect-conversion CMOS detectors for a sandwich detector with an imaging area of ~25 × 50 mm, thereby demonstrating its capability of single-shot DEI [20, 21]. To address the poor image quality caused by dominant Compton scattering, a similar detector-stacking approach was developed for radiotherapy imaging. Researchers combined four conventional *a*-Si FPDs with an imaging area of ~43 × 43 cm to improve the DQE [44] and demonstrated CNR-enhanced images [45]. This research group extended the application of multilayer detectors to megavoltage (MV) spectral imaging [46] and low-dose MV CBCT [47].

A three-layered *a*-Si FPD has also been suggested for DEI [48], with its commercial versions currently rendered by KA Imaging Inc. (Waterloo, ON, Canada). Conventional radiography images are reconstructed by summing all of the layers' images, whereas tissue-subtracted images are generated via logarithmic subtraction. Such a design does not require an interdetector metal filter layer; therefore, it may enable the potential for high dose efficiency while maintaining good spectral separation.

The author identified two studies on the development of prototype large-area flat-panel sandwich detectors and their applications. Varex Imaging Corp. (San Jose, CA) developed a prototype comprising a 0.2-mm-thick CsI:Tl-coupled *a*-Si panel, with a pitch of 0.15 mm for the front-detector layer, and a 0.55-mm-thick CsI:Tl-coupled same panel for the rear layer [49]. As the *a*-Si panel featured a pixel pitch of 0.15 mm in a 2880 × 2880 format, the corresponding imaging size was ~43 × 43 cm. They characterized the performance of single-shot DE radiography/fluoroscopy and DE CBCT along with their research group [22]. Using this prototype, researchers have demonstrated the feasibility of DE CBCT for contrast enhancement and material decomposition in orthopedics, including the development of elaborate model-based algorithms for high-resolution material decomposition [50, 51], measurement of bone mineral density [52], and quantification of bone marrow edema [53].

The other prototype, with an imaging size of ~38 × 29 cm, was developed by Philips Healthcare (Eindhoven, The Netherlands) using two CsI:Tl-coupled *a*-Si panels with a pixel pitch of 0.154 mm. The research group installed this prototype in a commercial interventional C-arm system and reported the performance of DE CBCT [54].

Using two indirect-conversion detectors as sandwich detectors is a common approach. However, to utilize the higher-spatial-resolution characteristics of direct-conversion detectors, a hybrid design employing a photoconductor, such as amorphous selenium, for the front layer has been reported [55, 56].

Acknowledgments The author thanks graduate students Seungjun Yoo and Hubeom Shin, who assisted in preparing figures and editing the manuscript and provided valuable advice. This work was supported by the National Research Foundation of Korea (NRF) grants funded by the Korea government (MSIT) (No. 2017M2A2A6A01019930 and No. 2021R1A2C1010161).

References

1. H.K. Kim, Development of energy-integrating detectors for large-area x-ray imaging, in *Radiation Detection Systems: Sensor Materials, Systems, Technology, and Characterization Measurements*, ed. by J.S. Iwanczyk, K. Iniewski (CRC Press, Boca Raton, Florida, 2022), pp. 163–224. https://doi.org/10.1201/9781003147633-7
2. S. Richard, J.H. Siewerdsen, Cascaded systems analysis of noise reduction algorithms in dual-energy imaging. Med. Phys. **35**(2), 586–601 (2008). https://doi.org/10.1118/1.2826556
3. D.W. Kim, J. Park, J. Kim, H.K. Kim, Noise-reduction approaches to single-shot dual-energy imaging with a multilayer detector. J. Instrum. **14**(1), C01021 (2019). https://doi.org/10.1088/1748-0221/14/01/c01021
4. International Commission on Radiation Units and Measurements: Medical Imaging–The Assessment of Image Quality. Tech. rep. (1995)
5. J.H. Siewerdsen, L.E. Antonuk, DQE and system optimization for indirect-detection flat-panel imagers in diagnostic radiology, in *Proc. SPIE*, vol. 3336 ed. by J.T. Dobbins III, J.M. Boone (SPIE, 1998), pp. 546–555. https://doi.org/10.1117/12.317057
6. J.H. Siewerdsen, Optimization of 2D and 3D radiographic imaging systems, in *The Handbook of Medical Image Perception and Techniques*, ed. by E. Samei, E.A. Krupinski (Cambridge University Press, Cambridge, 2018), pp. 417–439. https://doi.org/10.1017/9781108163781.028
7. M.J. Tapiovaara, R. Wagner, SNR and DQE analysis of broad spectrum x-ray imaging. Phys. Med. Biol. **30**(6), 519–529 (1985). https://doi.org/10.1088/0031-9155/30/6/002
8. R.K. Swank, Absorption and noise in x-ray phosphors. J. Appl. Phys. **44**(9), 4199–4203 (1973). https://doi.org/10.1063/1.1662918
9. R.N. Cahn, B. Cederström, M. Danielsson, A. Hall, M. Lundqvist, D. Nygren, Detective quantum efficiency dependence on x-ray energy weighting in mammography. Med. Phys. **26**(12), 2680–2683 (1999). https://doi.org/10.1118/1.598807
10. J. Tanguay, S. Yun, H.K. Kim, I.A. Cunningham, Cascaded-systems analyses of photon-counting x-ray detectors, in *Proc. SPIE*, vol. 8668, ed. by R.M. Nishikawa, B.R. Whiting, C. Hoeschen (SPIE, 2013), pp. 86680S. https://doi.org/10.1117/12.2008132
11. J. Tanguay, H.K. Kim, I.A. Cunningham, The role of x-ray Swank factor in energy-resolving photon-counting imaging. Med. Phys. **37**(12), 6205–6211 (2010). https://doi.org/10.1118/1.3512794
12. A.S. Wang, N.J. Pelc, Sufficient statistics as a generalization of binning in spectral x-ray imaging. IEEE Trans. Med. Imag. **30**(1), 84–93 (2011). https://doi.org/10.1109/TMI.2010.2061862
13. T. Thuering, S. Gkoumas, P. Zambon, P. Trueb, M. Rissi, F.D. Prima, T. Donath, S. Sawall, C. Amato, M. Kachelrieß, C. Broennimann, Towards large-area photon-counting detectors for spectral x-ray imaging, in *Proc. SPIE*, vol. 10948, ed. by T.G. Schmidt, G.H. Chen, H. Bosmans (SPIE, 2019), pp. 1094817. https://doi.org/10.1117/12.2507190
14. S. Leng, M. Bruesewitz, S. Tao, K. Rajendran, A.F. Halaweish, N.G. Campeau, J.G. Fletcher, C.H. McCollough, Photon-counting detector CT: System design and clinical applications of an emerging technology. Radiographics **39**(3), 729–743 (2019). https://doi.org/10.1148/rg.2019180115
15. M. Danielsson, M. Persson, M. Sjölin, Photon-counting x-ray detectors for CT. Phys. Med. Biol. **66**(3), 03TR01 (2021). https://doi.org/10.1088/1361-6560/abc5a5
16. A.S. Wang, N.J. Pelc, Spectral photon counting CT: Imaging algorithms and performance assessment. IEEE Trans. Radiat Plasma Med. Sci. **5**(4), 453–464 (2021). https://doi.org/10.1109/TRPMS.2020.3007380
17. R.E. Alvarez, J.A. Seibert, S.K. Thompson, Comparison of dual energy detector system performance. Med. Phys. **31**(3), 556–565 (2004). https://doi.org/10.1118/1.1645679
18. T. Ishigaki, S. Sakuma, Y. Horikawa, M. Ikeda, H. Yamaguchi, One-shot dual-energy subtraction imaging. Radiology **161**(1), 271–273 (1986). https://doi.org/10.1148/radiology.161.1.3532182

19. R. Carmi, G. Naveh, A. Altman, Material separation with dual-layer CT, in *IEEE Nuclear Science Symposium Conference Record*, vol. 4 (IEEE, 2005), pp. 1879–1878. https://doi.org/10.1109/NSSMIC.2005.1596697
20. S. Yun, J.C. Han, D.W. Kim, H. Youn, H.K. Kim, J. Tanguay, I.A. Cunningham, Feasibility of active sandwich detectors for single-shot dual-energy imaging, in *Proc. SPIE*, vol. 9033, ed. by B.R. Whiting, C. Hoeschen (SPIE, 2014), pp. 90335T. https://doi.org/10.1117/12.2043368
21. J.C. Han, H.K. Kim, D.W. Kim, S. Yun, H. Youn, S. Kam, J. Tanguay, I.A. Cunningham, Single-shot dual-energy x-ray imaging with a flat-panel sandwich detector for preclinical imaging. Cur. Appl. Phys. **14**(12), 1734–1742 (2014). https://doi.org/10.1016/j.cap.2014.10.012
22. L. Shi, M. Lu, N.R. Bennett, E. Shapiro, J. Zhang, R. Colbeth, J. Star-Lack, A.S. Wang, Characterization and potential applications of a dual-layer flat-panel detector. Med. Phys. **47**(8), 3332–3343 (2020). https://doi.org/10.1002/mp.14211
23. D.W. Kim, J. Kim, S. Yun, H. Youn, H.K. Kim, Cascaded-systems analysis of sandwich x-ray detectors. J. Instrum. **11**(2), C12022 (2016). https://doi.org/10.1088/1748-0221/11/12/c12022
24. J. Kim, D.W. Kim, S.H. Kim, S. Yun, H. Youn, H. Jeon, H.K. Kim, Linear modeling of single-shot dual-energy x-ray imaging using a sandwich detector. J. Instrum. **12**(1), C01029 (2017). https://doi.org/10.1088/1748-0221/12/01/c01029
25. H. Shin, S. Yoo, S. Oh, J. Lee, H.K. Kim, Detective quantum efficiency of a double-layered detector for dual-energy x-ray imaging. J. Instrum. **18**(11), C11005 (2023). https://doi.org/10.1088/1748-0221/18/11/C11005
26. S.H. Kim, D.W. Kim, D. Kim, H. Youn, S. Cho, H.K. Kim, Microtomography with sandwich detectors for small-animal bone imaging. J. Instrum. **11**(10), C10011 (2016). https://doi.org/10.1088/1748-0221/11/10/p10011
27. M.K. Cho, H.K. Kim, T. Graeve, S.M. Yun, C.H. Lim, H. Cho, J. Kim, Measurements of x-ray imaging performance of granular phosphors with direct-coupled CMOS sensors. IEEE Trans. Nucl. Sci. **55**(3), 1338–1343 (2008). https://doi.org/10.1109/TNS.2007.913939
28. J. Kim, D.W. Kim, S. Kam, J. Park, C.-S. Shon, S.K. Heo, H.K. Kim, Development of a large-area multilayer detector for single-shot dual-energy imaging. J. Instrum. **13**(12), C12018 (2018). https://doi.org/10.1088/1748-0221/13/12/c12018
29. S. Yun, H.K. Kim, C.H. Lim, M.K. Cho, T. Achterkirchen, I. Cunningham, Signal and noise characteristics induced by unattenuated x rays from a scintillator in indirect-conversion CMOS photodiode array detectors. IEEE Trans. Nucl. Sci. **56**(3), 1121–1128 (2009). https://doi.org/10.1109/TNS.2009.2014231
30. M.J. Flynn, S.M. Hames, S.J. Wilderman, J.J. Ciarelli, Quantum noise in digital x-ray image detectors with optically coupled scintillators. IEEE Trans. Nucl. Sci. **43**(4), 2320–2325 (1996). https://doi.org/10.1109/23.531897
31. H.K. Kim, Generalized cascaded model to assess noise transfer in scintillator-based x-ray imaging detectors. Appl. Phys. Lett. **89**(23), 233504 (2006). https://doi.org/10.1063/1.2398926
32. K.A. Fetterly, N.J. Hangiandreou, Effects of x-ray spectra on the DQE of a computed radiography system. Med. Phys. **28**(2), 241–249 (2001). https://doi.org/10.1118/1.1339883
33. J. Tanguay, D. Richtsmeier, C. Dydula, J.A. Day, K. Iniewski, M. Bazalova-Carter, A detective quantum efficiency for spectroscopic x-ray imaging detectors. Med. Phys. **48**(11), 6781–6799 (2021). https://doi.org/10.1002/mp.15194
34. H.H. Barrett, W. Swindell, *Radiological Imaging* (Academic Press, New York, 1981)
35. R.A. Sones, G.T. Barnes, Noise correlations in images acquired simultaneously with a dual-energy sandwich detector. Med. Phys. **16**(6), 858–861 (1989). https://doi.org/10.1118/1.596311
36. N. Zarif Yussefian, J. Tanguay, An experimental framework for assessing the detective quantum efficiency of spectroscopic x-ray detectors. Med. Phys. **50**(3), 1318–1335 (2022). https://doi.org/10.1002/mp.16114
37. J. Kim, S. Yoo, J. Kim, H.K. Kim, Towards subtraction angiography using a multi-layered x-ray detector. Phys. Medica. **95**, 32–40 (2022). https://doi.org/10.1016/j.ejmp.2022.01.001

38. Z. Farbman, R. Fattal, D. Lischinski, R. Szeliski, Edge-preserving decompositions for multi-scale tone and detail manipulation. ACM Trans. Graph. **27**(3), 1–10 (2008). https://doi.org/10.1145/1360612.1360666
39. M. Trentacoste, R. Mantiuk, W. Heidrich, F. Dufrot, Unsharp masking, countershading and halos: Enhancements or artifacts? Comput. Graph. Forum. **31**(2pt3), 555–564 (2012). https://doi.org/10.1111/j.1467-8659.2012.03056.x
40. D.L. Ergun, C.A. Mistretta, D.E. Brown, R.T. Bystrianyk, W.K. Sze, F. Kelcz, D.P. Naidich, Single-exposure dual-energy computed radiography: Improved detection and processing. Radiology **174**(1), 243–249 (1990). https://doi.org/10.1148/radiology.174.1.2294555
41. L.A. Lehmann, R.E. Alvarez, A. Macovski, W.R. Brody, N.J. Pelc, S.J. Riederer, A.L. Hall, Generalized image combinations in dual KVP digital radiography. Med. Phys. **8**(5), 659–667 (1981). https://doi.org/10.1118/1.595025
42. L.A. Feldkamp, L.C. Davis, J.W. Kress, Practical cone-beam algorithm. J. Opt. Soc. Am. A. **1**(6), 612–619 (1984). https://doi.org/10.1364/JOSAA.1.000612
43. S. Ha, J. Kim, J. Yun, D.W. Kim, S.H. Kim, H.K. Kim, Linear analysis of single-shot dual-energy computed tomography with a multilayer detector. J. Instrum. **14**(1), C01022 (2019). https://doi.org/10.1088/1748-0221/14/01/c01022
44. J. Rottmann, D. Morf, R. Fueglistaller, G. Zentai, J. Star-Lack, R. Berbeco, A novel EPID design for enhanced contrast and detective quantum efficiency. Phys. Med. Biol. **61**(17), 6297 (2016). https://doi.org/10.1088/0031-9155/61/17/6297
45. S.S.F. Yip, J. Rottmann, H. Chen, D. Morf, R. Füglistaller, J. Star-Lack, G. Zentai, R. Berbeco, Technical note: Combination of multiple EPID imager layers improves image quality and tracking performance of low contrast-to-noise objects. Med. Phys. **44**(9), 4847–4853 (2017). https://doi.org/10.1002/mp.12422
46. M. Myronakis, R. Fueglistaller, J. Rottmann, Y.-H. Hu, A. Wang, P. Baturin, P. Huber, D. Morf, J. Star-Lack, R. Berbeco, Spectral imaging using clinical megavoltage beams and a novel multi-layer imager. Phys. Med. Biol. **62**(23), 912 (2017). https://doi.org/10.1088/1361-6560/aa94f9
47. M. Myronakis, P. Huber, M. Lehmann, R. Fueglistaller, M. Jacobson, Y.-H. Hu, P. Baturin, A. Wang, M. Shi, T. Harris, D. Morf, R. Berbeco, Low-dose megavoltage cone-beam computed tomography using a novel multi-layer imager (MLI). Med. Phys. **47**(4), 1827–1835 (2020). https://doi.org/10.1002/mp.14017
48. S.L. Maurino, A. Badano, I.A. Cunningham, K.S. Karim, Theoretical and Monte Carlo optimization of a stacked three-layer flat-panel x-ray imager for applications in multi-spectral diagnostic medical imaging, in *Proc. SPIE*, vol. 9783, ed. by D. Kontos, T.G. Flohr (SPIE, 2016), pp. 1061–1074. https://doi.org/10.1117/12.2217085
49. M. Lu, A. Wang, E. Shapiro, A. Shiroma, J. Zhang, J. Steiger, J. Star-Lack, Dual energy imaging with a dual-layer flat panel detector, in *Proc. SPIE*, vol. 10948, ed. by T.G. Schmidt, G.H. Chen, H. Bosmans (SPIE, 2019), pp. 1094815. https://doi.org/10.1117/12.2513499
50. Y.Q. Ma, W. Wang, M. Tivnan, J. Li, M. Lu, J. Zhang, J. Star-Lack, R.E. Colbeth, W. Zbijewski, J.W. Stayman, High-resolution model-based material decomposition for multi-layer flat-panel detectors, in *International Conference on Image Formation in X-Ray Computed Tomography*, pp. 62–64 (2020)
51. W. Wang, Y. Ma, M. Tivnan, J. Li, G.J. Gang, W. Zbijewski, M. Lu, J. Zhang, J. Star-Lack, R.E. Colbeth, J.W. Stayman, High-resolution model-based material decomposition in dual-layer flat-panel CBCT. Med. Phys. **48**(10), 6375–6387 (2021). https://doi.org/10.1002/mp.14894
52. C. Zhao, S.Z. Liu, W. Wang, M. Herbst, T. Weber, S. Vogt, L. Ritschl, S. Kappler, J.W. Stayman, J.H. Siewerdsen, W. Zbijewski, Effects of x-ray scatter in quantitative dual-energy imaging using dual-layer flat panel detectors, in *Proc. SPIE*, vol. 11595, ed. by H. Bosmans, W. Zhao, L. Yu (SPIE, 2021), pp. 115952A. https://doi.org/10.1117/12.2581822
53. S.Z. Liu, C. Zhao, M. Herbst, T. Weber, S. Vogt, L. Ritschl, S. Kappler, J.H. Siewerdsen, W. Zbijewski, Feasibility of dual-energy cone-beam CT of bone marrow edema using dual-layer flat panel detectors, in *Proc. SPIE*, vol. 12031, ed. by W. Zhao, L. Yu (SPIE, 2022), pp. 120311J. https://doi.org/10.1117/12.2613211

54. F. Ståhl, D. Schäfer, A. Omar, P. van de Haar, F. van Nijnatten, P. Withagen, A. Thran, E. Hummel, B. Menser, Å. Holmberg, M. Söderman, A. Falk Delgado, G. Poludniowski, Performance characterization of a prototype dual-layer cone-beam computed tomography system. Med. Phys. **48**(11), 6740–6754 (2021). https://doi.org/10.1002/mp.15240
55. X. Duan, A. Howansky, H. Huang, W. Zhao, Evaluation of a direct-indirect dual layer detector for contrast enhanced breast imaging: Monte Carlo simulations and physical experiments, in *Proc. SPIE*, vol. 12031, ed. by W. Zhao, L. Yu (SPIE, 2022), pp. 120310A. https://doi.org/10.1117/12.2613056
56. A. Swaby, A.S. Wang, M.J. Willemink, S. Abbaszadeh, Dual-layer flat panel detector with a-Se top layer for opportunistic screening of coronary artery calcium: A simulation study, in *Proc. SPIE*, vol. 12031, ed. by W. Zhao, L. Yu (SPIE, 2022), pp. 1203143. https://doi.org/10.1117/12.2611831

A Novel Algorithm to Create Quantitative X-Ray Images Based on Precise Analysis of Polychromatic X-Ray Attenuation

Natsumi Kimoto, Hiroaki Hayashi, Rina Nishigami, Daiki Kobayashi, Tatsuya Maeda, and Akitoshi Katsumata

1 Introduction

When X-rays were discovered by Dr. W. Röntgen in 1895, the publication of an X-ray image of the human hand had a tremendous impact on the progress of medicine. In his first paper on the discovery of X-rays, he published experimentally derived properties of X-rays and X-ray photographs. Although he described abstract trends related to the attenuation of X-rays, much more time was needed to elucidate this physical phenomenon based on discoveries, such as the photoelectric and Compton scattering effects and the maturity of corresponding quantum theories. Fortunately, the development of medicine did not necessarily require the physics theory related to X-ray attenuation. This is because medical doctors were able to understand radiographs without having knowledge of the physics theory behind X-rays. At that time, X-ray image diagnosis was performed by a doctor who was looking at an X-ray image reflected on a fluorescent plate. Gradually this came to be replaced by radiographs. When reviewing the history of X-ray photography, we should pay tribute to the historical success of radiography in qualitative diagnosis, and at the same time, we should notice the fact that the theoretical consideration of X-rays had not been reflected in the diagnosis using X-ray photographs.

Since then, X-ray images have become increasingly essential for medical examination and noninvasive inspection. In particular, the digitization of X-ray

N. Kimoto (✉) · H. Hayashi
College of Medical, Pharmaceutical and Health Sciences, Kanazawa University, Ishikawa, Japan
e-mail: hayashi.hiroaki@staff.kanazawa-u.ac.jp

R. Nishigami · D. Kobayashi · T. Maeda
Graduate School of Medical Sciences, Kanazawa University, Ishikawa, Japan

A. Katsumata
Oral Radiology and Artificial Intelligence, Asahi University, Gifu, Japan

K. Iniewski, H. Gadey (eds.), *Emerging Radiation Detection*,
https://doi.org/10.1007/978-3-031-63897-8_5

images using various image processing technologies, such as computed radiography (CR) and digital radiography (DR) systems, has made it possible to generate X-ray images that are easier to use for quantitative diagnoses [1, 2]. In recent years, new attempts to incorporate artificial intelligence into diagnostic technology are also attracting attention [3]. Although these techniques focus on how much useful information can be extracted from a generated X-ray image, it should be pointed out that the amount of information which is derived from the original X-ray image is limited. If the detection and analysis technologies of X-rays are improved, we would be able to derive more information from X-ray images. We would like to develop a new image generation technique and investigate new possibilities for X-ray images.

The concept of the diagnostic system that we want to propose is fusion of conventional and novel methods. In conventional diagnosis as shown in the right panel of Fig. 1, a medical doctor interprets a gray-scale image reflecting the absorbed X-ray energy based on anatomical knowledge. It means that a biological perspective plays an important role in this diagnosis. On the other hand, our goal is to carefully analyze the interaction of X-rays related to image generation; during the production of X-ray images, X-rays penetrate the human body and further interact with detector materials as shown in the left panel of Fig. 1. The analysis

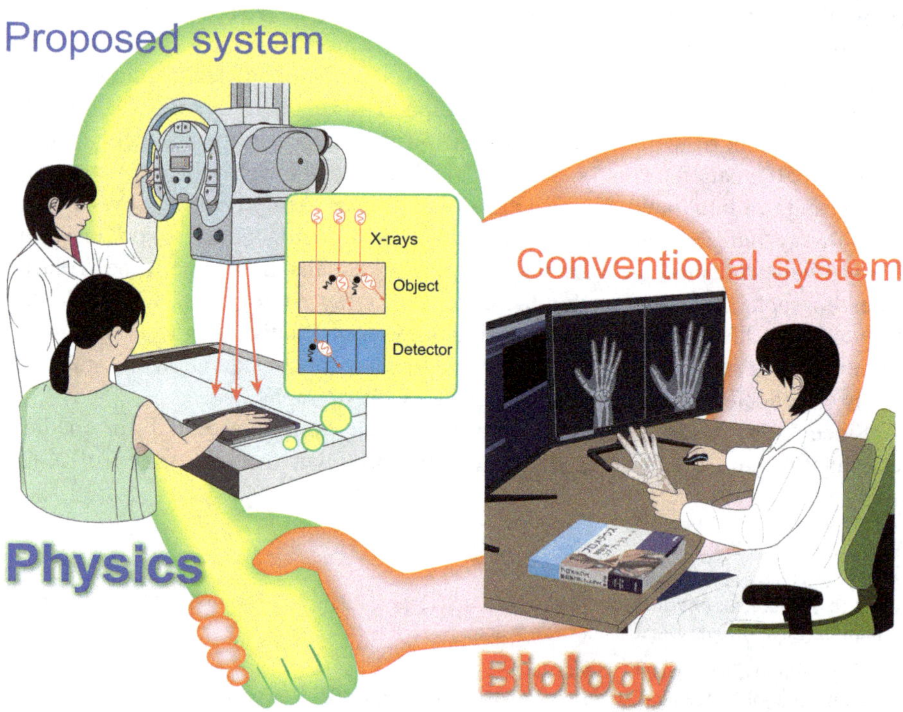

Fig. 1 Concept of the diagnostic system that we want to propose. The fusion of conventional qualitative diagnosis based on anatomical knowledge and quantitative diagnosis based on physics will lead to novel medical diagnosis

of the interaction in each process will lead to the generation of a novel concept X-ray image based on the perspective of physics. We expect that the fusion of conventional qualitative diagnosis based on biology and our proposed quantitative diagnosis based on physics will lead to innovations in diagnostic performance.

Currently, an energy-resolving photon-counting detector (ERPCD) which can generate images corresponding to the energy of each X-ray is being developed [4–6]. In our research, we focus on two-dimensional X-ray images and devise ways to make the best use of the X-ray attenuation information obtained from the ERPCD. In this chapter, we explain a novel method to identify materials using an ERPCD. First, we will explain the basic principle of material identification based on analysis of X-ray attenuation. Next, we will describe the issues which need to be considered when analyzing X-rays detected by ERPCD. Finally, we will present algorithms to derive an effective atomic number (Z_{eff}) image and to extract mass thickness (ρt) images related to the soft tissue and bone. Although the main topic of this chapter is to explain the novel method based on analysis of X-ray attenuation, the physics of X-ray attenuation is common for generating X-ray images using various systems such as plain X-ray and computed tomography (CT) examinations, and we are confident that the chapter will be useful for future research using ERPCDs. If readers are interested in these contents, we recommend reading our papers [7–16] and books [17–19] for more detailed information.

2 Energy-Resolving Photon-Counting Detector to Realize Quantitative Analysis of an Object

Next-generation-type ERPCD has been developed to realize novel analysis for such things as extracting properties of an object [4–6]. This section describes the processing procedure of the ERPCD and the principle of material identification based on the analysis of X-ray attenuation.

To clearly understand the mechanism of an ERPCD, we need to compare it with an energy-integrating detector (EID) [20, 21] which has been traditionally applied to medical and industrial X-ray examinations, as shown in Fig. 2. Figure 2a shows a schematic drawing of an EID. Assuming that four X-rays having different energies of E_1, E_2, E_3, and E_1 are incident to the pixel of the EID, corresponding charge clouds are generated in a pixel. Then, they are collected by an integrated circuit, and the corresponding totally absorbed energy ($\sum_{\mathrm{i}} E_{\mathrm{i}}$) values are digitalized and presented in a gray-scale image; this is the processing procedure for a conventional X-ray image. On the other hand, an ERPCD can analyze X-rays individually and discriminate X-ray energies corresponding to several energy thresholds [4–6]. Figure 2b shows a schematic drawing of an ERPCD. When setting threshold levels at $E_0^{'}$, $E_1^{'}$, $E_2^{'}$, and $E_3^{'}$, we can separate signals into three energy regions of $E_0^{'} < E_{\mathrm{i}} < E_1^{'}$, $E_1^{'} < E_{\mathrm{i}} < E_2^{'}$, and $E_2^{'} < E_{\mathrm{i}} < E_3^{'}$. In the case in which four X-rays having energies of E_1, E_2, E_3, and E_1 are incident to the pixel of ERPCD, E_1, E_2, and E_3 are classified into low-, middle-, and high-energy bins, respectively. This is a

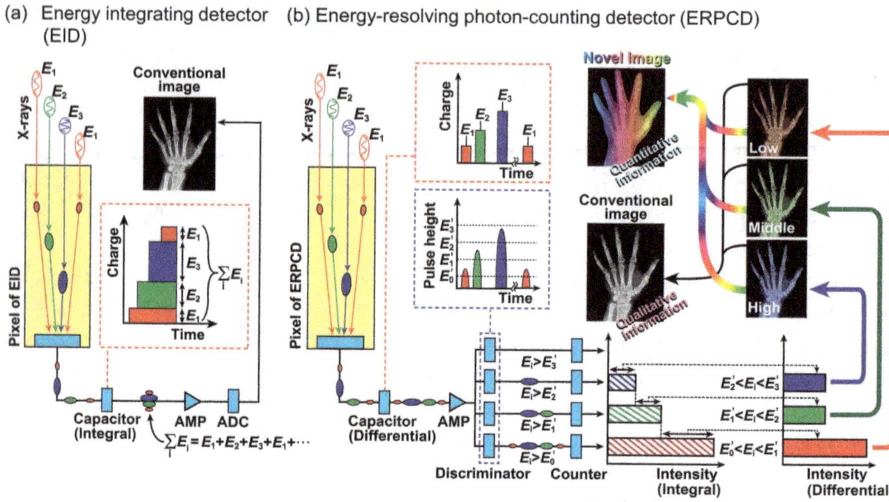

Fig. 2 Comparison between (**a**) energy-integrating detector (EID) and (**b**) energy-resolving photon-counting detector (ERPCD). The EID generates images with information related to the sum of the energies. On the other hand, the ERPCD which was the focus of our research can analyze each X-ray energy

main difference from the EID, which is used to measure not the energy of an X-ray but the number of X-rays. In this case, two counts are measured in the low-energy bin and one count is measured in each of the middle- and high-energy bins. We would like to create quantitative images that extract the properties of an object based on analysis using X-ray energy information. We can also produce a conventional X-ray image which provides qualitative information because the image related to the energy bin has information concerning intensity and effective energy.

Image generation using an ERPCD can quantitatively analyze the X-ray attenuation of an object. Here, we will introduce the method to extract the properties of an object from X-ray energy information. Figure 3 shows our concept for material identification. Assuming that an object having a certain Z_{eff} is measured, X-ray attenuation can be determined using the well-known formula [22]:

$$I = I_0 \times \exp(-\mu t), \tag{1}$$

where I_0 and I are intensities of the incident and penetrating X-rays, respectively, and μ and t are linear attenuation coefficient and material thickness, respectively. It is important to understand the fact that the probability of X-ray attenuation, μ, is determined by the atomic number and a function of the energy of the X-rays [22]. This means that if we can derive μ corresponding to a certain X-ray energy, we can determine the atomic number of an object. However, the physical quantity obtained from one piece of energy information is μt, and we cannot derive μ without using the information for t. To extract μ from μt, dual energy information is required,

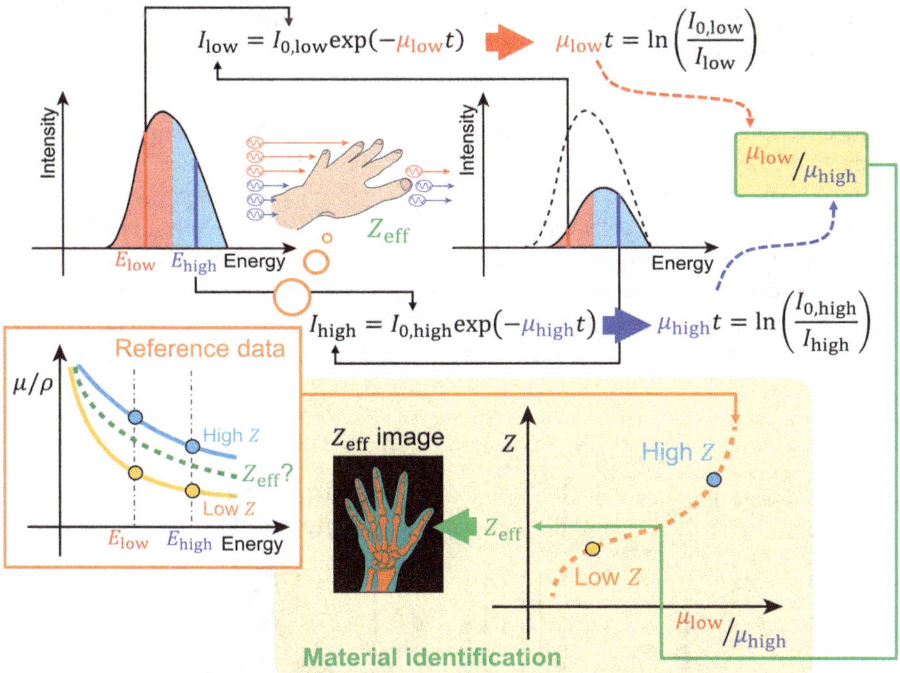

Fig. 3 Concept of material identification is based on the analysis of μ for determining differences in X-ray energies. We calculate μt values from the difference of intensities between the incident and penetrating X-rays and derive a $\frac{\mu_{low}}{\mu_{high}}$ value using $\mu_{low}t$ and $\mu_{high}t$ which correspond to low and high energies, respectively. Z_{eff} of an object can be determined from the experimentally derived $\frac{\mu_{low}}{\mu_{high}}$ value using a reference curve

and ERPCD can be used to provide information corresponding to multiple energies with a single irradiation. In this study, we focus our attention on the low energy E_{low} and high energy E_{high} and calculate the corresponding μt values as $\mu_{low}t$ and $\mu_{high}t$. Then, taking the ratio of these values, we can derive the value $\frac{\mu_{low}}{\mu_{high}}$ ($= \frac{\mu_{low}t}{\mu_{high}t}$) which can be uniquely determined by atomic number without information for t [23]. In order to determine the Z_{eff} from $\frac{\mu_{low}}{\mu_{high}}$, a unique relationship between $\frac{\mu_{low}}{\mu_{high}}$ and Z_{eff} is available, and these data can be derived from the reference data μ/ρ [24]. When using the theoretical relationship, experimentally derived $\frac{\mu_{low}}{\mu_{high}}$ can be converted into Z_{eff}. This means that a novel image that reflects the Z_{eff} of an object can be produced by analyzing X-ray energy information generated from an ERPCD. We should note that this description is based on ideal energy absorption resulting from the interaction between incident X-rays and detector materials and focuses on monochromatic X-ray energy. However, actual cases are far from the ideal condition, and therefore the value measured with an ERPCD needs to be corrected to a value related to monochromatic X-ray.

3 Issues That Need to Be Considered to Achieve Accurate Material Identification

In the description in the former section, we explained that X-ray energy information needs to be analyzed accurately in order to derive the properties of an object. In general, medical, and industrial applications, (1) polychromatic X-rays and (2) multipixel-type imaging detectors are widely applied, and therefore it is difficult to handle X-ray energy information ideally. In order to clarify the issues caused by these two restrictions, we show the differences in X-ray spectra obtained in each stage from incident to detection as shown in Fig. 4. The upper right panel shows the X-ray spectrum in which X-rays have various energies generated from the X-ray

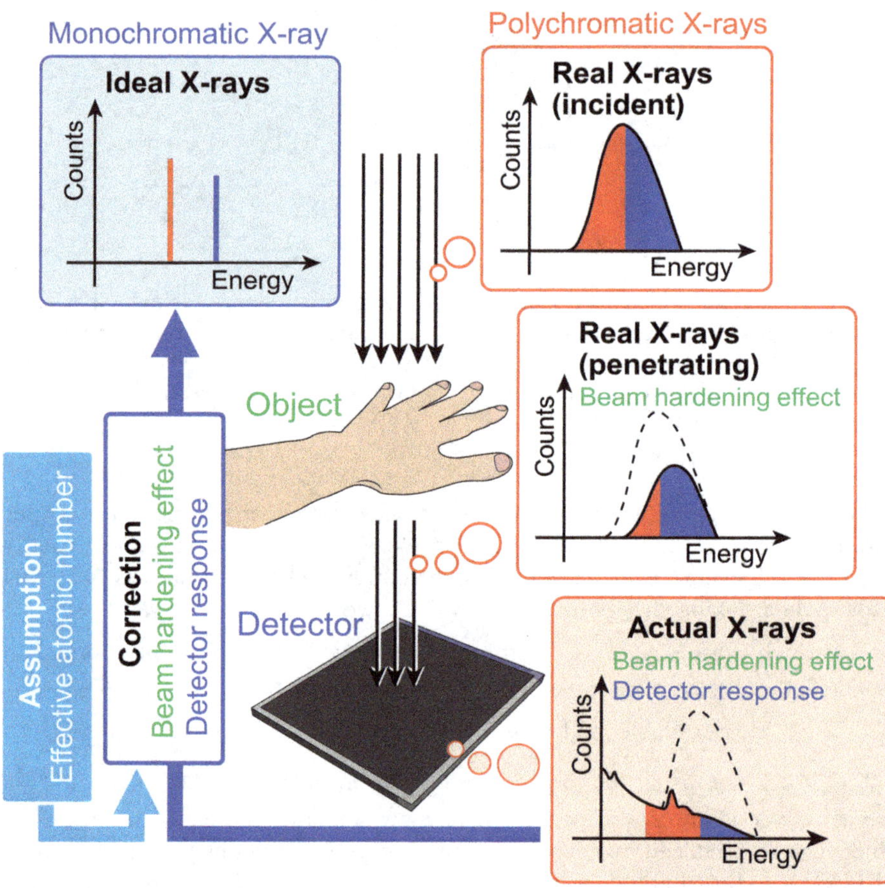

Fig. 4 Comparison of X-ray spectra obtained at each stage from incident to detection. The actual X-ray spectrum includes both the beam hardening effect and detector response. Our approach is to treat the measured polychromatic X-rays as if they are monochromatic X-rays by correcting for the beam hardening effect and detector response

equipment. When using an ERPCD having two energy bins, the low- and high-energy bins, presented by red and blue, respectively, are analyzed quantitatively. To perform the analysis, it is necessary to distinguish the energies, and we generally use the effective energies of each energy bin. In the spectrum obtained after penetrating an object, the degree of X-ray attenuation varies in the difference of Z_{eff} of an object, and as a result, the effective energy changes. This phenomenon is known as the "beam hardening effect" [25]. Furthermore, when X-rays are incident to the detector, the X-rays interact with detector materials, and the corresponding X-ray spectrum generated from the detector is much different than the incident X-ray spectrum. Although current commercial products use to take a practical approach for material decomposition based on the actual X-ray spectrum [26–29], our approach is much different. We apply the correction for the beam hardening effect and detector response to actual data by taking into consideration the Z_{eff} of an object [14–16, 19]. This method can process actual polychromatic X-rays as those equivalent to ideal monochromatic X-rays, which results in the accomplishment of what is considered "ideal" analysis based on monochromatic X-rays. In the following sections, we will carefully analyze the effects of the beam hardening effect and detector response and also introduce a novel method to correct these effects.

3.1 Beam Hardening Effect Depends on Effective Atomic Number

In this section, we will explain the beam hardening effect which is one of the issues to be considered for precise material identification. Figure 5 shows the relationship between the beam hardening effect and Z_{eff}. Figure 5a shows Z_{eff} values of the elements which compose the human body. The human body targeted for clinical examination is composed of various organs and tissue, and the Z_{eff} values of each organ and tissue are distributed at relatively low atomic number values. These Z_{eff} values can be calculated from the elemental composition of each organ [24] using a well-known formula [30]. As shown in this figure, adipose tissue is 6.33, muscle is 7.45, and cortical bone is 13.23; it is obvious that there is a slight difference in Z_{eff} between the soft tissues, and it can be roughly divided into soft tissue at around $Z_{\mathrm{eff}} = 7$ and bone around $Z_{\mathrm{eff}} = 13$. Next, we present the beam hardening effect depending on the difference of Z_{eff}. Figure 5b shows the changes in the spectra and the effective energies ($\overline{E}$) corresponding to objects having various Z_{eff} values. The black line is the incident X-ray spectrum at 50 kV which is reproduced by a semiempirical formula [31], and the $\overline{E}$ of the energy region 20–32 keV is 27.4 keV. We also present the X-ray spectra after penetrating objects of which Z_{eff} values and ρt value are set to be 5–15 and 1 g/cm^2, respectively. The distribution of X-rays is different depending on Z_{eff}, and it is important to especially focus on $Z_{\mathrm{eff}} = 7$ (blue line) and $Z_{\mathrm{eff}} = 13$ (red line). The corresponding $\overline{E}$ values are 27.8 keV and 29.5 keV, respectively. It is found that the difference between incident X-ray and penetrating X-ray shows the beam hardening effect, and the degree of effect

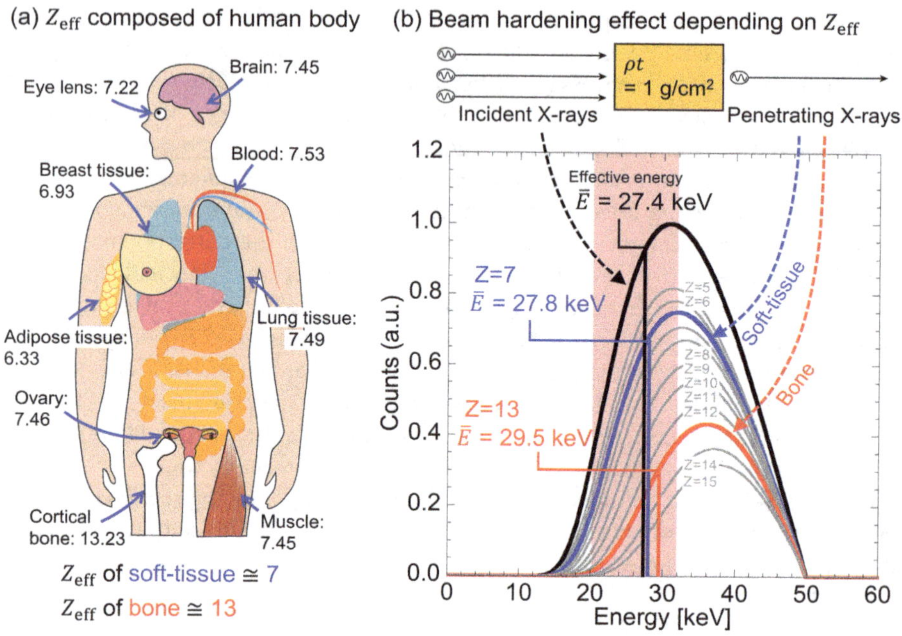

Fig. 5 Relationship between beam hardening effect and Z_{eff}. (**a**) shows the Z_{eff} values of various tissues which compose the human body. (**b**) shows the changes in the X-ray spectra and effective energies corresponding to an object having various Z_{eff} values

differs depending on the Z_{eff}. Unless these effects are properly corrected based on the information of an object, it is not possible to perform an analysis in which polychromatic X-rays are regarded as monochromatic X-rays.

3.2 Response of Multipixel-Type Energy-Resolving Photon-Counting Detector

Another issue to be considered for precise material identification is detector response. In this section, we will explain the concept of the response function of a multipixel-type ERPCD and the procedure to reproduce an X-ray spectrum using the response function. Figure 6 shows a concept of the various phenomena related to detector response when X-rays $\mathbf{\Phi}$ are incident to the detector. Figure 6a shows an ideal case where all of the energies of the incident X-rays are absorbed completely by the detector. Using the response function represented by an identity matrix $\mathbf{I}$, the corresponding X-ray spectrum can be calculated by the matrix operation $\mathbf{I\Phi}$ ($= \mathbf{\Phi}$). In actuality, the response function including phenomenon $\mathbf{R}^{(1)}$ which is related to radiation interactions and phenomenon $\mathbf{R}^{(2)}$ which is related to electric charge collecting processes should be considered to reproduce an X-ray spectrum

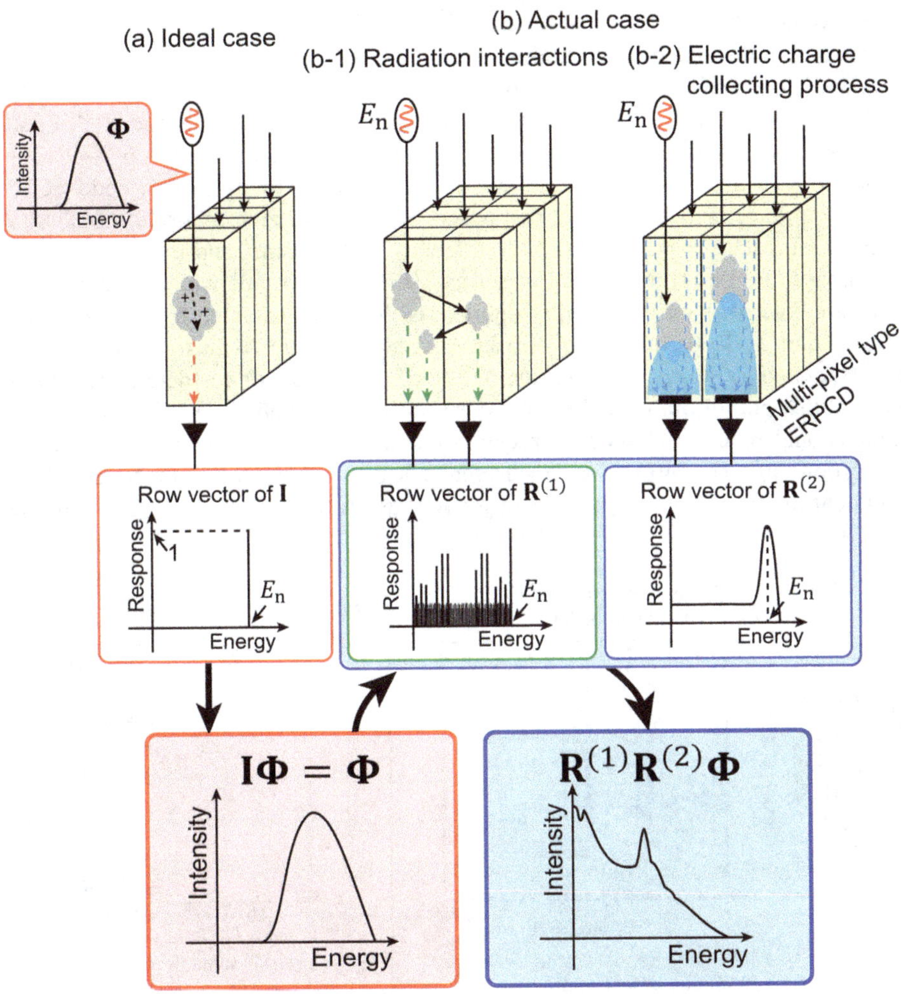

Fig. 6 Comparison between detector responses for ideal and actual cases. (**a**) in the ideal case, incident X-rays $\mathbf{\Phi}$ are totally absorbed within the pixel of the detector, and the detected X-ray spectrum ($\mathbf{I\Phi} = \mathbf{\Phi}$) is the same as the incident X-ray spectrum. (**b**) in actuality, incident X-rays are affected by two different phenomena: (**b**-1) radiation interactions and (**b**-2) electric charge collecting processes. Consequently, the detected X-ray spectrum $\mathbf{R}^{(1)}\mathbf{R}^{(2)}\mathbf{\Phi}$ is much different from that of the incident X-ray spectrum

measured with a multipixel-type ERPCD. Figure 6b-1 shows $\mathbf{R}^{(1)}$ in which we take into consideration the transportation of the characteristics X-rays caused by the photoelectric effect and scattered X-rays due to the Compton scattering effect [7, 9, 17, 19, 32–36]. Figure 6b-2 shows $\mathbf{R}^{(2)}$ in which the charge sharing effect [32–36] and energy resolution [32, 33, 35, 36] are considered. When considering both $\mathbf{R}^{(1)}$ and $\mathbf{R}^{(2)}$ operation of $\mathbf{\Phi}$, the X-ray spectrum under actual conditions can then be reproduced by matrix operation $\mathbf{R}^{(1)}\mathbf{R}^{(2)}\boldsymbol{\Phi}$. This model can predict the accurate detector response under the assumption in which a contamination of scattered X-rays and a generation of pulse pileup are not presented.

Here, we will look more closely at the response functions $\mathbf{R}^{(1)}$ and $\mathbf{R}^{(2)}$. First, the response function $\mathbf{R}^{(1)}$ is calculated by simulating the interaction between the incident X-rays and detector materials using Monte-Carlo simulation code EGS5 [37]. The detector materials are composed of cadmium zinc telluride (CZT) at a ratio of Cd:Zn:Te = 0.9:0.1:1.0 with a density of 5.8 g/cm^3 and the outer size of the monolithic detector material is set as 10 mm × 10 mm with a thickness of 1.5 mm. The response function is defined as the spectra related to a center pixel of 200 μm × 200 μm. The 0–140 keV monochromatic X-rays of 10^6 photons are irradiated to the center pixel, and the total irradiation area is determined so as to establish equilibrium for the secondary produced radiation [9, 17, 19]. A two-dimensional matrix $\mathbf{R}^{(1)}$ is constructed from the elements of response functions corresponding to 0–140 keV monochromatic X-rays. Here, we demonstrate 80 keV monochromatic X-ray vector element for $\mathbf{R}^{(1)}$ as presented in Fig. 7a. The red line

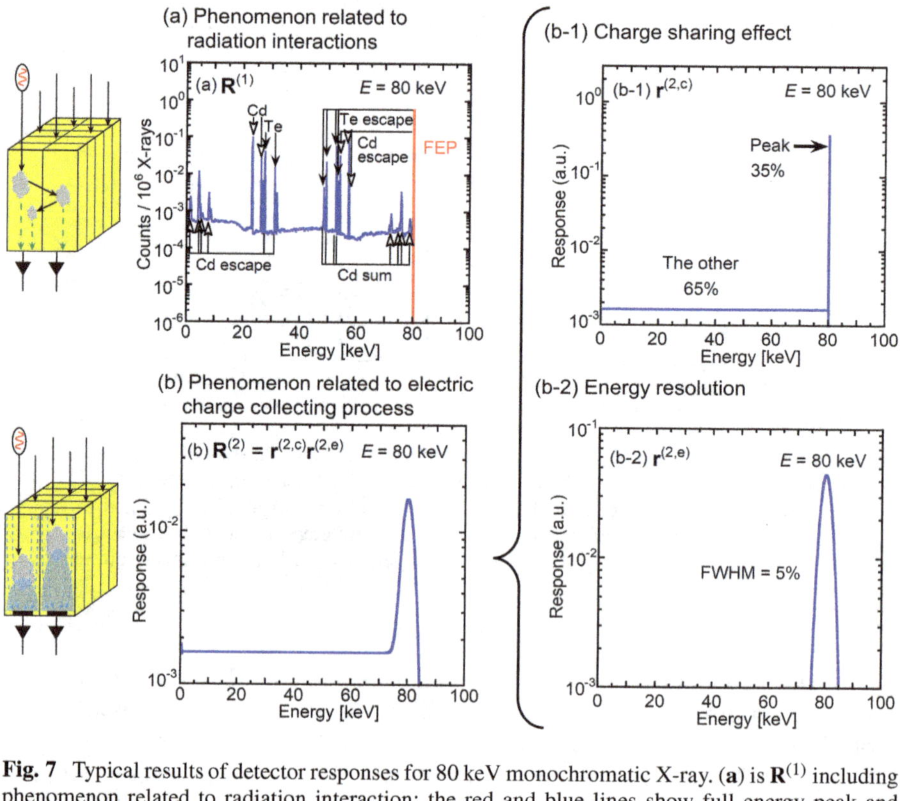

Fig. 7 Typical results of detector responses for 80 keV monochromatic X-ray. (**a**) is $\mathbf{R}^{(1)}$ including phenomenon related to radiation interaction; the red and blue lines show full energy peak and scattered X-rays, respectively. (**b**) is $\mathbf{R}^{(2)}$ including phenomenon related to the electric charge collection processes which are composed of (**b**-1) charge sharing effect $\mathbf{r}^{(2,c)}$ and (**b**-2) energy resolution $\mathbf{r}^{(2,e)}$

represents the full-energy peak (FEP) which appears when all of the incident X-ray energy is completely absorbed by the pixel of interest. On the other hand, the blue area relates to partially absorbed events, and some peaks are observed at 23–32 keV and 48–57 keV. It is important to understand these phenomena; when the X-rays are incident to the pixel, the photoelectric effect mainly occurs, and then characteristic X-rays of Cd (23–27 keV) and Te (27–32 keV) are generated. When focusing on the pixel of interest, there is a possibility of the characteristic X-rays escaping to adjacent pixels, which results in the existence of escape peaks (EPs) at 48–57 keV. On the other hand, when focusing on the pixel adjacent to the pixel of interest, the characteristic X-rays generating at pixel of interest are incident, which results in the existence of peaks at 23–32 keV. It should be noted that the characteristic X-ray peaks are constant regardless of the energy of the incident X-rays, whereas the EPs vary according to the energy of the incident X-rays. Next, the response function $\mathbf{R}^{(2)}$ is reproduced taking into consideration charge sharing effect $\mathbf{r}^{(2,\mathrm{c})}$ and energy resolution $\mathbf{r}^{(2,\mathrm{e})}$. During charge collection process in a detector, electrons undergo a diffusion process and drift due to an electric field. Consequently, not all of the charges are collected by the pixel of interest, and some charges spread into several adjacent pixels caused by diffusions and collisions of produced electrons. This phenomenon is called the "charge sharing effect." One vector element in $\mathbf{r}^{(2,\mathrm{c})}$ is exemplified in Fig. 7b-1. It consists of the peak component and the other part which is presented by a flat distribution. For our ERPCD, the respective proportions were optimized and the values were determined to be 35% and 65% so as to reproduce the X-ray spectrum measured with the ERPCD. In addition, as shown in Fig. 7b-2, energy resolution $\mathbf{r}^{(2,\mathrm{e})}$ was determined as 5% at 80 keV so as to reproduce characteristic X-ray peaks. Using $\mathbf{r}^{(2,\mathrm{c})}$ and $\mathbf{r}^{(2,\mathrm{e})}$, the 80 keV vector element in $\mathbf{R}^{(2)}$ is exemplified in Fig. 7b in which the peaks around 80 keV and flat distribution can be observed.

Next, we will explain the procedure to reproduce an X-ray spectrum using the response functions $\mathbf{R}^{(1)}$ and $\mathbf{R}^{(2)}$. To easily understand the procedure, we reproduce two different X-ray spectra obtained under the ideal and actual situations as shown in Fig. 6. First, we set the X-ray spectrum,

$$\boldsymbol{\Phi} = \begin{pmatrix} \Phi_1 \\ \Phi_2 \\ \vdots \\ \Phi_\mathrm{j} \\ \vdots \end{pmatrix}, \tag{2}$$

where the element of this vector is expressed as "Φ_j". Figure 8a shows $\boldsymbol{\Phi}$ when setting the tube voltage at 50 kV.

In the ideal case, the response function can be expressed as the identity matrix:

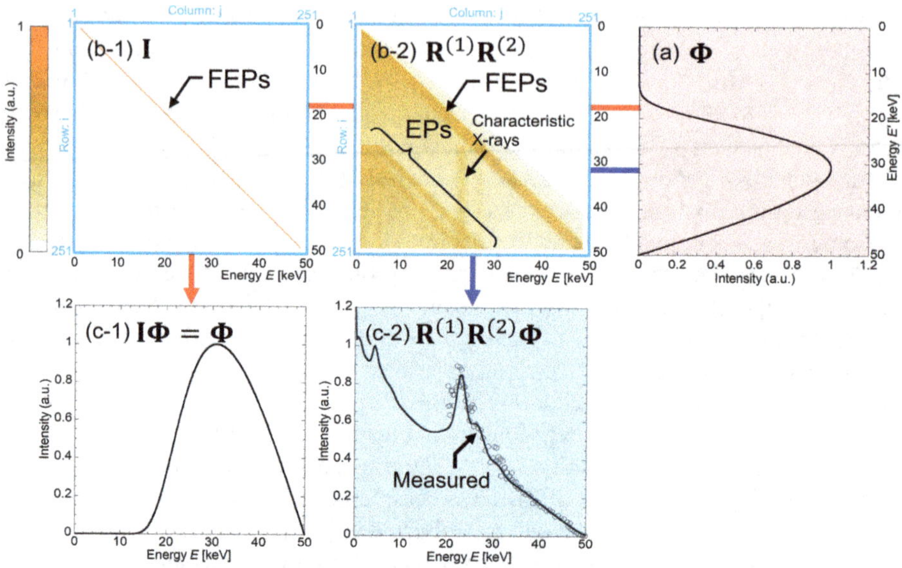

Fig. 8 The procedure for reproducing an X-ray spectrum taking into consideration detector response. (**a**) shows incident X-ray spectrum $\boldsymbol{\Phi}$. (**b**-1) and (**b**-2) show two-dimensional color maps of detector responses $\mathbf{I}$ and $\mathbf{R}^{(1)}\mathbf{R}^{(2)}$, respectively. (**c**-1) and (**c**-2) show reproduced X-ray spectra $\mathbf{I}\boldsymbol{\Phi} = \boldsymbol{\Phi}$ and $\mathbf{R}^{(1)}\mathbf{R}^{(2)}\boldsymbol{\Phi}$, respectively. By performing the mathematical calculations, we can reproduce X-ray spectra measured with an ERPCD

$$\mathbf{I} = \begin{pmatrix} 1 & 0 & \cdots & \cdots & 0 \\ 0 & 1 & \vdots & \vdots & \vdots \\ \vdots & \vdots & \ddots & \vdots & \vdots \\ \vdots & \vdots & \vdots & 1 & 0 \\ 0 & \cdots & \cdots & 0 & 1 \end{pmatrix}. \tag{3}$$

Figure 8b-1 shows a two-dimensional color map presented by $\mathbf{I}$. It can be seen that unit values corresponding to FEPs appear in diagonal line.

In actual circumstances, the response function can be calculated by the matrix operation for $\mathbf{R}^{(1)}$ and $\mathbf{R}^{(2)}$,

$$\mathbf{R}^{(1)}\mathbf{R}^{(2)} = \begin{pmatrix} R^{(1)}_{1,1} & R^{(1)}_{1,2} & \cdots & \cdots & \cdots \\ R^{(1)}_{2,1} & \ddots & \vdots & \vdots & \vdots \\ \vdots & \vdots & R^{(1)}_{\mathrm{i,k}} & \vdots & \vdots \\ \vdots & \vdots & \vdots & \ddots & \vdots \\ \cdots & \cdots & \cdots & \cdots & \cdots \end{pmatrix} \begin{pmatrix} R^{(2)}_{1,1} & R^{(2)}_{1,2} & \cdots & \cdots & \cdots \\ R^{(2)}_{2,1} & \ddots & \vdots & \vdots & \vdots \\ \vdots & \vdots & R^{(2)}_{\mathrm{k,j}} & \vdots & \vdots \\ \vdots & \vdots & \vdots & \ddots & \vdots \\ \cdots & \cdots & \cdots & \cdots & \cdots \end{pmatrix}$$

$$= \begin{pmatrix} R^{(1,2)}_{1,1} & R^{(1,2)}_{1,2} & \cdots & \cdots\cdots \\ R^{(1,2)}_{2,1} & \ddots & \vdots & \vdots & \vdots \\ \vdots & \vdots & R^{(1,2)}_{i,j} & \vdots & \vdots \\ \vdots & \vdots & \vdots & \ddots & \vdots \\ \cdots & \cdots & \cdots & \cdots\cdots \end{pmatrix}, \tag{4}$$

where $R^{(1,2)}_{i,j}$ is described as

$$R^{(1,2)}_{i,j} = \sum_{k} R^{(1)}_{i,k} R^{(2)}_{k,j}.$$

$\mathbf{R}^{(1)}$ consists of the elements of $\mathbf{R}^{(1)}_{E'}$ corresponding to various incident X-ray energies E'. Then, $\mathbf{R}^{(1)}_{E'}$ is expressed as $\left\{R^{(1)}_{i,1}, R^{(1)}_{i,2}, \ldots, R^{(1)}_{i,k}, \ldots\right\}$ where the corresponding response energy is the i-th element. In a similar way, $\mathbf{R}^{(2)}$ has elements of $\mathbf{R}^{(2)}_{E'}$. Then, the element $\mathbf{R}^{(1)}\mathbf{R}^{(2)}$ in the i-th row and j-th column is expressed as $R^{(1,2)}_{i,j}$. Figure 8b-2 shows a two-dimensional color map of $\mathbf{R}^{(1)}\mathbf{R}^{(2)}$. The response corresponding to FEPs is shown with the diagonal line and the sharpness is affected by the energy resolution. The characteristic X-rays are observed at around 23 keV. EPs are represented by diagonal lines where the highest value is around 27 keV. The charge sharing effect causes an increase in intensities in the low-energy region.

Finally, we will reproduce X-ray spectra using response functions **I**·and $\mathbf{R}^{(1)}\mathbf{R}^{(2)}$. In ideal case, X-ray spectrum can be obtained by folding $\mathbf{\Phi}$ with **I**,

$$\mathbf{I\Phi} = \begin{pmatrix} 1 & 0 & \cdots\cdots & 0 \\ 0 & 1 & \vdots & \vdots & \vdots \\ \vdots & \vdots & \ddots & \vdots & \vdots \\ \vdots & \vdots & \vdots & 1 & 0 \\ 0 & \cdots\cdots & & 0 & 1 \end{pmatrix} \begin{pmatrix} \Phi_1 \\ \Phi_2 \\ \vdots \\ \Phi_j \\ \vdots \end{pmatrix} = \begin{pmatrix} \Phi_1 \\ \Phi_2 \\ \vdots \\ \Phi_j \\ \vdots \end{pmatrix}. \tag{5}$$

Figure 8c-1 shows $\mathbf{I\Phi}$. This is the same as incident X-ray spectrum $\mathbf{\Phi}$.

In the actual case, we can reproduce the X-ray spectrum by replacing **I** in Eq. (5) by $\mathbf{R}^{(1)}\mathbf{R}^{(2)}$:

$$\mathbf{R}^{(1)}\mathbf{R}^{(2)}\mathbf{\Phi} = \begin{pmatrix} R^{(1,2)}_{1,1} & R^{(1,2)}_{1,2} & \cdots & \cdots\cdots \\ R^{(1,2)}_{2,1} & \ddots & \vdots & \vdots & \vdots \\ \vdots & \vdots & R^{(1,2)}_{i,j} & \vdots & \vdots \\ \vdots & \vdots & \vdots & \ddots & \vdots \\ \cdots & \cdots & \cdots & \cdots\cdots \end{pmatrix} \begin{pmatrix} \Phi_1 \\ \Phi_2 \\ \vdots \\ \Phi_j \\ \vdots \end{pmatrix} = \begin{pmatrix} \sum_j R^{(1,2)}_{1,j}\Phi_j \\ \sum_j R^{(1,2)}_{2,j}\Phi_j \\ \vdots \\ \sum_j R^{(1,2)}_{i,j}\Phi_j \\ \vdots \end{pmatrix}, \tag{6}$$

where each element is described as

$$\sum_{\rm j} R^{(1,2)}_{\rm i,j}\Phi_{\rm j} = \sum_{\rm j}\left(\sum_{\rm k} R^{(1)}_{\rm i,k} R^{(2)}_{\rm k,j}\right)\Phi_{\rm j} = \sum_{E'}\left(\sum_{\rm k} R^{(1)}_{\rm i,k} R^{(2)}_{{\rm k},E'}\right)\Phi\left(E'\right),$$

where j is the j-th element of $\boldsymbol{\Phi}$, for the sake of clarification, j is expressed as E'. Figure 8c-2 shows the reproduced X-ray spectrum $\mathbf{R}^{(1)}\mathbf{R}^{(2)}\boldsymbol{\Phi}$ plotted with the X-ray spectrum measured using our prototype ERPCD [13, 15, 38]. $\mathbf{R}^{(1)}\mathbf{R}^{(2)}\boldsymbol{\Phi}$ is in good agreement with the experimental data. We can see that the $\mathbf{R}^{(1)}\mathbf{R}^{(2)}\boldsymbol{\Phi}$ has two major features: relatively large intensities in the low-energy region and the presence of the characteristic X-ray peaks of Cd and Te. The $\mathbf{R}^{(1)}\mathbf{R}^{(2)}\boldsymbol{\Phi}$ is much different than that of $\boldsymbol{\Phi}$.

Here, in order to simplify the mathematical notation, $\boldsymbol{\Phi}$ is redefined as following formations: $\boldsymbol{\Phi} = \{\Phi(0\ \text{keV}), \Phi(0.2\ \text{keV}), \cdots, \Phi(E_{\rm k}), \cdots\}$ where $E_{\rm k}$ is X-ray energy at intervals of 0.2 keV (subscript of "k" means k-th vector element) for matrix formation and $\Phi(E)$ for continuous function. We also redefine $\mathbf{R}^{(1)}\mathbf{R}^{(2)}\boldsymbol{\Phi}$ as matrix formation: $\mathbf{R}^{(1)}\mathbf{R}^{(2)}\boldsymbol{\Phi} = \{R^{(1)}R^{(2)}\Phi(0\ \text{keV}), R^{(1)}R^{(2)}\Phi(0.2\ \text{keV}), \cdots, R^{(1)}R^{(2)}\Phi(E_{\rm k}), \cdots\}$. These formation roles are applied to all equations in the following description.

3.3 Procedure to Correct for the Beam Hardening Effect and Detector Response

To accomplish highly accurate material identification, the biggest challenges that need to be solved are the issues of the beam hardening effect and detector response. In this section, we will introduce the correction procedure for the beam hardening effect and detector response. Figure 9 shows a schematic drawing that explains the concept of the correction. The correction is to convert $(\mu t)_{\rm meas}$ which is calculated from polychromatic X-rays detected by the ERPCD into $(\mu t)_{\rm cor}$ calculated from monochromatic X-rays. To easily understand the correction of both effects, we focus on (a) monochromatic X-rays in $\boldsymbol{\Phi}$, (b) polychromatic X-rays in $\boldsymbol{\Phi}$, and (c) polychromatic X-rays in $\mathbf{R}^{(1)}\mathbf{R}^{(2)}\boldsymbol{\Phi}$. (b) includes the beam hardening effect and (c) includes both beam hardening effect and detector response. As shown, common for conditions (a) through (c), we set the lower and upper energies of each energy bin as E_1 and E_2, respectively, and $\overline{E}$ is calculated. Assuming that an object with certain $Z_{\rm eff}$ and ρt values is measured, we simulate the X-ray spectra and calculate corresponding μt values related to conditions (a) through (c). Then, we derive unique relationships between $\rho t \times \mu/\rho$ and μt which were proposed in a previous study [12, 14–16, 19].

Here, we explain the relationship of (a) monochromatic X-ray energy $\overline{E}$. The $(\mu t)_{\rm cor}$ is calculated from the equation, $\ln\left(\frac{\Phi_{\rm i}(\overline{E})}{\Phi_{\rm p}(\overline{E})}\right)$. Because the $\overline{E}$ is always constant even when ρt and/or $Z_{\rm eff}$ are varied, $(\mu t)_{\rm cor}$ values can be plotted as a

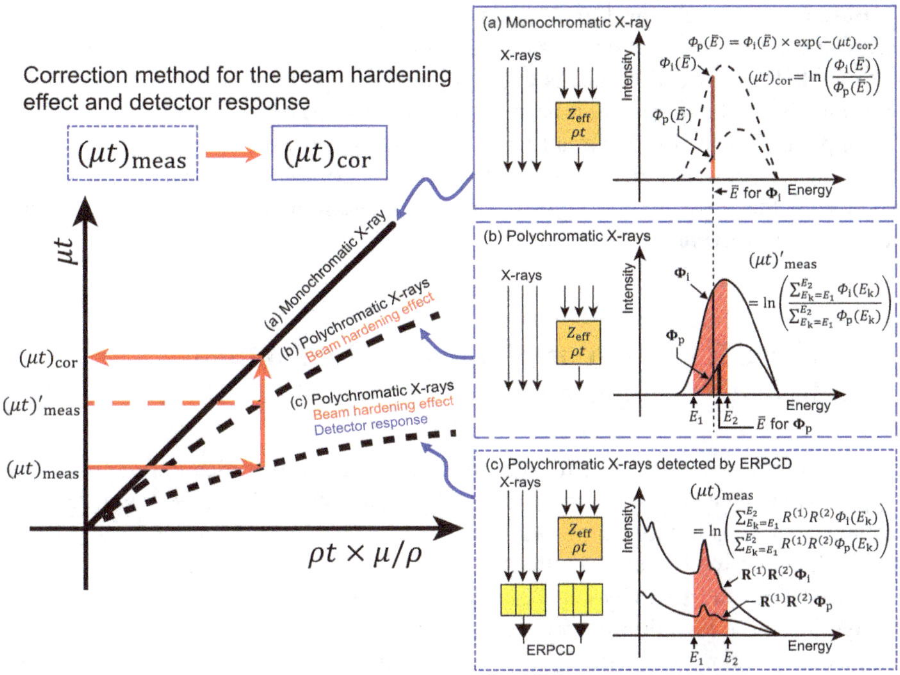

Fig. 9 Concept of the procedure to correct both the beam hardening effect and detector response. The correction is to convert $(\mu t)_{meas}$ calculated from polychromatic X-rays measured with an ERPCD to $(\mu t)_{cor}$ which is related to monochromatic X-rays. The former case consists of energy information disruption caused by the beam hardening effect and detector response, and the latter case is the retention of original energy information reflecting X-ray attenuation of the object. To clearly show the correction for both effects, we present each step of the correction: (**a**) $(\mu t)_{cor}$ calculated from monochromatic X-rays, (**b**) $(\mu t)'_{meas}$ calculated from polychromatic X-rays, and (**c**) $(\mu t)_{meas}$ calculated from polychromatic X-rays measured with the ERPCD

straight line Y = X; X and Y axes are $\rho t \times \mu/\rho$ and μt, respectively. Next, for (b) polychromatic X-rays, because the $\overline{E}$ of $\mathbf{\Phi}_p$, which is calculated from the equation, $\frac{\sum_{E_k=E_1}^{E_2} \Phi_p(E_k)E_k}{\sum_{E_k=E_1}^{E_2} \Phi_p(E_k)}$, is higher due to the beam hardening effect, the $(\mu t)'_{meas}$ calculated from the equation, $\ln\left(\frac{\sum_{E_k=E_1}^{E_2} \Phi_i(E_k)}{\sum_{E_k=E_1}^{E_2} \Phi_p(E_k)}\right)$, differs from the $(\mu t)_{cor}$; the $(\mu t)'_{meas}$ shows a curve which goes away from the Y = X line. Finally, when using (c) polychromatic X-rays detected using an ERPCD, $(\mu t)_{meas}$ calculated from the equation, $\ln\left(\frac{\sum_{E_k=E_1}^{E_2} R^{(1)}R^{(2)}\Phi_i(E_k)}{\sum_{E_k=E_1}^{E_2} R^{(1)}R^{(2)}\Phi_p(E_k)}\right)$, is much different from the Y = X line due to the beam hardening effect and detector response. The curve was calculated at intervals of 0.1 g/cm^2 for virtual objects having Z_{eff} = 5.0–15.0 with ρt = 0–150 g/cm^2.

Based on the relationship between (a) and (c), the $(\mu t)_{\text{meas}}$ measured with ERPCD can be converted into $(\mu t)_{\text{cor}}$ as shown by the direction of red arrows in the left panel. This correction means that both effects can be corrected simultaneously and polychromatic X-rays can be treated as monochromatic X-rays. Although we can apply the correction accurately under the assumption that the Z_{eff} of an object is known, it is a rare case that we know the Z_{eff} of an object in advance. To make the best use of the correction curve, we propose to determine Z_{eff} while estimating the amount of correction in later sections.

4 Material Identification Method Leading to Innovations in Imaging Technology

In this section, we would like to introduce our material identification method. The biggest advantage of our method is that the beam hardening effect and detector response can be corrected appropriately by taking into consideration the Z_{eff} of an object. This correction realizes the ideal analysis in which polychromatic X-rays measured with the ERPCD can be treated as those equivalent to monochromatic X-rays. As an application example showing the effectiveness of this analysis, we succeeded not only in identifying the Z_{eff} of an object but also in extracting ρt related to the soft tissue and bone. In the following section, we introduce the proposed method to derive Z_{eff} and ρt values of the soft tissue and bone. We then show the feasibility of our method visually using images produced with prototype ERPCD.

4.1 Novel Method to Derive Effective Atomic Number Image and Extract Mass Thickness Images Related to the Soft Tissue and Bone

We will introduce a method to derive the Z_{eff} and extract the ρt values related to the soft tissue and bone. Figure 10 shows a diagram of the method. Assuming that a bilayer structure consisting of "a" and "b" is measured using an ERPCD having several energy bins, we perform the experiments without and with the presence of objects, as shown in the upper center panel. In the current system, we adopted to use two energy bins. Steps shown in Fig. 10a–f are applied to derive Z_{eff} value and each ρt value. First, we prepare count values which are measured in the experiments without and with objects as shown in Fig. 10a. The count values without and with objects correspond to mathematical expressions $\sum_{E_k=E_1}^{E_2} R^{(1)} R^{(2)} \Phi_i\,(E_k)$ and $\sum_{E_k=E_1}^{E_2} R^{(1)} R^{(2)} \Phi_p\,(E_k)$, respectively. Next, using these values, attenuation factor $(\mu t)_{\text{meas}}$ is calculated as

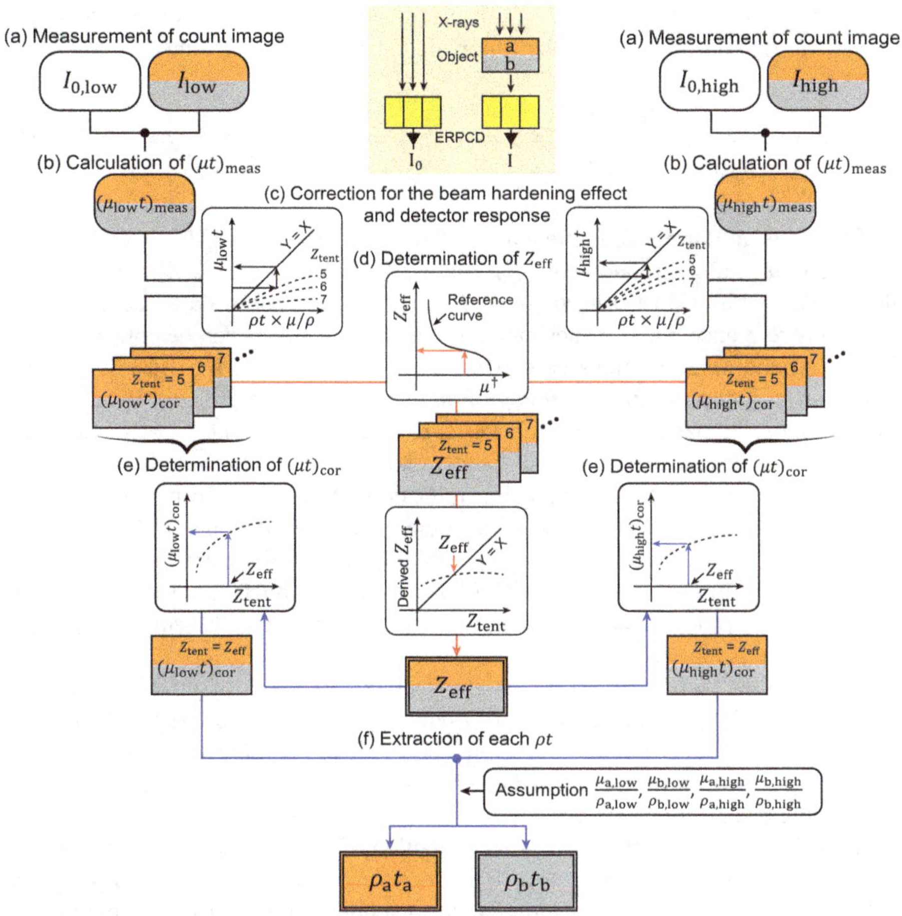

Fig. 10 Diagram of the proposed algorithm for determining the Z_{eff} and ρt values. Assuming that a bilayer structure consisting of elements "a" and "b" is measured using an ERPCD, we perform the experiments without and with the presence of the object. The six steps (**a**) through (**f**) are applied to measurement data from two energy bins to derive the Z_{eff} value and the ρt values corresponding to the soft tissue and bone

$$(\mu t)_{\mathrm{meas}} = \ln\left(\frac{\sum_{E_{\mathrm{k}}=E_1}^{E_2} R^{(1)} R^{(2)} \Phi_{\mathrm{i}}(E_{\mathrm{k}})}{\sum_{E_{\mathrm{k}}=E_1}^{E_2} R^{(1)} R^{(2)} \Phi_{\mathrm{p}}(E_{\mathrm{k}})}\right). \tag{7}$$

This calculation is applied to low- and high-energy bins, and $(\mu_{\mathrm{low}}t)_{\mathrm{meas}}$ and $(\mu_{\mathrm{high}}t)_{\mathrm{meas}}$ are derived as shown in Fig. 10b. Because $(\mu t)_{\mathrm{meas}}$ is affected by the beam hardening effect and detector response, corrections related to various Z_{tent} values which are the values tentatively assigned as the Z_{eff} of an object are performed in each energy bin; namely, $(\mu t)_{\mathrm{meas}}$ is converted into $(\mu t)_{\mathrm{cor}}$ values related to Z_{tent} values 5, 6, 7, …, 15, as shown in Fig. 10c. Then, to extract μ using $(\mu_{low}t)_{cor}$ and $(\mu_{high}t)_{cor}$, the following calculation is performed:

$$\mu^{\dagger} \equiv \frac{(\mu_{high}t)_{cor}}{\sqrt{(\mu_{low}t)_{cor}^2 + (\mu_{high}t)_{cor}^2}} = \frac{(\mu_{high})_{cor}}{\sqrt{(\mu_{low})_{cor}^2 + (\mu_{high})_{cor}^2}}, \tag{8}$$

where $\mu^{\dagger}$ is a normalized attenuation factor [8, 10–12, 14–16, 18, 19], and we obtain $\mu^{\dagger}$ values related to Z_{tent} values. In order to derive Z_{eff} from $\mu^{\dagger}$, a theoretical relationship between $\mu^{\dagger}$ and Z_{eff} is created using database μ/ρ [24]. Using this reference curve, experimentally derived $\mu^{\dagger}$ values can be converted into the corresponding Z_{eff} values as shown in Fig. 10d. The derived Z_{eff} values are plotted as a function of Z_{tent}, and Z_{eff} of an object can be determined by searching for the intersectional point between the derived Z_{eff} curve and Y = X line. This algorithm to determine the Z_{eff} of an object takes advantage of the following trend. When Z_{tent} is the same as Z_{eff} of an object, the correction can perform appropriately, and the derived Z_{eff} becomes Z_{eff} of an object. On the other hand, when Z_{tent} is different from Z_{eff} of an object, the derived Z_{eff} is also different from Z_{eff} of an object due to improper correction. This means that the intersectional point of the Z_{tent} versus Z_{eff} plot indicates the Z_{eff} of an object, and we can determine Z_{eff} of the object using this algorithm.

Furthermore, taking advantage of the fact that the Z_{eff} of an object can be determined and simultaneously appropriate correction can be applied, we can perform further analysis and extract other properties of an object. We focus on the use of $(\mu t)_{cor}$ with the correction when obtaining the Z_{eff} of an object. Fortunately, when going back to Fig. 10c, a database of $(\mu t)_{cor}$ values related to Z_{tent} values has been obtained. Now plotting $(\mu t)_{cor}$ as a function of Z_{tent}, the $(\mu t)_{cor}$ related to Z_{eff} can be determined as shown in Fig. 10e. Because the $(\mu t)_{cor}$ related to Z_{eff} shows that the beam hardening effect and detector response can be corrected completely, ideal analysis based on monochromatic X-ray energy can be carried out. As one valuable application using $(\mu t)_{cor}$, we can extract ρt values of each element by solving simultaneous equations:

$$(\mu_{low}t)_{cor} = \frac{\mu_{a,low}}{\rho_a}\rho_a t_a + \frac{\mu_{b,low}}{\rho_b}\rho_b t_b, \tag{9-1}$$

$$(\mu_{high}t)_{cor} = \frac{\mu_{a,high}}{\rho_a}\rho_a t_a + \frac{\mu_{b,high}}{\rho_b}\rho_b t_b, \tag{9-2}$$

where $\frac{\mu_{a,low}}{\rho_a}$ and $\frac{\mu_{b,low}}{\rho_b}$ are μ/ρ values at low energy for "a" and "b," respectively, $\frac{\mu_{a,high}}{\rho_a}$ and $\frac{\mu_{b,high}}{\rho_b}$ are μ/ρ values at high energy for "a" and "b," respectively, additionally $\rho_a t_a$ and $\rho_b t_b$ are ρt values of "a" and "b," respectively. Because we have information related to monochromatic X-ray energy ($\overline{E}$ of $\boldsymbol{\Phi}_i$) under the assumption that bilayer structure consisting of elements "a" and "b" is measured, the μ/ρ values can be assigned by using a well-known database [24]. Consequently,

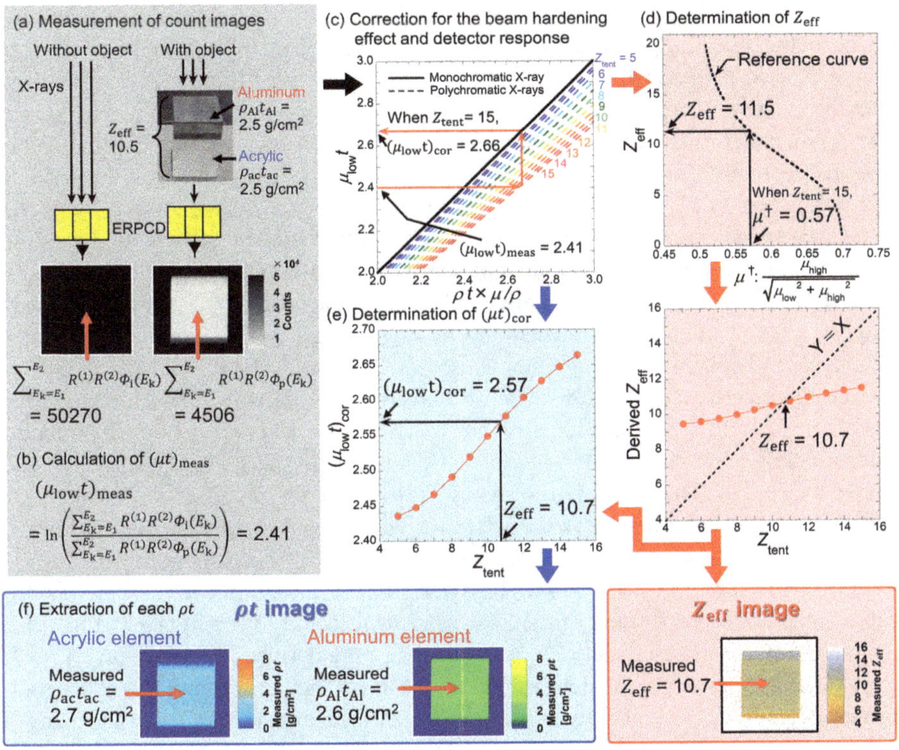

Fig. 11 Typical results to explain each process to derive Z_{eff} and ρt images. Using the data measured for a bilayer structure consisting of acrylic and aluminum, the Z_{eff} and ρt images can be derived by performing the six steps (**a**) through (**f**)

the unknown values are $\rho_a t_a$ and $\rho_b t_b$, and therefore they can be derived by solving the simultaneous equations as shown in Fig. 10f. This is one analysis that makes full use of our method in which monochromatic X-rays can be virtually obtained through the correction of beam hardening effect and detector response considering the identification of Z_{eff}. We are convinced that there are additional possibilities of discovering novel values using our method.

Next, to more clearly understand the procedure to determine Z_{eff} and ρt, we explain each process using typical results. Figure 11 shows typical results obtained when measuring a bilayer structure (Z_{eff} = 10.5) of acrylic ($\rho_{ac}t_{ac}$ = 2.5 g/cm^2) and aluminum ($\rho_{Al}t_{Al}$ = 2.5 g/cm^2). As shown in Fig. 11a, we measure the count images obtained in the experiments without and with the presence of an object. Although we use count images from the low-energy bin (32–40 keV) and high-energy bin (40–50 keV) for analysis, for simplicity, we show the results related to a certain pixel of images from the low-energy bin. The pixel values of interest in count images are 50,270 and 4506 which correspond to the calculations $\sum_{E_k=E_1}^{E_2} R^{(1)}R^{(2)}\Phi_i(E_k)$ and $\sum_{E_k=E_1}^{E_2} R^{(1)}R^{(2)}\Phi_p(E_k)$, respectively. Substituting these values into Eqs. (7), $(\mu_{\text{low}}t)_{\text{meas}}$ is calculated as 2.41 as shown in Fig. 11b. Because the $(\mu_{\text{low}}t)_{\text{meas}}$ is affected by the beam hardening effect and detector response, correction is performed as shown in Fig. 11c. Figure 11c shows the correction curves presented by the relationship between $\rho t \times \mu/\rho$ and $\mu_{\text{low}}t$. The solid and broken lines show monochromatic X-rays and polychromatic X-rays, respectively. In this graph, monochromatic X-rays are plotted on the Y = X line regardless of Z_{tent}, and the polychromatic X-ray trend depends on Z_{tent}. The difference between monochromatic X-rays and polychromatic X-rays indicates the amount of correction; it is seen that at higher Z_{tent}, the amount of correction is larger. By assuming the Z_{tent} values are between 5 and 15, the $(\mu_{\text{low}}t)_{\text{meas}}$ can be converted into $(\mu_{\text{low}}t)_{\text{cor}}$ values which correspond to Z_{tent} values. In this demonstration of a typical case shown in Fig. 11c, when we set the $Z_{\text{tent}} = 15$, the $(\mu_{\text{low}}t)_{\text{meas}} = 2.41$ is converted into $(\mu_{\text{low}}t)_{\text{cor}} = 2.66$, as represented by the red arrows. These corrections are also applied to the high-energy bin to obtain $(\mu_{\text{high}}t)_{\text{cor}}$ values. In order to determine Z_{eff}, $\mu^{\dagger}$ values are calculated by substituting $(\mu_{\text{low}}t)_{\text{cor}}$ and $(\mu_{\text{high}}t)_{\text{cor}}$ values to Eq. (8). The upper figure in Fig. 11d shows the relationship between $\mu^{\dagger}$ and Z_{eff}. The broken line shows the reference curve determined using a well-known database [24]. As shown by the arrows, the $\mu^{\dagger} = 0.57$ corresponding to $Z_{\text{tent}} = 15$ can be converted to $Z_{\text{eff}} = 11.5$. To determine Z_{eff} of an object, the derived Z_{eff} values are plotted as a function of Z_{tent} as shown in the lower figure in Fig. 11d. By searching for the intersectional point between the derived Z_{eff} curve and Y = X line, the Z_{eff} of an object is determined to be 10.7. The determined Z_{eff} is in good agreement with $Z_{\text{eff}} = 10.5$ which is predetermined as a reference value using a well-known formula [30].

Furthermore, we want to determine ρt by taking advantage of the Z_{eff}. Because $(\mu_{\text{low}}t)_{\text{cor}}$ values related to Z_{tent} values are already obtained in the correction process as shown in Fig. 11c, the $(\mu_{\text{low}}t)_{\text{cor}}$ is plotted as a function of Z_{tent} as shown in Fig. 11e. Then, the $(\mu_{\text{low}}t)_{\text{cor}}$ related to an $Z_{\text{eff}} = 10.7$ can be determined as 2.57. Because the determined $(\mu_{\text{low}}t)_{\text{cor}}$ has completely corrected the beam hardening effect and detector response, we can perform the following analysis which can be carried out under the limited conditions of being able to use monochromatic X-rays. Substituting $(\mu_{\text{low}}t)_{\text{cor}}$ and $(\mu_{\text{low}}t)_{\text{cor}}$ values into Eqs. (9-1) and (9-2) under the assumption that the object consists of acrylic and aluminum, namely, μ/ρ values are known, the $\rho_{\text{ac}}t_{\text{ac}}$ and $\rho_{\text{Al}}t_{\text{Al}}$ can be determined to be 2.7 g/cm^2 and 2.6 g/cm^2, respectively, as shown in Fig. 11f. The determined ρt values are in good agreement with reference ρt values of 2.5 g/cm^2. These procedures are applied pixel by pixel to create Z_{eff} and ρt images.

To confirm that our method can determine Z_{eff} and ρt values, we present images produced using our prototype ERPCD. Figure 12 shows images of acrylic, aluminum, and bilayer structures of acrylic and aluminum having (a) $\rho t = 1.0$ g/cm^2 and (b) $\rho t = 5.0$ g/cm^2. As shown in the photograph, acrylic, aluminum, and bilayer structures are arranged in order from left to right. In our system, the conventional X-ray image, Z_{eff} image, and ρt image can be obtained. In conventional X-ray images which are equivalent to the image generated by EID, the image density

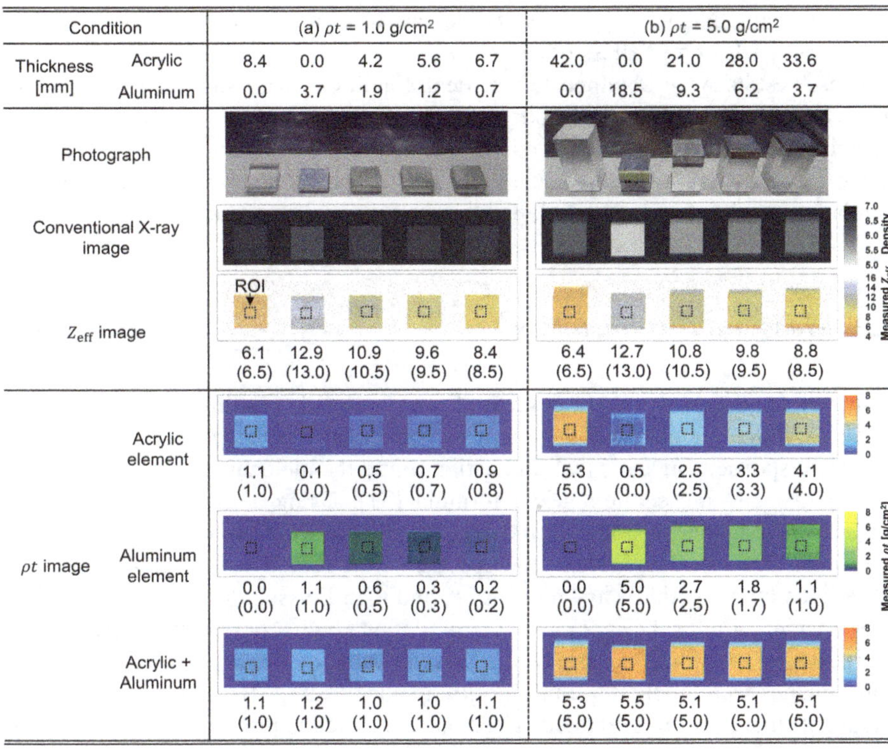

Fig. 12 The results for verifying our novel method to derive Z_{eff} and ρt images using a prototype ERPCD system. The upper row of images shows photographs of samples having (**a**) $\rho t = 1.0$ g/cm^2 and (**b**) $\rho t = 5.0$ g/cm^2. The acrylic, aluminum, and bilayer structures of acrylic and aluminum are arranged in order left to right. The second row of images shows conventional X-ray images. The third row of images shows Z_{eff} images, and the ROI of 50 $\times$ 50 pixels is set in each image. The mean Z_{eff} value measured in ROI is presented below each image, and the theoretical Z_{eff} value is also presented in parentheses. The last three rows of images show ρt images for the acrylic element, the aluminum element, and both the acrylic and aluminum elements. The mean ρt value and theoretical value are presented below. In the Z_{eff} and ρt images, the measured values are in good agreement with theoretical values

varies depending on Z_{eff} and ρt. When the Z_{eff} is higher with the same ρt, the image density is smaller. Also, when the ρt is larger with the same Z_{eff}, the image density is smaller. It means that the property of an object cannot be identified because the image density changes depending on the two parameters, Z_{eff} and ρt. The Z_{eff} image is presented in color scale, and the region of interest (ROI) of 50 $\times$ 50 pixels is set in the center of each sample. The mean Z_{eff} value measured in ROI is presented below each image, and theoretical Z_{eff} is provided in parentheses. The measured Z_{eff} value agrees with the theoretical Z_{eff} with an accuracy of $Z_{eff} \pm 0.4$. In the ρt images, elements for acrylic, aluminum, and acrylic and aluminum are presented. The ρt value for each sample is in good agreement with theoretical ρt with an accuracy of

$\rho t \pm 0.5$ g/cm^2. Namely, these results indicate that our method can determine $Z_{\rm eff}$ and extract each ρt regardless of sample.

These results were obtained under ideal conditions with sufficient statistics to prove the principle of the methodology. In order to open the application of our method to current diagnosis, it is necessary to evaluate the accuracy of measurement. In the next section, we will discuss the feasibility of our method based on the simulation results showing the relationship between accuracy and statistics.

4.2 Measurement Accuracy

To investigate the accuracy of our method, we simulated two-dimensional images in which the following physical phenomena were considered: attenuation of an object, detector response, and statistical fluctuation. First, the incident X-ray spectrum was calculated using the semiempirical formula [31]. Using a μ/ρ database [24], a virtual sample having a certain ρt was created, and the spectrum penetrating through the object was calculated. Next, the spectrum was folded with detector response $\mathbf{R}^{(1)}\mathbf{R}^{(2)}$ as the procedure shown in Fig. 8, and then Poisson distribution was taken into consideration for the spectrum. Accumulating counts for two energy regions of the spectrum, the value was assigned as a pixel value of a count image. Finally, the count images were applied to our algorithm shown in Fig. 10, and the accuracy of $Z_{\rm eff}$ and ρt images were evaluated. The conditions used in the simulation were set to be similar to that of the experiment as shown in Fig. 12: tube voltage 50 kV, energy bins 32–40 keV and 40–50 keV, and acrylic, aluminum, and bilayer structure objects having ρt of 1 and 5 g/cm^2. In addition, referring to the fact that the diagnostic reference level (DRL) for general radiography is 0.4 mGy [39], the total incident X-ray counts were adjusted so as to calculate the exposure doses of 0.04, 0.4, and 4 mGy.

Figure 13 shows the results of $Z_{\rm eff}$ and ρt images when simulating the conditions at three different exposure doses. The upper left panel shows the objects; the upper and lower rows correspond to $\rho t = 1$ g/cm^2 and $\rho t = 5$ g/cm^2, respectively, and the arranged order of acrylic, aluminum, and three bilayer structures is similar to that shown in Fig. 12. To clearly show the difference from the true value, reference images are also presented. As the exposure dose increases from 0.04 mGy to 4 mGy, the variation in pixel values is reduced gradually, and images closer to that of the reference are obtained. In order to investigate the difference quantitatively, we plot the mean value and standard deviation measured in the ROI (10×10 pixels) of each image as shown in Fig. 14. The upper and lower figures correspond to the results of $Z_{\rm eff}$ and ρt, respectively. The data for $\rho t = 1$ g/cm^2 are plotted, and (a), (b), (c), (d), and (e) correspond to acrylic, aluminum, and three bilayer structures, respectively. For each object, (a) through (e), the reference value, measured values for 0.04 mGy, 0.4 mGy, and 4 mGy are presented on the left to right. For the ρt results, the

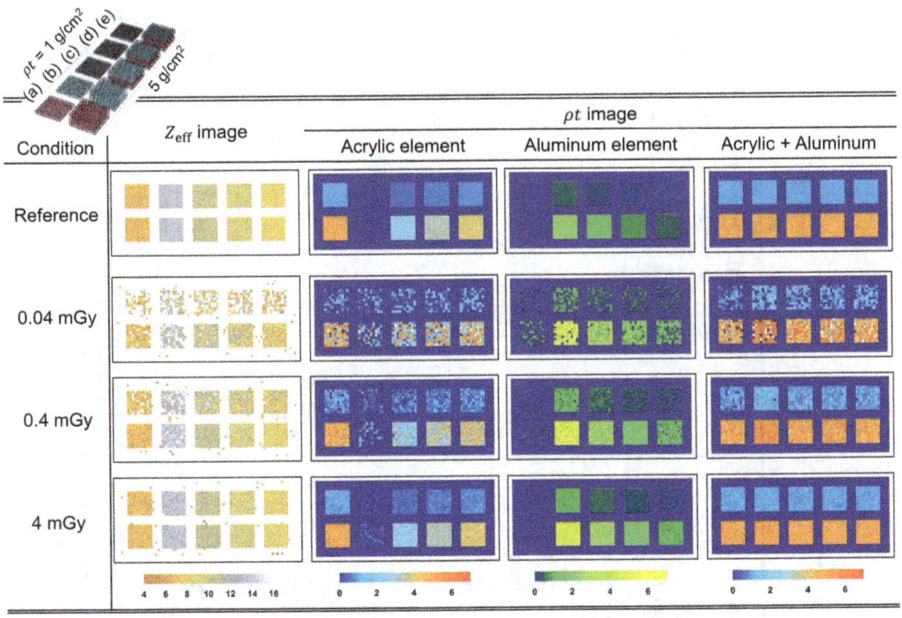

Fig. 13 Simulation results of Z_{eff} and ρt images related to setting exposure doses of 0.04 mGy, 0.4 mGy, and 4 mGy. The acrylic, aluminum, and bilayer structures are arranged in order from left to right. The upper and lower rows correspond to $\rho t = 1$ g/cm^2 and 5 g/cm^2, respectively. As the exposure dose increases, the coloration is close to that of the reference image

composition ratio of acrylic and aluminum is presented in areas filled with black and shaded, respectively. As the exposure dose increases, the measured Z_{eff} value agrees with the reference value and the standard deviation rapidly decreases. This trend is similar to the ρt results. To further analyze the accuracy of our method, the relationship between the standard deviations of Z_{eff} and ρt is presented in Fig. 15. The data for $\rho t = 1$ g/cm^2 and $\rho t = 5$ g/cm^2 are plotted. The red, green, and blue circles correspond to 0.04 mGy, 0.4 mGy, and 4 mGy, respectively. The standard deviations of Z_{eff} and ρt correlate; as indicated by the markers, a dose of 4 mGy is required to determine Z_{eff} and ρt with an accuracy of 0.5. This is far above DRL, and it is difficult to apply this method without modification being made to be suitable for clinical diagnosis. These results were derived from raw images obtained under various conditions such as the tube voltage and energy bins which are not optimized. We have the prospect to achieve high accuracy by optimizing these conditions and applying proper image filtration. For clinical applications, we would like to design a system that achieves both applicable dose and high-precision measurement.

If our method can realize medical applications, what clinical value could we get from the proposed Z_{eff} and ρt images? In the next section, we discuss the meaning of functional images by presenting images of measured biological samples.

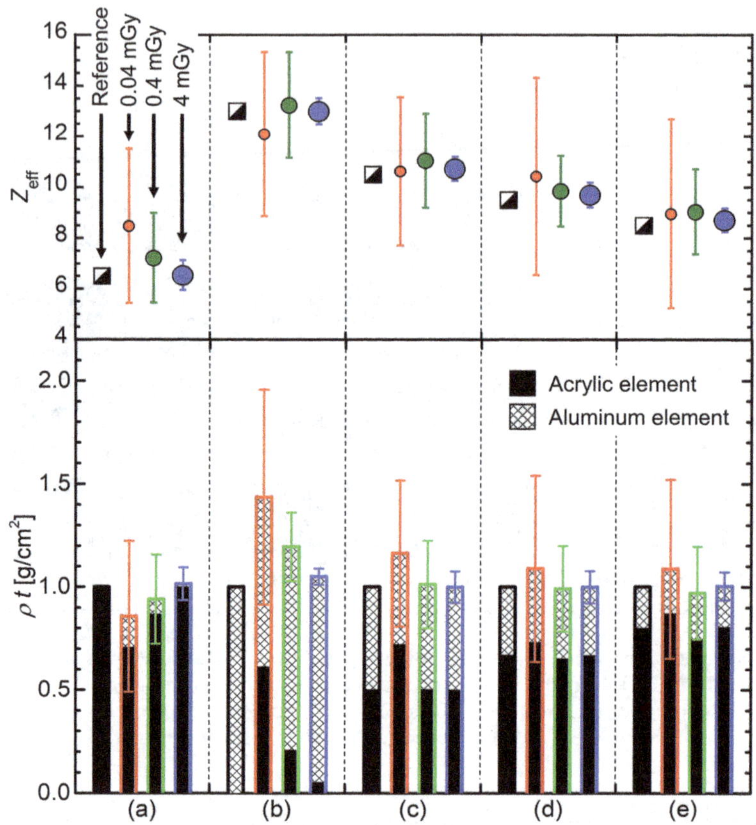

Fig. 14 Results of quantitative analysis of images shown in Fig. 13. For each image, ROI were set and the mean values were plotted. The upper and lower figures correspond to the results of Z_{eff} and ρt, respectively. With increasing exposure of doses (0.04, 0.4, and 4 mGy), the measured value agrees more with the reference value

4.3 *Spillover Effect on Applications*

In this section, we will discuss the usefulness of the images by visually presenting images measured with our prototype ERPCD. Figure 16 shows images produced when measuring fish samples. Figure 16a shows photographs of freshwater and saltwater fish. Our system can generate four different images as shown in Fig. 16b–e. Figure 16b shows the conventional X-ray image. We can clearly see the bone structure presented by white and the gas bladder shown in black. The X-ray path including bone has a large amount of X-ray attenuation, and therefore image density decreases and it is expressed by white color. On the other hand, the X-ray path including the gas bladder has a relatively small amount of attenuation when compared with that of the surrounding area filled with soft tissue, and therefore the image density increases and is expressed by black color. Although the value

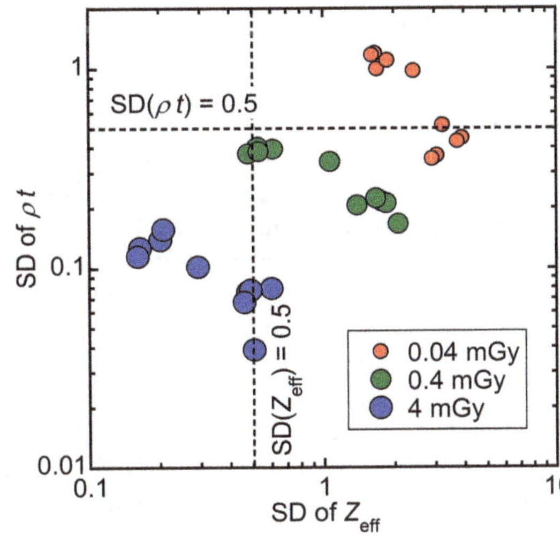

Fig. 15 Relationship between the standard deviations of Z_{eff} and ρt. To derive Z_{eff} and ρt within an accuracy of 0.5, an exposure dose of 4 mGy is needed

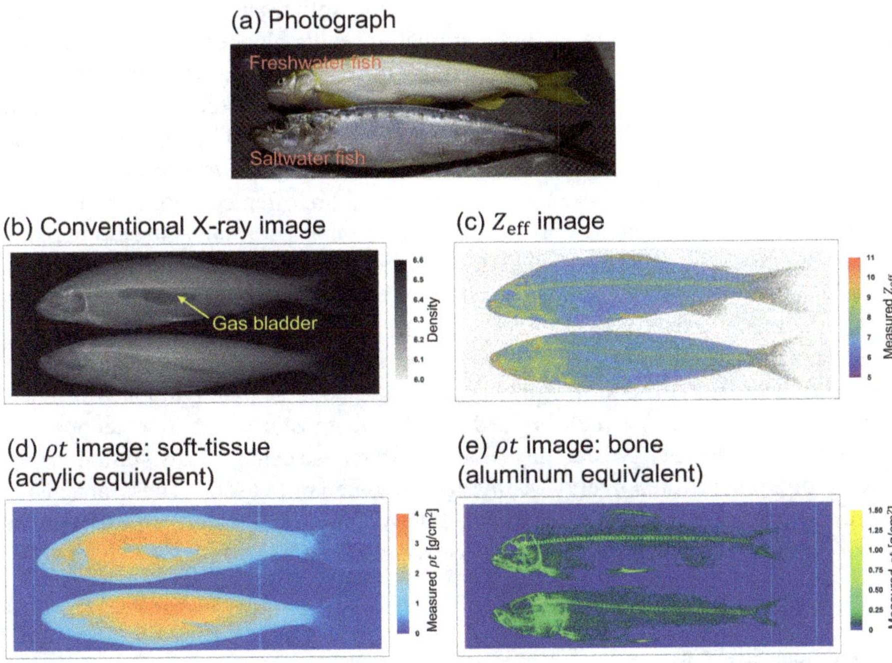

Fig. 16 Produced images of fish samples to demonstrate the feasibility of our method. (**a**) shows a photograph of a freshwater fish and saltwater fish. In our system, we can obtain (**b**) conventional X-ray image, (**c**) Z_{eff} image, (**d**) ρt image for soft tissue, and (**e**) ρt image for the bone. We hope that these novel images lead to novel diagnosis techniques in the near future

of image density does not directly represent the property of an object, we can identify the object by comparing the appearance presented by the contrast between the target and surrounding material with the appearance assumed in advance. This approach for image analysis is similar to that of medical diagnosis; normal/abnormal judgment can be performed qualitatively by analyzing thc appearance aspect which can be read from the image based on anatomical knowledge. On the other hand, our system can produce quantitative images to express the functional aspect of an object. Figure 16c shows the Z_{eff} image. The bone structure can be seen as high Z_{eff} values of 9–11, with high Z_{eff} accumulations observed around the head, fins, and tail. On the other hand, the gas bladder is not observed because the small X-ray attenuation is not reflected in the calculation of Z_{eff}. The coloration of the Z_{eff} image is different from that of a conventional X-ray image and reflects the physical phenomenon of X-ray attenuation. To extract various useful information from the Z_{eff} image, a new evaluation based on physical perspective is needed. Figure 16d shows a ρt image related to soft tissue; in the analytical procedure, we assume that soft tissue is equivalent to acrylic. By comparing the conventional X-ray image shown in Fig. 16b, we can see that the structure of only soft tissue is extracted clearly. The ρt increases in the thick area of the soft tissue near the center and decreases in the area with the gas bladder; therefore, the bladder is easily identified as an area where soft tissue does not exist. Figure 16e shows a ρt image related to bone which is assumed to be equivalent to aluminum. The structure of only bone can be extracted completely. By comparing to the Z_{eff} image shown in Fig. 16c, it can be seen that the high Z_{eff} component is extracted more sharply. As you know, biological tissue such as the human body, food materials, etc. are often composed of the soft tissue and bone, and images corresponding to their separation will help achieve various diagnoses through examination. In the radiography of the lung area, images with the bone removed are in great demand to confirm the presence of lesions [40–42]. In the mammography examination, identification of calcification buried in soft tissue is also necessary [43]. Furthermore, our system can accurately measure the ρt of the bone from a bone image. The ρt of the bone is synonymous with bone mineral density (BMD) which is used for the diagnostic index of osteoporosis, which indicates that our system may be able to diagnose osteoporosis without using specific equipment (dual-energy X-ray absorptiometry) [44–46]. These are just a few examples of how images can be useful. We think there are various fields in which Z_{eff} and ρt images can be used effectively.

5 Summary

In this chapter, an algorithm for extracting the properties of an object using an ERPCD was described. Based on the principle in which the information of the object can be derived from the differences in the X-ray attenuation depending on X-ray energy, we analyzed polychromatic X-rays measured with an ERPCD. The most important issue of the analysis is that the measured information related to each energy bin is distorted by the beam hardening effect and detector response. We described these effects from a viewpoint based on physics and presented a methodology to correct these effects simultaneously. By applying this correction, we can treat polychromatic X-rays as those equivalent to monochromatic X-rays. This results in the accomplishment of correct analysis in principle. In order to demonstrate our method, we succeeded in the derivation of a Z_{eff} image and extraction of soft tissue and bone images by applying our algorithm to each pixel of an image measured with a prototype ERPCD.

It should be noted that the experiments in this chapter were performed under ideal conditions, and our conclusion may not necessarily apply to actual CT systems. This is because the phantoms were measured at a low dose rate and the effect of pulse pileup was negligibly small. Furthermore, since the object and the detector distance are sufficiently separated, the influence of scattered X-rays can be almost ignored. When considering the actual equipment that the examination is being performed under clinical condition in which pulse pileup and/or contamination of scattered X-rays occur, these effects must be included in the estimation process of the detector's response. The algorithm we presented is very simple in which information related to only two energy bins was used. If scientists want to apply our algorithm to other systems in which three or more energy bins are used, we recommend to modify the measured data; by summing the information of different energy bins, a two-energy-bin system can be easily achieved using a software approach.

We hope that the content described in this chapter opens up possibilities for the use of photon-counting techniques in medical and industrial fields.

Acknowledgments The description in this chapter was partially supported by collaborative research between the Kanazawa University and JOB CORPORATION (https://www.job-image.com/), Japan. The authors wish to express gratitude to members of the JOB CORPORATION, Dr. Shuichiro Yamamoto, Mr. Masahiro Okada, Mr. Fumio Tsuchiya, Mr. Daisuke Hashimoto, Mr. Yasuhiro Kuramoto, and Mr. Masashi Yamasaki for their valuable contributions. We wish to thank Mr. Takumi Asakawa who belongs to the GE Healthcare Japan for his important research results when he belonged to graduate school in Kanazawa University, Japan. We would like to thank Dr. Yuki Kanazawa who belongs to Tokushima University, Japan, for discussing the feasibility of our research from a clinical point of view. We would also like to thank Dr. Yoshie Kodera and Dr. Shuji Koyama who belong to Nagoya University, Japan, for discussing the clinical application of a photon-counting detector.

References

1. E.D. Pisano et al., Image processing algorithms for digital mammography: a pictorial essay. Radiographics **20**(5), 1479–1491 (2000). https://doi.org/10.1148/radiographics.20.5.g00se311479
2. M. Prokop et al., Principles of image processing in digital chest radiography. J. Thorac. Imaging. **18**(3), 148–164 (2003)
3. D.S. Kermany et al., Identifying medical diagnoses and treatable diseases by image-based deep learning. Cell **172**(5), 1122–1131 (2018). https://doi.org/10.1016/j.cell.2018.02.010

4. S. Leng et al., Photon-counting detector CT: system design and clinical applications of an emerging technology. Radiographics **39**(3), 729–743 (2019). https://doi.org/10.1148/rg.2019180115
5. K. Taguchi, J.S. Iwanczyk, Vision 20/20: single photon counting x-ray detectors in medical imaging. Med. Phys. **40**(10), 100901–1–19 (2013). https://doi.org/10.1118/1.4820371
6. M.J. Willemink et al., Photon-counting CT: technical principles and clinical prospects. Radiology **289**(2), 293–312 (2018). https://doi.org/10.1148/radiol.2018172656
7. T Asakawa et al., Importance of considering the response function of photon counting detectors with the goal of precise material identification. IEEE Nuclear Science Symposium and Medical Imaging Conference (NSS/MIC), 1–7 (2019). https://doi.org/10.1109/NSS/MIC42101.2019.9059844
8. H Hayashi et al., A fundamental experiment for novel material identification method based on a photon counting technique: Using conventional X-ray equipment. IEEE Nuclear Science Symposium and Medical Imaging Conference (NSS/MIC), 1–4 (2015). https://doi.org/10.1109/NSSMIC.2015.7582027
9. H. Hayashi et al., Response functions of multi-pixel-type CdTe detector -toward development of precise material identification on diagnostic X-ray images by means of photon counting. Proc. SPIE. Med. Imaging **10132**, 1–18 (2017). https://doi.org/10.1117/12.2251185
10. N. Kimoto et al., Precise material identification method based on a photon counting technique with correction of the beam hardening effect in X-ray spectra. Appl. Radiat. Isotopes **124**, 16–26 (2017). https://doi.org/10.1016/j.apradiso.2017.01.049
11. N. Kimoto et al., Development of a novel method based on a photon counting technique with the aim of precise material identification in clinical X-ray diagnosis. Proc. SPIE. Med. Imaging **10132**, 1–11 (2017). https://doi.org/10.1117/12.2253564
12. N Kimoto et al., Novel material identification method using three energy bins of a photon counting detector taking into consideration Z-dependent beam hardening effect correction with the aim of producing an X-ray image with information of effective atomic number. IEEE Nuclear Science Symposium and Medical Imaging Conference (NSS/MIC), 1–4 (2017). https://doi.org/10.1109/NSSMIC.2017.8533059
13. N Kimoto et al., Reproduction of response functions of a multi-pixel-type energy-resolved photon counting detector while taking into consideration interaction of X-rays, charge sharing and energy resolution. IEEE Nuclear Science Symposium and Medical Imaging Conference (NSS/MIC), 1–4 (2018). https://doi.org/10.1109/NSSMIC.2018.8824417.
14. N Kimoto et al., Feasibility study of photon counting detector for producing effective atomic number image. IEEE Nuclear Science Symposium and Medical Imaging Conference (NSS/MIC). 1–4 (2019). https://doi.org/10.1109/NSS/MIC42101.2019.9059919
15. N. Kimoto et al., Effective atomic number image determination with an energy-resolving photon-counting detector using polychromatic X-ray attenuation by correcting for the beam hardening effect and detector response. Appl. Radiat Isotopes **170**, 109617 (2021). https://doi.org/10.1016/j.apradiso.2021.109617
16. N. Kimoto et al., A novel algorithm for extracting soft-tissue and bone images measured using a photon-counting type X-ray imaging detector with the help of effective atomic number analysis. Appl. Radiat. Isotopes **176**, 109822 (2021). https://doi.org/10.1016/j.apradiso.2021.109822
17. S. Reza et al., *Semiconductor Radiation Detectors: Technology And Applications* (CRC Press, New York, 2017), pp. 85–108
18. H. Hayashi et al., *Advances in Medicine and Biology* (Nova Science Publishers, Inc., New York, 2019), pp. 1–46
19. H. Hayashi et al., *Photon Counting Detectors for X-ray Imaging: Physics and Applications* (Springer, Switzerland, 2021), pp. 1–119. https://doi.org/10.1007/978-3-030-62680-8
20. E. Samei, J.M. Flynn, An experimental comparison of detector performance for direct and indirect digital radiography systems. Med. Phys. **30**(4), 608–622 (2003). https://doi.org/10.1118/1.1561285

21. G. Spekowius, T. Wendler, *Advances in Healthcare Technology: Shaping the Future of Medical Care* (Springer, Dordrecht, 2006), pp. 49–64
22. G.F. Knoll, *Radiation Detection and Measurement* (Wiley, New York, 2000), pp. 1–802
23. B.J. Heismann et al., Density and atomic number measurements with spectral x-ray attenuation method. J. Appl. Phys. **94**(3), 2073–2079 (2003). https://doi.org/10.1063/1.1586963
24. J.H. Hubbell, Photon mass attenuation and energy-absorption coefficients. Int. J. Appl. Radiat. Isotopes **33**(11), 1269–1290 (1982). https://doi.org/10.1016/0020-708X(82)90248-4
25. R.A. Brooks, G. Di Chiro, Beam hardening in x-ray reconstructive tomography. Phys. Med. Biol. **21**(3), 390–398 (1976). https://doi.org/10.1088/0031-9155/21/3/004
26. M.M. Good et al., Accuracies of the synthesized monochromatic CT numbers and effective atomic numbers obtained with a rapid kVp switching dual energy CT scanner. Med. Phys. **38**(4), 2222–2232 (2011). https://doi.org/10.1118/1.3567509
27. T.R.C. Johnson, Dual-energy CT: general principles. Am. J. Roentgenol. **199**(5), S3–S8 (2012). https://doi.org/10.2214/AJR.12.9116
28. F. Tatsugami et al., Measurement of electron density and effective atomic number by dual-energy scan using a 320-detector computed tomography scanner with raw data-based analysis: a phantom study. J. Comput. Assist. Tomogr. **38**(6), 824–827 (2014). https://doi.org/10.1097/RCT.0000000000000129
29. X. Wang et al., Material separation in x-ray CT with energy resolved photon-counting detectors. Med. Phys. **38**(3), 1534–1546 (2011). https://doi.org/10.1118/1.3553401
30. F.W. Spiers, Effective atomic number and energy absorption in tissues. Br. J. Radiol. **19**(218), 52–63 (1946). https://doi.org/10.1259/0007-1285-19-218-52
31. R. Birch, M. Marshall, Computation of bremsstrahlung X-ray spectra and comparison with spectra measured with a Ge(Li) detector. Phys. Med. Biol. **24**(3), 505–517 (1979). https://doi.org/10.1088/0031-9155/24/3/002
32. S.S. Hsieh et al., Spectral resolution and high-flux capability tradeoffs in CdTe detectors for clinical CT. Med. Phys. **45**(4), 1433–1443 (2018). https://doi.org/10.1002/mp.12799
33. P. Otfinowski, Spatial resolution and detection efficiency of algorithms for charge sharing compensation in single photon counting hybrid pixel detectors. Nucl. Inst. Methods Phys. Res. A **882**, 91–95 (2018). https://doi.org/10.1016/j.nima.2017.10.092
34. K. Taguchi et al., Spatio-energetic cross-talk in photon counting detectors: numerical detector model (PcTK) and workflow for CT image quality assessment. Med. Phys **45**(5), 1985–1998 (2018). https://doi.org/10.1002/mp.12863
35. P. Trueb et al., Assessment of the spectral performance of hybrid photon counting x-ray detectors. Med. Phys. **44**(9), e207–e214 (2017). https://doi.org/10.1002/mp.12323
36. P. Zambon et al., Spectral response characterization of CdTe sensors of different pixel size with the IBEX ASIC. Nucl. Inst. Methods Phys. Res. A. **892**, 106–113 (2018). https://doi.org/10.1016/j.nima.2018.03.006
37. H Hirayama et al., The EGS5 code system. KEK Report. 2005-8 SLAC-R-730. 1–418 (2005)
38. M. Sasaki et al., A novel mammographic fusion imaging technique: the first results of tumor tissues detection from resected breast tissues using energy-resolved photon counting detector. Proc. SPIE **109486**, 1094864 (2019). https://doi.org/10.1117/12.2512271
39. J-RIME (2020) National Diagnostic Reference Levels in Japan (2020) -Japan DRLs 2020
40. J.E. Kuhlman et al., Dual-energy subtraction chest radiography: what to look for beyond calcified nodules. Radiographics **26**(1), 79–92 (2006). https://doi.org/10.1148/rg.261055034
41. H. MacMahon et al., Dual energy subtraction and temporal subtraction chest radiography. J. Thorac. Imaging **23**(2), 77–85 (2008). https://doi.org/10.1097/RTI.0b013e318173dd38
42. H.P. McAdams et al., Recent advances in chest radiography. Radiology **241**(3), 663–683 (2006). https://doi.org/10.1148/radiol.2413051535
43. S.C. Kappadath, C.C. Shaw, Quantitative evaluation of dual-energy digital mammography for calcification imaging. Phys. Med. Biol. **49**(12), 2563–2576 (2004). https://doi.org/10.1088/0031-9155/49/12/007
44. M.G. Blake, I. Fogelman, Technical principles of dual energy X-ray absorptiometry. Semin. Nucl. Med. **27**(3), 210–228 (1997). https://doi.org/10.1016/S0001-2998(97)80025-6

45. D.I. Cullum et al., X-ray dual-photon absorptiometry: a new method for the measurement of bone density. Br. J. Radiol. **62**(739), 587–592 (1989). https://doi.org/10.1259/0007-1285-62-739-587
46. J.D. Theodorou et al., Dual-energy X-ray absorptiometry in diagnosis of osteoporosis: Basic principles, indications, and scan interpretation. Compr. Ther. **28**(3), 190–200 (2002). https://doi.org/10.1007/s12019-002-0028-6

Characterization of Noise and Noise Correlation in Photon-Counting CT for Multimaterial Decomposition-Based Spectral Imaging

Xiangyang Tang, Yan Ren, Huiqiao Xie, Xiaofeng Yang, and Arthur E. Stillman

1 Introduction

In the late 1970s, Alvarez and Macovski discovered that attenuation—the foundation for computed tomography (CT) and other X-ray imaging modalities—could be readily decomposed into the photoelectric and Compton scattering interactions via the data acquisition at two distinct peak X-ray tube voltages (termed as dual-energy X-ray or CT since then). Through a linear transform, the interaction decomposition could be converted into material decomposition while the data acquisition (dual energy scan) was maintained unchanged. In their milestone papers, Alvarez and Macovski were well aware of the fact that the noise in the projection data acquired at the two distinct peak tube voltages are relevant to determining spectral CT's imaging performance [1, 2]. In presentation of their derivation under the framework of maximum likelihood estimation, Alvarez and Macovski showed that the Jacobian of the transform from the projection space to the so-called A-space [3] determines the noise property that is governed by the Cramér-Rao lower bound (CRLB) [9]. Alternatively, by taking noise as the uncertainty in estimating the basis materials' mass density, Kelcz et al. investigated the same problem [4]. Through first-order Taylor expansion, Kelcz et al. arrived at a basically identical conclusion and found that the so-called R-quantity [4], which is the ratio of the mass attenuation coefficients associated with the two interested materials at the two distinct peak voltages, plays the role that is almost identical to the Jacobian of the transform from projection space to A-space [5].

X. Tang (✉) · Y. Ren · H. Xie · X. Yang · A. E. Stillman
Emory University School of Medicine, Atlanta, GA, USA
e-mail: xiangyang.tang@emory.edu

K. Iniewski, H. Gadey (eds.), *Emerging Radiation Detection*,
https://doi.org/10.1007/978-3-031-63897-8_6

In their earliest work, Alvarez and Macovski assumed that the noise correlation in projection data vanishes, that is, no correlation in noise between the projection data acquired at the two distinct peak voltages (correlation-vanishing henceforth), which is the case in the data acquisition implemented by photon counting under ideal spectral response or voltage switching (termed as fast kV switching nowadays) [1, 2, 6], even though they showed that there existed a negative correlation in the noise corresponding to each basis material [1, 2, 6, 7]. It is interesting to note that the assumption of correlation vanishing initially seemed invalid in the case of data acquisition by layered (sandwiched) X-ray detector. However, it turned out that the noise correlation in the data acquired by the first and second layers was negligibly small and thus can be virtually treated as another case of correlation vanishing [8]. Also notable is the fact that the correlation-vanishing assumption is actually not valid in the data acquisition by photon counting if realistic detector response is considered, since the spectral distortion inevitably induces correlation in the projection data acquired at the low and high peak voltage [9–11].

Focused on the case of spectral imaging based on two-material decomposition (2-MD), Roessl et al. carried out an investigation to study the effect of noise correlation in the projection data on the noise and its correlation in the material-specific images [12]. Analytically, they derived a set of equations and arrived at the conclusions that the correlation of noise in the basis images (material-specific images) is negative. They also showed that, if the correlation in the projection data vanishes, the analytics degenerates to the case that was initially derived by Alvarez and Macovski [1, 2, 6]. Though they took the counting-integrating mode of data acquisition as the case of their work and only plotting of analytic profiles were given, their derivation was in fact general and readily extendable to other cases of 2-MD-based spectral imaging implemented in different schemes of data acquisition [12].

Contrast agent has been indispensable in conventional CT since the very beginning and will continue to be the case in photon-counting CT [13–15], especially in light of the prospect that novel contrast agents in the form of biodegradable nanoparticles may play more important roles [15, 16]. The spectral imaging in photon-counting CT with administration of one or multiple contrast agents demands in-depth studies of multimaterial decomposition (m-MD). With increasing number of spectral channels and basis materials, the dimension of Jacobian corresponding to the transform from the projection space to the A-space increases accordingly, which in turn complicates the mechanism at which the noise in the projection data acquired in each spectral channel impacts the noise and its correlation in the material-specific images. Notably, the basics of linear algebra tell us that the sign of the off-diagonal elements in the inverse of an $M\times$ ($M > 2$) matrix may vary, which implies that the correlation in the noise corresponding to each basis image in the m-MD-based spectral imaging may not be always negative, as has been the case in 2-MD-based spectral imaging.

Recently, the research and development in m-MD-based spectral imaging in photon-counting CT is gaining momentum. Based on an in-depth revisiting of the physics underlying photon-counting CT, the conditioning of basis materials (functions) and spectral channelization (energy binning) for spectral imaging has been studied. Using the approach based on singular value decomposition (SVD) and analysis [10, 11], it has been shown how the conditioning of basis materials and spectral channelization can affect the performance of spectral imaging in photon-counting CT based on multimaterial decomposition (*m*-MD). In this chapter, we briefly present our recent work on the noise and its correlation in the material-specific (basis) images for spectral imaging in photon-counting CT based on multimaterial decomposition [17, 18]. Via analytical derivation, statistical analysis, and phantom studies supported by computer simulation, our work is focused on *m*-MD that is carried out in the projection domain [9–11, 17, 18], with the goal to provide the information for further understanding of the basics of spectral imaging in photon-counting CT, energy-integration CT, and other X-ray imaging modalities that are based on multimaterial decomposition. Notably, the work is designed to focus on phantom study supported by computer simulation, so that any adverse effects that may exist due to the imperfection in photon-counting CT's imaging chain on the investigation and its conclusion can be excluded.

2 Signal Detection and Material Decomposition in Photon-Counting CT

The signal and its detection in photon-counting CT have been modeled in our recent publication [9–11] and are briefly reiterated here for reader's convenience. If the noise in data acquisition is taken into account, the detection of signal in photon-counting CT can be analytically expressed as

$$I_k(L) = \text{Poisson}\left(\int_{E_{\min}}^{E_{\max}} D_k(E)\, N_0(E) \exp\left(-\int_L \mu\,(\boldsymbol{x};\, E)\, dl\right) dE\right), \tag{1}$$

where $N_0(E)$ denotes the spectrum of X-ray source and L is the path penetrated by the X-ray beam. Subscript k indexes the spectral channel in which the raw data are acquired, while $D_k(E)$ is defined in the k-th spectral channel $[E_{\min}^k E_{\max}^k]$ that takes detector's efficiency $\eta(E)$ and spectral response into account [9–11]. *Poisson*$(\cdot)$ denotes the numerical operation that generates random numbers to observe the Poisson distribution [19].

In material decomposition, one can write

$$\mu\,(\boldsymbol{x};\, E) = \textstyle\sum_{i=1}^{K} a_i\,(\boldsymbol{x})\, \mu_i(E)\,. \tag{2}$$

Subsequently, Eq. (1) is converted into

$$I_k(L) = \text{Poisson}\Big(\textstyle\int_{E_{\min}}^{E_{\max}} D_k(E)\, N_0(E) \exp\left(-\sum_{p=1}^{P} A_p(L) \mu_p(E)\right) dE\,. \tag{3}$$

Given $p = 1, 2, \ldots, P$, $A_p(L) = \int_L a_p(\boldsymbol{x})dl$ denotes the line integral of $a_p(\boldsymbol{x})$ associated with the mass distribution of material p, which can be obtained from $I_k(L)$ $(k = 1, 2, \ldots, K)$ by solving the set of integral equations specified in Eq. (3) through numerical methods, for example, the iterative Newton-Raphson algorithm [2, 20], in which the integral equations' initial conditions are obtained by system calibration via polynomial data fitting [2, 20, 21].

3 Data Acquisition and Image Formation in Photon-Counting CT

In simulation [22], the X-ray technique is set at 1000 mA/140 keV, and the gantry of photon-counting CT rotates at 1 rot/sec. The spectral channels are defined as [1 ~ 58, 59 ~ 140] keV in 2-MD (Fig. 1a–a′), [1 ~ 51, 52 ~ 68, 69 ~ 140] keV in 3-MD (Fig. 1b–b′), and [1 ~ 43, 44 ~ 58, 59 ~ 72, 73 ~ 140] keV in 4-MD (Fig. 1c–c′). The goal in spectral channelization is to ensure that the photon counts in each channel be about the same at the incident surface of the object to be imaged, despite of other criteria that have been used in practice [23]. For illustration, the spectral channels under ideal detector response are plotted in Fig. 1a–c while their counterparts under realistic detector response are in Fig. 1a′–c′. The material in the photon-counting detector assumed in the simulation is cadmium zinc telluride (CZT) or cadmium telluride, which is associated with Compton scattering, charge sharing, and fluorescent escaping that may distort the detector's spectral response [10, 11]. The photon-counting detector is designed as a curved two-dimensional

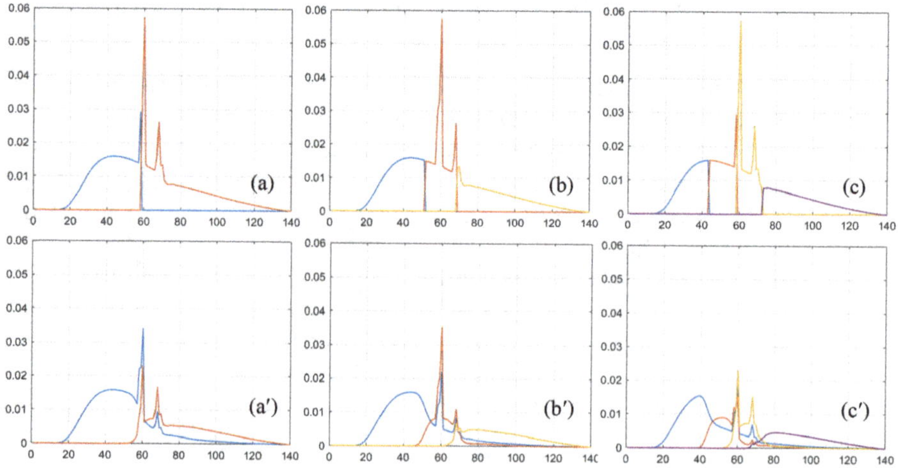

Fig. 1 Two (**a**), three (**b**), and four (**c**) spectral channels under ideal detector spectral response and their counterparts (**a′**–**c′**) under realistic detector response (keV is the unit of abscissa)

array at 864 × 16 dimension and 1.024 × 1.092 mm^2 pitch. The source-to-iso and source-to-detector distances are 541.0 mm and 949.0 mm, respectively, and consistent to one of the multidetector row CT scanners in the clinic. After material decomposition, each basis image $a_p(\boldsymbol{x})$ is in turn reconstructed from $A_p(L)$ ($p = 1, 2, \ldots, P$) via an FBP image reconstruction algorithm [24].

4 Noise and Its Propagation in M-MD-Based Spectral Imaging

By treating $I_k(L)$ ($k = 1, \ldots, K$) and $A_p(L)$ ($p = 1, \ldots, P$) as random variables in the projection space and A-space, respectively, the mapping from the former to the latter can be written as

$$I_k(L) = f_k\left(A_1(L), \ldots, A_p(L)\right) \quad (k = 1, \ldots, K) \tag{4}$$

Then, one can have the covariance matrix of A as [12, 25]

$$V[A] = \left(F^{-1}\right) \cdot \mathrm{V}[I] \cdot \left(F^{-1}\right)^T \tag{5}$$

where each entry of the mapping matrix F is defined as $F_{kp} = \partial I_k / \partial A_p$.

5 Noise and Its Propagation in 3-MD-Based Spectral Imaging and Beyond

In 3-MD, one can have the Jacobian

$$\begin{aligned}\Delta_{3\times3} &= \mu_{11}\mu_{22}\mu_{33} + \mu_{12}\mu_{23}\mu_{31} + \mu_{21}\mu_{32}\mu_{13} - \mu_{31}\mu_{22}\mu_{13} \\ &\quad - \mu_{32}\mu_{23}\mu_{11} - \mu_{11}\mu_{21}\mu_{33} \\ &\triangleq \frac{1}{A_1 A_2 A_3}\det(F),\end{aligned} \tag{6}$$

Again starting from Eq. (5), the covariance matrix $V[A]$ becomes

$$V_{3\times3}[A] = \begin{bmatrix} \sigma_{A_1}^2 & \sigma_{A_1A_2} & \sigma_{A_1A_3} \\ \sigma_{A_2A_1} & \sigma_{A_2}^2 & \sigma_{A_2A_3} \\ \sigma_{A_3A_1} & \sigma_{A_3A_2} & \sigma_{A_3}^2 \end{bmatrix} \tag{7}$$

with the entries being derived in the appendix of Ref. [18]. The interchannel spectral overlap may induce positive correlation between the data acquired in different channels, which in general may increase the magnitude of noise and noise correlation in the data denoted by A_1, A_2, and A_3 after material decomposition (see Ref. [18] for more detail).

As shown in the appendix of Ref. [18], the analytic expression of each entry of matrix $V_{3\times 3}[A]$ becomes exhaustively complicated in *3*-MD, though their derivation can be carried out according to Eq. (5). Hence, an analysis of the noise and noise correlation via simulation study makes more sense in terms of feasibility in the case of *4*-MD-based spectral imaging (and beyond) in photon-counting CT.

6 Noise Correlation and Its Impact on the Performance of Spectral Imaging: Simulation Study

The noise correlation and its impact on the performance of spectral imaging are assessed via simulation study, in which the imaging chain of photon-counting CT is modeled and simulated using a software kit [9–11, 22].

6.1 Analysis of Noise Correlation

A cylinder of water at 20 cm diameter, which consists of three rods at 7 cm diameter, is designed as the phantom to assess the correlation of noise in material-specific images in photon-counting CT (henceforth the noise correlation phantom). The materials that make rods 1–3 corresponding to the cases of 2-MD, 3-MD, and 4-MD are listed in Table 1. Specifically, two materials (soft tissue and cortical bone) are assumed in the 2-MD case, while iodine and gadolinium are added into the cases of 3-MD and 4-MD, respectively. The fraction of each material in the rods are chosen for quantitative purpose without any implication toward clinical or preclinical applications, as such the difference in material composition is adequate to ensure the accuracy in quantitative measurement. A sectional view of the phantom is presented in Fig. 2a, in which a region of interest (ROI) at 6 cm diameter within rod 3 is defined for illustration of noise and noise correlation measurement.

A number of basis materials have been investigated in our previous studies [9–11]. In the study of analyzing noise correlation, we choose soft tissue and cortical bone as the basis materials for 2-MD and soft tissue, 10 mg/ml gadolinium, and cortical bone for 3-MD. In the case of 4-MD, the basis materials are chosen as soft tissue, 10 mg/ml iodine, 10 mg/ml gadolinium, and cortical bone. Once the material-specific images are generated, ROIs are drawn in each of them as illustrated in Fig. 2a. The Pearson correlation coefficient (R-coefficient) is adopted to quantitatively assess the correlation of noise in those images.

Table 1 Material configuration (fraction in weight) of the three rods in the phantom used for evaluation and verification of noise correlation in the material-specific (basis) images in 2-MD, 3-MD, and 4-MD, respectively

	2-MD	3-MD	4-MD
Rod 1	1/3 soft tissue 2/3 cortical bone	2/7 soft tissue 4/7 iodine (10 mg/ml) 1/7 cortical bone	2/7 soft tissue 2.5/7 iodine (10 mg/ml) 2/7 gadolinium (10 mg/ml) 0.5/7 cortical bone
Rod 2	1/2 soft tissue 1/2 cortical bone	3/7 soft tissue 3/7 iodine (10 mg/ml) 1/7 cortical bone	3/7 soft tissue 2/7 iodine (10 mg/ml) 1.5/7 gadolinium (10 mg/ml) 0.5/7 cortical bone
Rod 3	2/3 soft tissue 1/3 cortical bone	4/7 soft tissue 2/7 iodine (10 mg/ml) 1/7 cortical bone	3/7 soft tissue 2.5/7 iodine (10 mg/ml) 1/7 gadolinium (10 mg/ml) 0.5/7 cortical bone

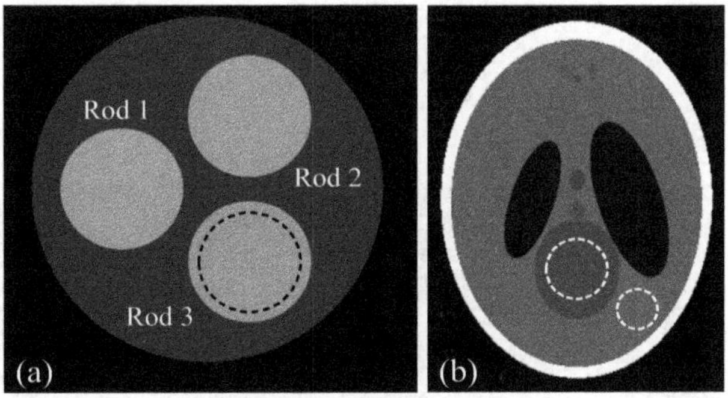

Fig. 2 Sectional view of the phantoms for gauging the noise correlation (**a**) and imaging performance (**b**)

6.2 Assessment of Spectral Imaging Performance

A humanoid head phantom (henceforth head phantom) is used to study the noise correlation's impact on the performance of virtual monochromatic imaging (a form of spectral imaging) in photon-counting CT. Shown in Fig. 2b is a sectional view of the phantom wherein two ROIs are defined as the signal (big circle) and background (small circle) for gauging contrast, noise, and contrast-to-noise ratio (CNR). The mass attenuation coefficients of the materials in the head phantom are set according to authoritative publications, for example, Refs. [26, 27] and the EPDL library [28]. The readers who are interested in the phantom's details are referred to the Refs. [10, 11]. The study of noise correlation's impact on the performance of spectral imaging is focused on the 3-MD case, in which four sets of basis materials are selected:

(*i*) BM-1: adipose, PMMA, and Teflon; (*ii*) BM-2: adipose, PMMA, and 20 mg/ml iodine; (*iii*) BM-3: soft tissue, 20 mg/ml iodine, and cortical bone; and (*iv*) BM-4: adipose, 20 mg/ml iodine, and Teflon.

6.3 Results: Noise and Noise Correlation in 3-MD Material-Specific Images

The material-specific images corresponding to basis material soft tissue, 10 mg/ml iodine, and cortical bone acquired via 3-MD under ideal detector spectral response are presented in Fig. 3a–c. Note that the interchannel overlap in detector's response increases the magnitude of noise in each material-specific image significantly (Fig. 5), which is within anticipation according to Eqs. (A-1–A-3) in the appendix of Ref. [18].

The correlation of noise in the material-specific images corresponding to basis material soft tissue, 10 mg/ml iodine, and cortical bone under ideal detector spectral response is presented in the scatterplots in Fig. 4a–c. The polarity of noise correlation over the material-specific images in 3-MD-based spectral imaging alternates between ± 1. The interchannel overlap induced by the spectral distortion in a realistic detector's spectral response increases the magnitude of noise correlation modestly, which is again within expectation according to Eqs. (A-1′–A-3′) in the appendix of Ref. [18].

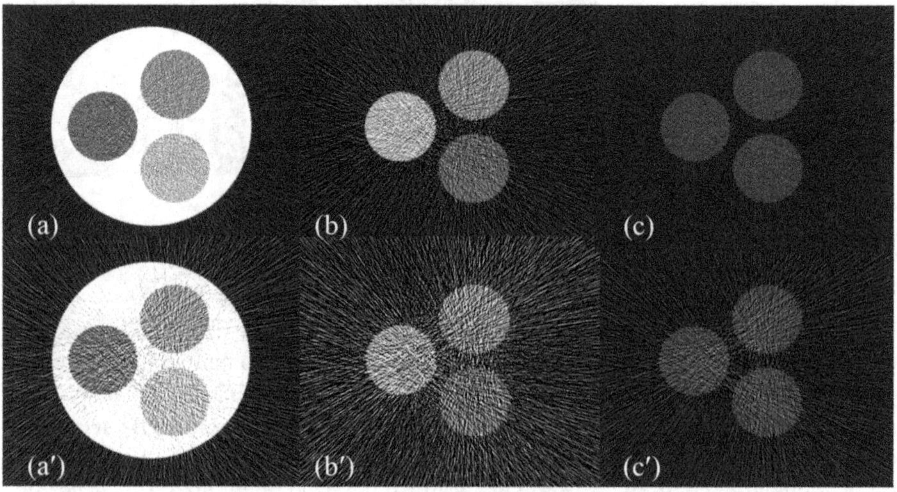

Fig. 3 Material-specific (tissue (**a**–**a′**), 10 mg/ml I (**b**–**b′**), and bone (**c**–**c′**)) images of the noise correlation phantom acquired via 3-MD under ideal (**a**–**c**) and realistic (**a′**–**c′**) detector response

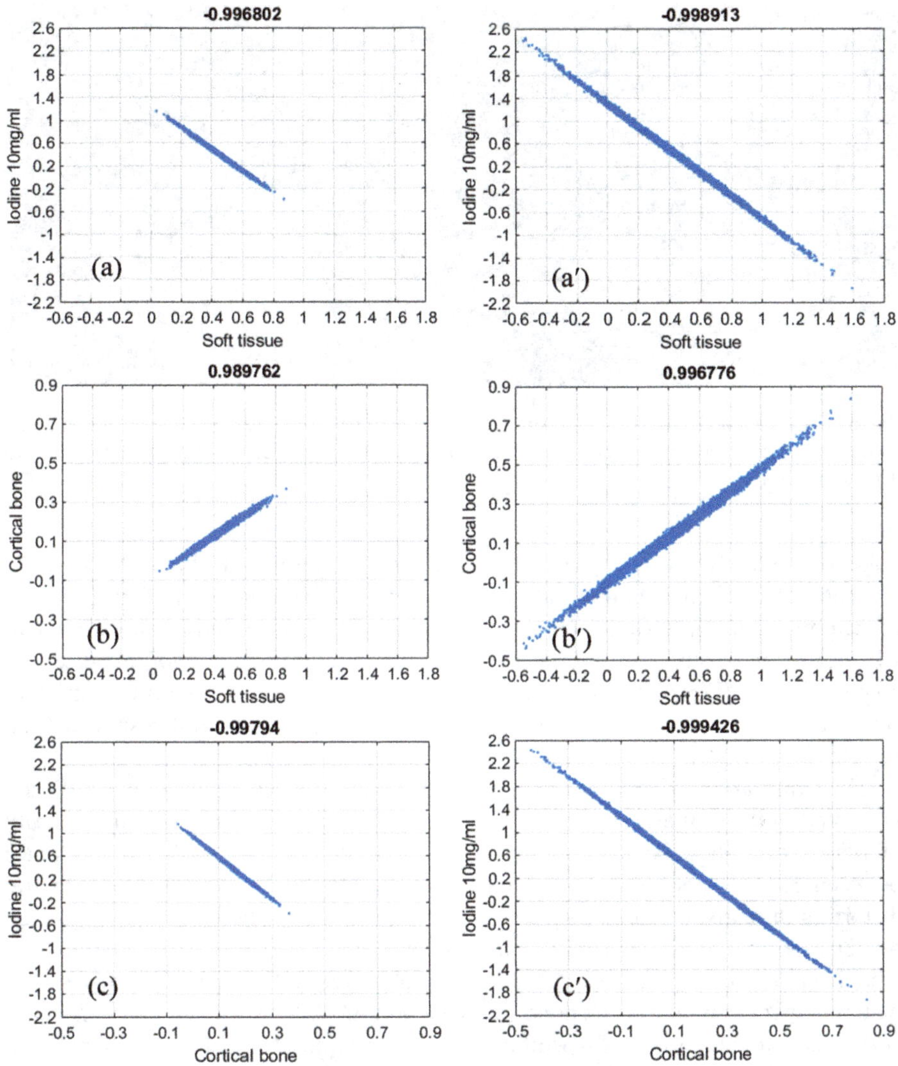

Fig. 4 Correlation of noise between the material-specific images acquired via 3-MD under both ideal (**a–c**) and realistic (**a′–c′**) detector responses

6.4 *Results: Noise and Noise Correlation in 4-MD Material-Specific Images*

The material-specific images corresponding to basis material soft tissue, 10 mg/ml gadolinium, 10 mg/ml iodine, and cortical bone via 4-MD under ideal detector spectral response are displayed in Fig. 5a–c. Overall, given same X-ray dose, the

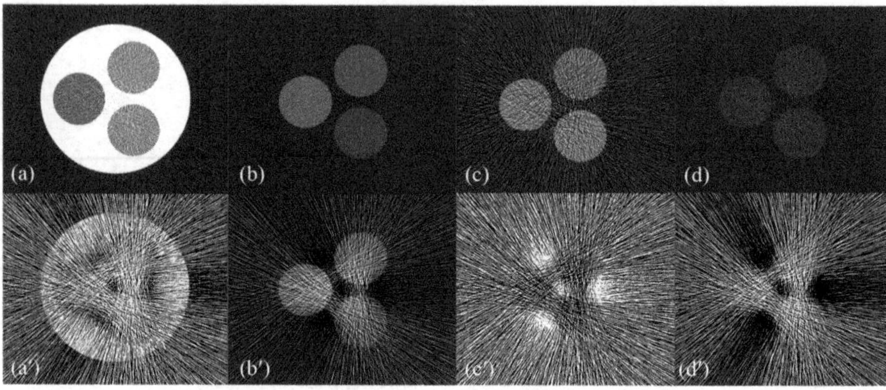

Fig. 5 Material-specific (left to right: soft tissue, (10 mg/ml) I, (10 mg/ml) Gd, and cortical bone) images of the noise correlation phantom acquired via 4-MD under ideal (**a**–**d**) and realistic (**a**′–**d**′) detector response

noise in the 4-MD basis images is much stronger than that in the 3-MD case, as the number of X-ray photons falling into each spectral channel is further fewer, as demonstrated by visual comparison between Figs. 3a–c and 5a–d. Similar to what has been observed in the case of 3-MD, the interchannel overlap in detector's response in 4-MD makes the noise in each material-specific image much stronger (Fig. 5a′–d′) over their counterparts under ideal detector spectral response (Fig. 5a–d).

The correlation of noise between the material-specific images in 4-MD are presented in the scatterplots in Fig. 6a–f. Similar to the cases in 3-MD, the polarity of noise correlation between those material-specific images in 4-MD alternates between ±1. Notably, the interchannel spectral overlap induced by the spectral distortion in realistic detector's spectral response in 4-MD increases the magnitude of noise correlation considerably (compare Fig. 6a–f and Fig. 6a′–f′), similar to what happens in the case of 3-MD. Also, it is observed in Fig. 5 that the noise in the material-specific image corresponding to gadolinium (Fig. 5c, c′) is the strongest in comparison to other material-specific images. Consequently, the Pearson correlation coefficient decreases significantly whenever gadolinium is involved, as illustrated in Fig. 6a, d, e in comparison to Fig. 6b, c, f and Fig. 6a′, d′, e′ compared to Fig. 6b, c, f as well. We believe that the underlying reason is that gadolinium is much more attenuating than other materials in the phantom. In particular, the K-edge of gadolinium is at 50.2 keV, which is very close to the effective energy of the polychromatic X-ray source utilized in the simulation study.

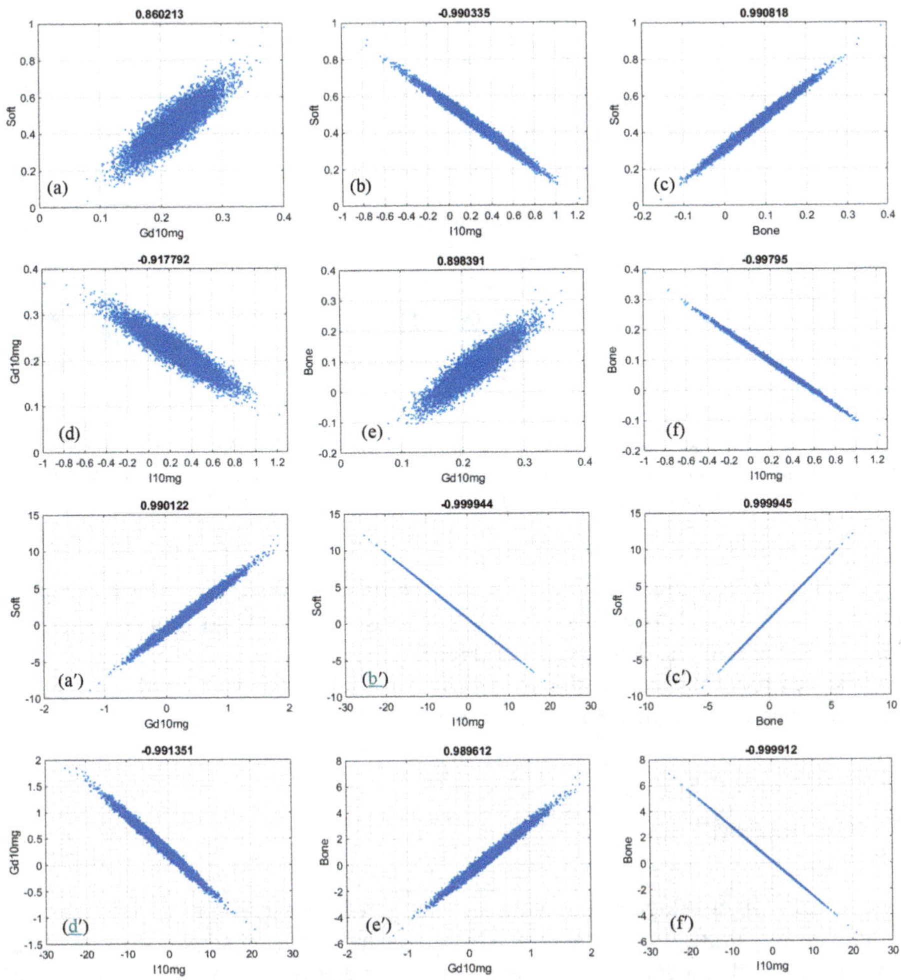

Fig. 6 Noise correlation in 4-MD under ideal detector spectral response (**a**) soft tissue vs gadolinium, (**b**) soft tissue vs iodine, (**c**) soft tissue vs bone, (**d**) gadolinium vs iodine, (**e**) gadolinium vs bone, and (**f**) iodine vs bone and their counterparts under realistic detector spectral response (**a′**–**f′**)

6.5 *Results: Noise Correlation and Its Impact on the Performance of Spectral Imaging*

Taking 3-MD as an example, the noise gauged in each of the material-specific images under ideal detector spectral response is charted in Fig. 7a, showing that the variation of noise within the basis material sets BM-1 and BM-2 is drastic. Similar phenomenon is observed under realistic detector spectral response (Fig. 7a′), except for the markedly increased magnitude in noise. Moreover, an inspection of Fig. 7a,

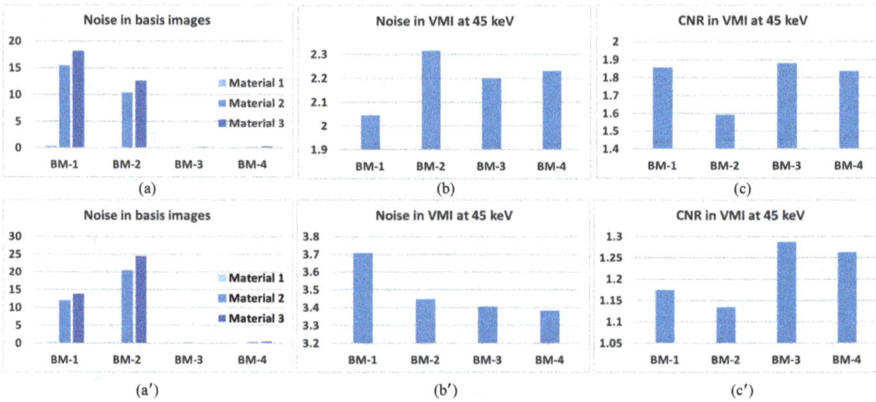

Fig. 7 Noise in the material-specific images of each basis material set (column 1) and the resultant noise (column 2) and contrast-to-noise (column 3) in virtual monochromatic images at 45 keV under ideal (top row) and realistic (bottom row) detector spectral response

a′ reveals that the variation of noise in the material-specific images across the basis material sets BM-1, BM-2, BM-3, and BM-4 is drastic too. On the other hand, under both ideal and realistic detector spectral responses, the noise in the VMIs at 45 keV only varies modestly across the four basis material sets (Fig. 7b, b′), leading to moderate variation in the CNR (Fig. 7c, c′).

7 Discussion

The research in development of photon-counting CT and exploration of its clinical potentials is gaining the momentum. In this chapter, we presented the equations governing the behavior of noise and noise correlation in the material-specific images for m-MD-based spectral imaging in photon-counting CT. Since the analytic expression for noise and noise correlation in the cases beyond 2-MD is complicated, we conducted a simulation study to investigate the noise and its correlation between the material-specific images and its impact on the performance of *m*-MD-based spectral imaging in photon-counting CT. Supported by quantitative data, we characterized the relation between the noises in material-specific (basis) images in m-MD-based spectral imaging. As shown in Figs. 4 and 6, the polarity of noise correlation in the material-specific images of 3-MD and 4-MD alternates between ± 1, which makes the noise property in *m*-MD-based spectral imaging in photon-counting CT complicated. Potentially, such a noise behavior may bring about the opportunity for us to design and implement sophisticated algorithms for denoising in *m*-MD-based spectral imaging in photon-counting CT, for example, similar to the ones reported in the literature [29] or in particular using neural network or deep learning (DL)-based approaches [30, 31].

In the case of 3-MD, Eqs. (A-1)–(A-6) in the appendix of Ref. [18] show that the magnitude of noise in each material-specific image is jointly determined by the SNR of the data acquired in each spectral channel, the interchannel data correlation $\rho_{I_{k1k2}}$ ($k_1 = 1, 2, 3$; $k_2 = 1, 2, 3$), the effective attenuation coefficients μ_{kp} ($k = 1, 2, 3$; $p = 1, 2, 3$), and the Jacobian $\Delta^2_{3\times3}$ that is dependent on μ_{kp} ($k = 1, 2, 3$; $p = 1, 2, 3$). Notably, the distortion in realistic detector's spectral response (detector spectral distortion) induces interchannel data correlation, that is, $\rho_{I_{k1k2}} > 0$ ($k_1 = 1,2,3; k_2 = 1,2,3$). On the other hand, the detector spectral distortion also alters the effective attenuation coefficient μ_{11}, μ_{12}, μ_{21}, and μ_{22} or μ_{kp} ($k = 1, 2, 3$; $p = 1, 2, 3$) and thus $\Delta^2_{3\times3}$. In reality, the role played by the Jacobian in the denominator is dominant over that by the interchannel data correlation in the nominator. Hence, the material-specific images under realistic detector spectral response are noisier than their counterparts under ideal detector spectral response, as we have observed visually in Fig. 3 and quantitatively in Fig. 7a–a′. Similar observation can be made on the noise in 4-MD (see Fig. 5a–d, a′–d′), though the analytic equations that characterize the noise behavior in 4-MD are not presented.

The polarity of noise correlation in the material-specific images is determined by the corresponding material's order of attenuating property at the low energy end among the basis materials (see Fig. 7 and its related context and Ref. [10] as well). In turn, the order decides the sign of the co-factors in getting the inverse matrixes according to Eq. (5). Meanwhile, the magnitude of noise correlation between the 3-MD-based material-specific images seemingly approaches unity and implies that the correlation is actually very strong. As shown in Eqs. (A-1)–(A-6) in the appendix of Ref. [18], the interchannel data correlation ($\rho_{I_{k1k2}}$) and the effective attenuation coefficient μ_{kp} (p-th material in k-th energy bin) play important roles in determining the magnitude of noise correlation in 3-MD, in a manner that is similar to their influence on the magnitude of noise as just explained above. A close inspection of Fig. 4 shows that the detector's spectral distortion, which may be induced by charge sharing, Compton scattering, and florescence escaping [11], increases the magnitude of noise correlation between the material-specific images in comparison to their counterparts under ideal detector spectral response. Hence, further investigation to characterize the magnitude of noise correlation is deemed necessary, especially its variation over the conditioning of basis materials. Similar observation can be made on the noise correlation in the case of 4-MD if one compares the results given in Fig. 6a–f, a′–f′, despite the analytic equations that characterize the behavior of noise correlation in 4-MD that is not explicitly given.

The strong variation of noise in each of the material-specific images in 3-MD (Fig. 7a–a′) is attributed to the similarity in attenuation between the basis material adipose and PMMA and accordingly poor conditioning in the basis material sets BM-1 and BM-2 [9, 10]. By improving the conditioning of basis materials in BM-3 and BM-4, the variation of noise in the material-specific images becomes substantially more tractable. Hence, for the material-specific imaging, attention should be paid to the selection of basis materials [9, 10] (see Fig. 7a–a′). On the other hand, though the noise varies radically in the material-specific images over the basis material sets BM-1, BM-2, BM-3, and BM-4 (Fig. 7a–a′), the noise variation in virtual monochromatic images is still moderate (Fig. 7b–b′), leading to moderate variation in the imaging performance assessed by CNR (Fig. 7c–c′). It is believed that the tolerance to the conditioning of basis materials in the virtual monochromatic imaging is mainly due to the negative correlation in the noise between material-specific images.

Only the data related to the cases of 3-MD and 4-MD are presented in the chapter as the examples to evaluate and verify the property of noise and noise correlation in m-MD. Literally, it should be straightforward to apply Eq. (5) and its derivatives in the cases wherein more than four basis materials are engaged in material decomposition. Fortunately, as indicated in our study on the conditioning of basis materials and spectral channelization and their impacts on the performance of photon-counting spectral imaging in CT [10, 11], the likelihood for more than four basis materials (and accordingly four spectral channels) is low, in light of the fact that more basis materials (and thus spectral channels) may radically impact the imaging performance, leading to the consequence of severe noise or even photon starvation, as we have observed in the cases presented in Fig. 5a′–d′. We hope that the data presented herein would be informative to the clinical medical physicists who are interested in spectral imaging in either photon-counting or energy-integration spectral CT and also beneficiary to the clinical medical physicists who are working in other X-ray imaging modalities, such as digital radiography and tomosynthesis, as essentially there is no difference in the underlying physics and implementation of material decomposition between CT and those X-ray-related imaging modalities.

References

1. R.E. Alvarez, Extraction of energy-dependent information in radiography. (Doctoral dissertation) Stanford University, Stanford, CA (1976)
2. R.E. Alvarez, A. Macovski, Energy-selective reconstructions in x-ray computerised tomography. Phys. Med. Biol. **21**(5), 733–744 (1976)
3. R.E. Alvarez, Dimensinality and noise in energy selective x-ray imaging. Med. Phys. **40**(11), 111909 (2013)
4. F. Kelcz, P.M. Joseph, S.K. Hilal, Noise considerations in dual energy CT scanning. Med. Phys. **6**(5), 418–425 (1979)
5. A.N. Primak, J.C. Ramirez Giraldo, X. Liu, L. Yu, C.H. McCollough, Improved dual-energy material discrimination for dual-source CT by means of additional spectral filtration. Med. Phys. **36**(4), 1359–1369 (2009)
6. R. Alvarez, E. Seppi, A comparison of noise and dose in conventional and energy selective computed tomography. IEEE Trans. Nucl. Sci. **26**(2), 2853–2856 (1979)
7. W.A. Kalender, E. Klotz, L. Kostaridou, Algorithm for noise suppression in dual energy CT material density images. IEEE Trans. Med. Imaging. **7**(3), 218–224 (1988)
8. R.A. Sones, G.T. Barnes, Noise correlation in image acquired simultaneously with a dual-energy sandwich detector. Med. Phys. **16**(6), 858–861 (1989)
9. Y. Ren, H. Xie, W. Long, X. Yang, X. Tang, Optimization of basis material selection and energy binning in three material decomposition for spectral imaging without contrast agents in photon-counting CT. SPIE Proc.**11312**, 113124X. (8 pages) (2020). https://doi.org/10.1117/12.2549678

10. X. Tang, Y. Ren, On the conditioning of basis materials and its impact on multi-material decomposition based spectral imaging in photon-counting CT. Med Phys **48**(3), 1100–1116 (2021)
11. Y. Ren, H. Xie, W. Long, X. Yang, X. Tang, On the conditinaing of spectral channelization (energy binning) and its impact on multi-material decomposition based spectral imaging in photon-counting CT. IEEE TBME **68**(9), 2678–2688 (2021)
12. E. Roessl, A. Ziegler, R. Proksa, On the influence of noise correlations in measurement data on basis image noise in dual-energylike x-ray imaging. Med. Phys. **34**(3), 959–966 (2007)
13. National Council on Radiation Protection and Measurements, Report No. 184 – Medical Radiation Exposure of patients in the United States, Bethasda, MD, 2019
14. K.T. Bae, Intravenous contrast medium administration and scan timing at CT: considerations and approaches. Radiology **256**(1), 32–61 (2010)
15. H. Lusic, M.W. Grinstaff, X-ray computed tomography contrast agents. Chem Rev **113**, 1641–1666 (2013)
16. Z. Zhou, L. Ren, K. Rajendran, F.E. Diehn, J.G. Fletcher, C.H. McCollough, L. Yu, Simultaneous dual-contrast imaging using energy-integrating detector multi-energy CT: An in vivo feasibility study. Med. Phys **49**(3), 1458–1467 (2022)
17. X. Tang, Y. Ren, Noise correlation in multi-material decomposition-based spectral imaging in photon-counting CT, in *Proceedings of the 16th Virtual International Meeting on Fully 3D Image Reconstruction in Radiology and Nuclear Medicine*, (2021), pp. 414–418. Fully 3D
18. T. Tang, Y. Ren, H. Xie, Noise correlation and its impact on theperformance of multi-materialdecomposition-based spectral imaging inphoton-counting CT. J. Appl. Clin. Med. Phys. **e13830** (2023). https://doi.org/10.1002/acm2.13830
19. J.H. Ahrens, U. Dieter, Computer generation of poisson deviates, from modified normal distributions. ACM Trans. Math. Software **8**(2), 163–179 (1982)
20. C. Keh-Shih, H.K. Huang, Comparison of four dual energy image decomposition methods. Phys. Med. Biol. **33**(4), 455 (1988)
21. M.M. Goodsitt, E.G. Christodoulou, S.C. Larson, Accuracies of the synthesized monochromatic CT numbers and effective atomic numbers obtained with a rapid kVp switching dual energy CT scanner. Med. Phys. **38**(4), 2222–2232 (2011)
22. B. De Man, S. Basu, C. Naveen, B. Dunham, P. Edic, M. Iatrou, S. McOlash, P. Sainath, C. Shaughnessy, T. Brenton, E. Williams, CatSim: a new computer assisted tomography simulation environment. SPIE Proc **6510**, 65102G. (8 pages) (2007). https://doi.org/10.1117/12.710713
23. H. Chen, C. Xu, M. Persson, M. Danielsson, Optimization of beam qualty for photon-counting spectral computed tomoraphy in head imaging: simulation study. SPIE J. Med. Imaging **2**(4), 043504 (2015) (16 pages)
24. T.M. Buzug, *Computed Tomography: From Photon Statistics to Modern Cone-Beam CT*, 1st edn. (Springer, Berlin/Heidelberg, 2008). https://doi.org/10.1007/978-3-540-39408-2
25. G. Cowan, *Statistical Data Analysis* (Oxford University Press, Oxford, 1998)
26. H.Q. Woodard, D.R. White, The composition of body tissues. Br. J. Radiol **59**(12), 1209–1219 (1986)
27. D.R. White, R.V. Griffith, I.J. Wilson, ICRU Report 46: Photon, eletron, proton and neutron interaction data for body tissues. J. Int. Comm. Radiat. Units Meas. **os24**(1) (1992)
28. D.E. Cullen, J.H. Hubbell, L. Kissel, EPDL97: the evaluated photon data library Lawrence Livermore National Laboratory Report UCRL-50400 **6**, rev 5 (1997)
29. S. Zhang, D. Han, D.G. Politte, J.F. Williamson, J.A. O'Sullivan, Impact of joint statistical dual-energy CT reconstruction of proton stopping power images: comparison to image- and sinogram-domain material decompositin approaches. Med. Phys. **45**(5), 2129–2142 (2018)
30. Z. Gao, X. Wang, S. Sun, W. Dan, J. Bai, Y. Yin, X. Liu, H. Zhang, V.H.C. de Albuquerque, Learning physical properties in complex visual scenes: an intelligent machine for perceiving blood flow dynamics from static CT angiography imaging. Neural Netw **123**(3), 82–93 (2020)
31. J.-P. Charbonnier, E.M. van Rikxoort, A.A.A. Setio, C.M. Schaefer-Prokop, B. van Ginneken, F. Ciompi, Improving airway segmentation in computed tomography using leak detection with convolutional networks. Medical Image Analysis **36**(2), 52–60 (2017)

Photon-Counting Computed Tomography (PCCT) for Radiation Therapy

Guyue Hu, Katharina Niepel, Franka Risch, Christopher Kurz, Matthias Würl, Thomas Kröncke, Florian Schwarz, Katia Parodi, and Guillaume Landry

1 Introduction

Protons passing through tissue deposit a dose along their path, which reaches a maximum at the Bragg peak, immediately before the protons stop. This characteristic makes it possible to create a spread-out Bragg peak by superimposing proton beams of different energies and weights. Compared to conventional photon therapy, proton

---

Katia Parodi, Florian Schwarz, Guillaume Landry contributed equally to this work.

G. Hu (✉) · K. Niepel · M. Würl · K. Parodi
Department of Medical Physics, Faculty of Physics, Ludwig-Maximilians-Universität München, München, Germany
e-mail: Guyue.Hu@physik.uni-muenchen.de

F. Risch · T. Kröncke
Department of Diagnostic and Interventional Radiology, Universitätsklinikum Augsburg, Augsburg, Germany

C. Kurz
Department of Radiation Oncology, LMU Klinikum, Munich, Germany

F. Schwarz
Department of Diagnostic and Interventional Radiology, Universitätsklinikum Augsburg, Augsburg, Germany

Medical Faculty, Ludwig-Maximilians-Universität München, Munich, Germany

G. Landry
Department of Medical Physics, Faculty of Physics, Ludwig-Maximilians-Universität München, München, Germany

Department of Radiation Oncology, LMU Klinikum, Munich, Germany

German Cancer Consortium (DKTK), Partner Site Munich, Munich, Germany

K. Iniewski, H. Gadey (eds.), *Emerging Radiation Detection*,
https://doi.org/10.1007/978-3-031-63897-8_7

therapy allows a highly conformal dose to the tumour region, sparing surrounding healthy tissue [19]. To produce the high dose at just the right place, the challenge is the accurate prediction of proton range, which requires knowledge of proton relative (to water) stopping power (RSP) in matter.

In current clinical practice, RSP is typically estimated using single-energy computed tomography (SECT) with a stoichiometric calibration between CT numbers and RSP [15]. In this method, linear fits between RSP and Hounsfield units (HU) are made for different groups of tissues, but the relation between CT numbers and RSP is not unique, leading to range uncertainties up to 3.5% [27].

In contrast to single-energy CT, dual-energy CT (DECT) uses two different energy spectra. As the linear attenuation coefficient is energy dependent and can be parameterised in terms of electron density and atomic number of the material, it is then possible to directly estimate the relative electron density (RED) and the effective atomic number (EAN) using scans at two different energies. Subsequently, RSP can be directly calculated using the Bethe–Bloch formula [8, 9, 20, 21, 25]. This way, the range uncertainties are reduced below 2% [10].

As one of the latest developments in X-ray CT, photon-counting CT (PCCT) [17, 23, 24], which provides CT data with intrinsic spectral information, is a topic of increasing interest in radiation therapy. PCCT images reconstructed at different virtual monoenergies allow us to estimate RSP with the same method as described above for DECT, but offering flexibility to choose the best energy combination of different virtual monoenergetic reconstructions.

Early studies with prototype PCCT scanners have compared these different methods of RSP estimation, showing similar or slightly better RSP accuracy for PCCT compared to methods based on DECT and SECT [22].

The first-generation clinical PCCT scanners have been recently introduced. Their performance for evaluation of quantitative information for radiation therapy treatment planning such as RED, EAN, and RSP has been recently investigated. We aimed at assessment of RED, EAN, and RSP relying on the virtual monoenergetic images (VMIs) provided by a first-generation clinical PCCT scanner [4]. Here, we summarise the main findings.

2 PCCT Measurements at a NAEOTOM Alpha Scanner

2.1 NAEOTOM Alpha PCCT Scanner

In a conventional energy-integrating detector, X-rays are first converted into visible light in the scintillator, which is then converted into electrical signals at the photodiodes. Light pulses generated by thousands of X-ray photons are accumulated and measured altogether. In contrast, in a photon-counting detector (PCD), X-rays generate electron–hole pairs in the semiconductor, which are separated by a strong electric field between the cathode and the pixelated anode. This direct conversion from X-ray to electrical signal enables individual measurements of X-ray photons.

In our work, a NAEOTOM Alpha CT scanner (Siemens Healthineers, Forchheim, Germany) with PCD technology was used. For the NAEOTOM Alpha PCDs, each anode pixel is further divided into subpixels of 0.15×0.176 mm^2, which can be read out as 2×2 or 1×1 groups for the normal- and the high-resolution modes, respectively. For each subpixel, there are four manufacturer-specified energy thresholds. Data from each threshold are weighted to yield two basis images, which are combined to generate VMIs [11].

2.2 Phantom Scans and Image Reconstruction

In our experiments, customised phantoms housing tissue-equivalent inserts were scanned at the PCCT scanner. A cylindrical PMMA phantom with a diameter of 15 cm housing twelve Gammex 467 inserts (Gammex, Inc., Middleton, WI, USA) was used as the calibration phantom. A similar phantom with a diameter of 13 cm containing CIRS 062M inserts (CIRS, Inc., Norfolk, VA, USA) was used for evaluation. For both phantoms, the elemental composition and the mass density of the inserts provided by the vendor are listed in Table 1.

For all scans in this chapter, we used a tube voltage of 120 kVp and the normal-resolution mode at a dose level of $\mathrm{CTDI_{vol}} = 40\,\mathrm{mGy}$. For each phantom, VMIs with a slice thickness of 1 mm and a pixel size of 0.38×0.38 mm^2 were reconstructed at virtual monoenergies ranging from 40 keV to 190 keV in 10 keV increments using the Qr36f kernel.

3 Conversion of CT Numbers into RED/EAN/RSP

To estimate quantitative information for radiation therapy from PCCT data, we used the approach adopted by [7], converting measured CT numbers into RED, EAN, and mean excitation energy, using a calibration based on a pair of low-energy and high-energy CT numbers for a set of inserts with known mass density and elemental composition.

As mentioned in Sect. 2.2, VMIs were reconstructed at energies ranging from 40 keV to 190 keV in 10 keV increments. We exploited various combinations of a low-energy and a high-energy VMI to estimate RED, EAN, and RSP. The VMI pairs were chosen in such a way that, for each low energy E_L ranging from 40 keV to 180 keV, we used a sequence of high energy ranging from $E_\mathrm{L} + 10$ keV to 190 keV. For each VMI pair, three calibration procedures were conducted. The PCCT data of the twelve Gammex inserts were used for calibration.

Table 1 Mass percent elemental composition and mass density of insert materials [4]

Element	ρ	H	C	N	O	Mg	Si	P	Cl	Ca
Z	[g cm^{-3}]	1	6	7	8	12	14	15	17	20
A		1.008	12.01	14.01	16.00	24.31	28.09	30.97	25.45	40.08
I		19.2	81	82	106	176	195	195	180	216
Gammex										
BR-12 Breast	0.98	8.59	70.11	2.33	17.90	0	0	0	0.13	0.95
AP6 Adipose	0.94	9.06	72.30	2.25	16.27	0	0	0	0.13	0
LN-450 Lung	0.45	8.47	59.79	1.97	18.11	11.21	0.58	0	0.10	0
LN-300 Lung	0.30	8.46	59.38	1.96	18.14	11.19	0.78	0	0.10	0
LV1 Liver	1.10	8.06	67.01	2.47	20.01	0	0	0	0.14	2.31
BRN-SR2 Brain	1.05	10.83	72.54	1.69	14.86	0	0	0	0.08	0
Muscle	1.05	8.10	67.17	2.42	19.85	0	0	0	0.14	2.32
CT Solid Water	1.02	8.02	67.23	2.41	19.91	0	0	0	0.14	2.31
CB2-50% CaCO3	1.56	4.77	41.63	1.52	32.00	0	0	0	0.08	20.02
CB2-30% CaCO3	1.34	6.68	53.48	2.12	25.61	0	0	0	0.11	12.01
B200 Bone Mineral	1.15	6.65	55.52	1.98	23.64	0	0	3.24	0.11	8.87
IB Inner Bone	1.14	6.67	55.64	1.96	23.52	0	0	3.23	0.11	8.86
CIRS										
Liver	1.07	9.00	69.40	2.10	17.10	0	0	0	0.10	2.20
Cortical Bone	1.91	3.30	25.37	0.91	35.28	3.36	0	8.82	0.03	22.91
Muscle	1.06	9.10	69.70	2.10	16.80	0	0	0	0.10	2.20
Adipose	0.96	10.00	71.30	1.80	16.40	0	0	0	0.20	0.30
Trabecular Bone 200	1.16	7.00	56.30	2.00	22.70	0	0	3.30	0.20	8.50
Breast 50/50	0.99	9.60	70.30	1.90	17.00	0	0	0	0.20	0.90

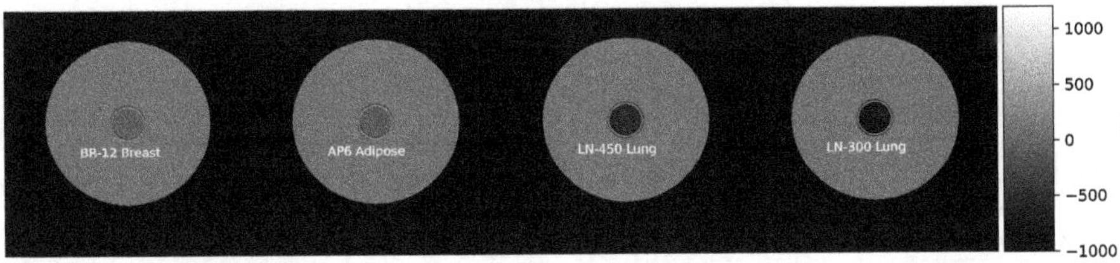

Fig. 1 HU maps of four Gammex (calibration) inserts. ROIs for the four inserts are shown in red

3.1 Three Calibration Procedures

HU-RED Calibration

As a first step, the mean measured CT numbers, $\mathrm{HU_H}$ and $\mathrm{HU_L}$, were calculated over a cylindrical region of interest (ROI) within each calibration insert for the high- and low-energy VMIs, respectively. Each ROI has a radius 5 pixels (i.e., 1.9 mm) smaller than the known radius of the insert, as illustrated in Fig. 1.

Next, the RED values, denoted by ρ_e, were calculated from vendor-provided elemental composition of the inserts:

$$\rho_e = \rho \cdot \frac{\sum_i \omega_i Z_i / A_i}{\omega_H Z_H / A_H + \omega_O Z_O / A_O}, \tag{1}$$

where ω_i is the mass percent of the composing element with an atomic number of Z_i and a mass number of A_i. The values used for RED calculation are listed in Table 1. Following [12], RED was calibrated against a weighted difference between the low- and the high-energy CT number:

$$\rho_e = a \frac{(1+\alpha)\mathrm{HU_H} - \alpha \mathrm{HU_L}}{1000} + b, \tag{2}$$

where a, b, and α are the fit parameters, which were obtained by least-squares fit.

HU-EAN Calibration

A second calibration was performed to fit the ratio between insert EAN and that of water, denoted by $Z_{\mathrm{eff}}/Z_{\mathrm{eff,w}}$, to the mean measured CT numbers and RED values of the inserts [13]:

$$\left(\frac{Z_{\mathrm{eff}}}{Z_{\mathrm{eff,w}}}\right)^{3.3} - 1 = \gamma_L \left(\frac{u_L}{\rho_e} - 1\right) + \gamma_0, \tag{3}$$

where $u_L \equiv \frac{\mathrm{HU_L}}{1000} + 1$ is the reduced CT number, Z_{eff} the EAN, calculated from the known elemental composition of the inserts [6, 14]:

$$Z_{\text{eff}} = \left(\frac{\sum_i \frac{\omega_i Z_i}{A_i} Z_i^{3.3}}{\sum_i \frac{\omega_i Z_i}{A_i}} \right)^{1/3.3}. \tag{4}$$

With least-squares fit, we obtained the fit parameters γ_{L} and γ_0.

EAN-*I* Calibration

A third calibration was needed to convert EAN into mean excitation energy [14]:

$$\ln \frac{I}{I_{\text{w}}} = C_1^{\text{soft/bone}} \left[\left(\frac{Z_{\text{eff}}}{Z_{\text{eff,w}}} \right)^{3.3} - 1 \right] - C_0^{\text{soft/bone}}, \tag{5}$$

where I/I_{w} is the ratio of mean excitation energy of an insert to that of water. The mean excitation energy of each insert was calculated by [14]

$$\ln I = \frac{\sum_i \frac{\omega_i Z_i}{A_i} \ln I_i}{\sum_i \frac{\omega_i Z_i}{A_i}}. \tag{6}$$

Here, the mean excitation energy values of the composing elements, denoted by I_i, are listed in Table 1. A mean excitation energy value of $I_{\text{w}} = 78$ eV for water was used according to the ICRU report 90 [16, 18].

For this calibration, the inserts were divided into two groups, soft-tissue inserts (EAN<8.8) and bone-tissue inserts (EAN>8.8). Two sets of fit parameters $C_1^{\text{soft/bone}}$ and $C_2^{\text{soft/bone}}$ were obtained by performing least-squares fit for the two groups of inserts.

3.2 Evaluation of RED, EAN, and RSP

With the obtained fit parameters a, b, α, γ_{L}, γ_0, $C_1^{\text{soft/bone}}$, and $C_2^{\text{soft/bone}}$, the RED (ρ_{e}), EAN (Z_{eff}), and mean excitation energy (I) maps were converted from measured CT numbers of a pair of low-energy and high-energy VMIs using the Eqs. (2), (3), and (5).

RSP was calculated voxel by voxel based on the estimated RED (ρ_{e}) and mean excitation energy (I) maps using a simplified version of the Bethe–Bloch formula [14]:

$$\text{RSP} = \rho_{\text{e}} \frac{\ln \frac{2m_{\text{e}}c^2\beta^2}{I(1-\beta^2)} - \beta^2}{\ln \frac{2m_{\text{e}}c^2\beta^2}{I_{\text{w}}(1-\beta^2)} - \beta^2}, \tag{7}$$

where m_{e} is electron mass, c the speed of light, and β the velocity relative to the speed of light.

4 Accuracy Evaluation of RED/EAN/RSP Estimation

To evaluate the accuracy of the estimation, we first calculated the mean estimated RED/EAN/RSP values within an ROI covering the central volume of each insert. For both calibration inserts (CIRS, diameter 30 mm) and evaluation inserts (Gammex, diameter 29 mm), the cylindrical ROIs have a diameter 10 pixels smaller than the insert diameter. The mean estimated values ($\text{value}_{\text{est}}$) were compared to reference values ($\text{value}_{\text{ref}}$):

$$\text{residuals} = \frac{\text{value}_{\text{est}} - \text{value}_{\text{ref}}}{\text{value}_{\text{ref}}} \times 100\%. \tag{8}$$

Here, the reference values for RED and EAN were calculated from either vendor information using Eqs. (1) and (4). For RSP, they were obtained through water column measurements with a carbon ion beam [5].

Root-mean-squared errors (RMSEs) of these estimated quantities were calculated over the phantom inserts:

$$\text{RMSE} = \sqrt{\frac{1}{N}\sum_{i=1}^{N} \text{residuals}^2}, \tag{9}$$

where N is the number of inserts for each phantom.

5 Experimental Results

Various pairs of low- and high-energy VMIs were selected for data evaluation. We compared the achieved accuracy of RED, EAN, and RSP accuracy. Results for the optimal VMI pair with a low energy of 60 MeV and a high energy of 180 keV, which showed overall best accuracy, are presented in the following sections.

5.1 Calibration

Following the calibration procedures described in Sect. 3.1, Fig. 2 and Table 2 show the fit curves and the obtained fit parameters, respectively. The results were obtained based on the mean measured CT numbers of the calibration inserts, which are listed in Table 3.

RED was well fitted with a coefficient of determination of $R^2 = 0.9999$ (Fig. 2a). For fitting EAN, the R^2 was 0.9934. As shown in Fig. 2b, the outlier with the largest residuals was a lung-mimicking insert. For conversion of EAN into mean excitation energy, the inserts were grouped into soft and bone tissues. The two

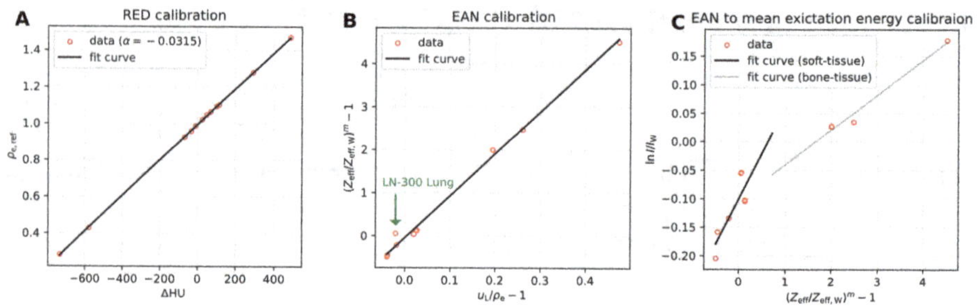

Fig. 2 Calibration curves for (**a**) RED calibration, (**b**) EAN calibration, and (**c**) RSP calibration based on measured CT numbers of the Gammex inserts at virtual monoenergies of 60 MeV and 180 MeV [4]

Table 2 Fit parameters for the three calibration procedures obtained using PCCT data at virtual monoenergies of 60 MeV and 180 MeV [4]

	Soft tissue	Bone tissue
a	0.9737	
b	0.9955	
α	−0.0315	
γ_L	9.7653	
γ_0	−0.0448	
C_1	0.1620	0.0612
C_0	0.0999	0.1011

corresponding fits yielded $R^2 = 0.6974$ and $R^2 = 0.9804$, respectively. The coefficient of determination for the soft-tissue inserts is relatively low, possibly due to the suboptimal tissue equivalence of the inserts in terms of the relation between $\ln I$ and Z_{eff}. Higher R^2 when performing the fit on tabulated human tissues was reported in [26].

5.2 *RED/EAN/RSP Accuracy*

With the fit parameters shown in Table 2, we first converted the measured CT numbers into RED (Fig. 3a). The estimated RED values, along with the measured CT numbers, were used to evaluate the EAN (Fig. 3b), from which we obtained the mean excitation energy values (Fig. 3c). Subsequently, RSP (Fig. 3d) was directly calculated from the estimates RED and mean excitation energy values using the Bethe–Bloch formula. Figure 3 shows the resulting images of three slices of the evaluation (CIRS) phantom. The ROIs for calculating mean estimated values are shown in red in Fig. 3a.

The same evaluation procedures were also conducted for the calibration (Gammex) phantom. The mean estimated RED, EAN, and RSP were evaluated within the same ROIs as chosen for the mean CT number calculation.

Table 3 Mean and standard deviation of CT numbers for Gammex and CIRS inserts from 60 keV (HU_L) and 180 keV (HU_H) VMIs, along with the mean and standard deviation of the estimated RED ($\rho_{e,est}$), EAN ($Z_{eff,est}$), and RSP values for all inserts. Reference and residual values as well as the resulting RMSEs are also given. (Adapted from [4])

	HU_L	HU_H	$\rho_{e,ref}$	$\rho_{e,est}$	ρ_e res. [%]	$Z_{eff,ref}$	$Z_{eff,est}$	Z_{eff} res. [%]	RSP_{ref}	RSP_{est}	RSP res. [%]
Gammex											
BR-12 Breast	-61 ± 5	-34 ± 5	0.957	0.961 ± 0.004	0.43	6.9	6.8 ± 0.2	−2.28	0.973	0.979 ± 0.006	0.63
AP6 Adipose	-111 ± 4	-69 ± 4	0.922	0.927 ± 0.004	0.48	6.2	6.3 ± 0.2	0.45	0.943	0.947 ± 0.005	0.42
LN-450 Lung	-563 ± 29	-579 ± 27	0.439	0.433 ± 0.026	−1.57	7.6	7.6 ± 0.5	0.07	0.436	0.438 ± 0.027	0.36
LN-300 Lung	-723 ± 14	-737 ± 13	0.292	0.278 ± 0.012	−4.79	7.6	7.1 ± 1.0	−6.20	0.272	0.283 ± 0.013	3.87
LV1 Liver	89 ± 6	68 ± 5	1.069	1.062 ± 0.005	−0.69	7.8	7.9 ± 0.2	2.00	1.079	1.071 ± 0.006	−0.74
BRN-SR2 Brain	6 ± 5	49 ± 4	1.046	1.042 ± 0.004	−0.42	6.1	6.4 ± 0.2	5.61	1.064	1.064 ± 0.005	−0.04
Muscle	48 ± 5	23 ± 5	1.021	1.019 ± 0.005	−0.23	7.8	8.0 ± 0.2	2.78	n. a.	1.027 ± 0.006	n. a.
CT Solid Water	11 ± 5	-13 ± 5	0.991	0.984 ± 0.005	−0.78	7.7	8.0 ± 0.2	2.63	1.000	0.992 ± 0.006	−0.85
CB2-50% CaCO3	1165 ± 10	463 ± 9	1.471	1.468 ± 0.008	−0.18	12.5	12.6 ± 0.1	0.45	1.434	1.434 ± 0.010	0.02
CB2-30% CaCO3	611 ± 7	279 ± 8	1.286	1.277 ± 0.008	−0.70	10.9	10.9 ± 0.1	0.29	1.279	1.269 ± 0.009	−0.82
B200 Bone Mineral	314 ± 6	107 ± 7	1.102	1.106 ± 0.006	0.28	10.4	10.2 ± 0.1	−2.05	1.100	1.104 ± 0.007	0.39
IB Inner Bone	306 ± 6	95 ± 6	1.093	1.095 ± 0.006	0.18	10.4	10.3 ± 0.1	−1.57	1.092	1.093 ± 0.007	0.10
RMSE (excluding lung)					**0.49**			**2.50**			**0.54**
RMSE					**1.52**			**2.90**			**1.27**
CIRS											
Liver	69 ± 5	46 ± 4	1.048	1.041 ± 0.004	−0.69	7.6	7.9 ± 0.1	4.48	1.060	1.049 ± 0.005	−1.02
Cortical Bone	1893 ± 8	749 ± 8	1.770	1.760 ± 0.008	−0.53	13.6	13.6 ± 0.1	−0.05	1.690	1.697 ± 0.009	0.40
Muscle	52 ± 6	39 ± 6	1.040	1.034 ± 0.005	−0.60	7.6	7.8 ± 0.2	2.01	1.057	1.044 ± 0.007	−1.20
Adipose	-81 ± 3	-40 ± 3	0.950	0.955 ± 0.003	0.57	6.5	6.3 ± 0.2	−1.86	0.970	0.976 ± 0.003	0.60
Trabecular Bone 200	316 ± 6	116 ± 6	1.116	1.114 ± 0.006	−0.13	10.3	10.1 ± 0.1	−2.05	1.112	1.114 ± 0.006	0.15
Breast 50/50	-51 ± 3	-26 ± 3	0.975	0.969 ± 0.003	−0.62	6.9	6.9 ± 0.1	−0.67	0.988	0.986 ± 0.004	−0.15
RMSE					**0.56**			**2.32**			**0.71**

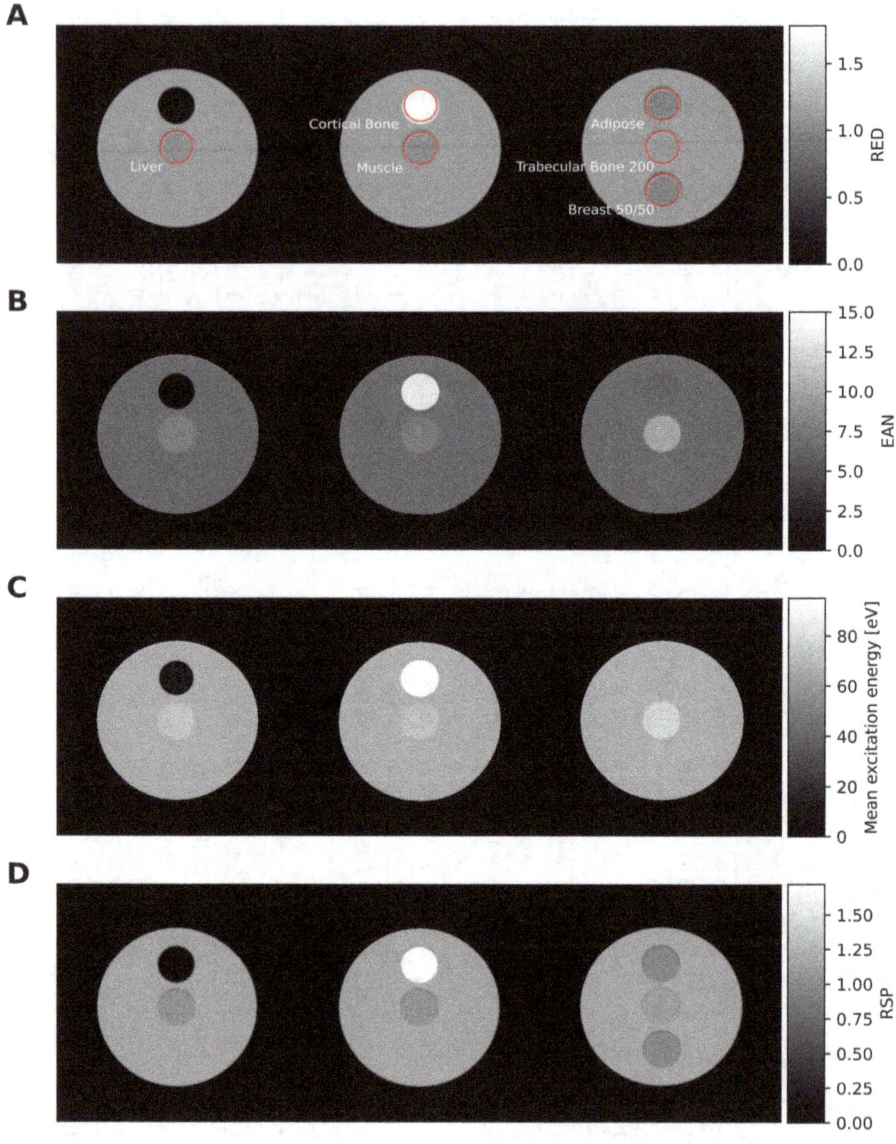

Fig. 3 Slices at three different positions of the evaluation (CIRS) phantom for the estimated (**a**) RED, (**b**) EAN, (**c**) mean excitation energy, and (**d**) RSP. ROIs for the six CIRS inserts are shown in red [4]

The residuals of RED estimation for the phantom inserts are presented in Fig. 4a. Most of the inserts showed an accuracy better than 1%, except the two lung-mimicking inserts, which showed much higher errors, leading to much higher RMSE for the Gammex phantom (1.52%) compared to the CIRS phantom (0.56%).

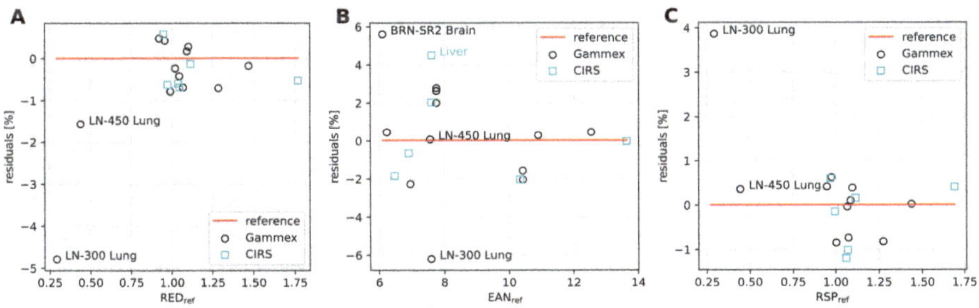

Fig. 4 Residuals of the estimated (**a**) RED, (**b**) EAN, and (**c**) RSP for the phantom inserts [4]

The Gammex RMSE was reduced to 0.49% when excluding the lung-mimicking inserts.

Similar to RED estimation, the measured CT numbers were converted into EAN through a linear relation. We can observe in Fig. 4b that the estimated EAN agrees with the ground truth within 2.8% for most of the inserts. The RMSEs calculated for both phantoms are listed in Table 3.

Figure 4c shows residuals of the estimated RSP for the phantom inserts. Comparing the estimation results to RSP measurements, the majority of the inserts showed accuracy better than 1%. For *Liver* and *Muscle*, the errors were slightly above 1%. The RSP estimation for *Lung* showed highest errors. This resulted in RMSE of 1.27% for the Gammex phantom, which was reduced to 0.54% when excluding lung-mimicking inserts, and the RSP accuracy for the CIRS phantom was 0.71%. This accuracy was found comparable to that obtained with dual-energy CT from previous studies [1, 5]. In [5], Gammex and CIRS inserts were scanned at a dual-source DECT scanner (SOMATOM Definition Force, Siemens Healthineers, Forchheim, Germany), and RMSEs of 1.1% and 0.8% were achieved for the Gammex and CIRS inserts, respectively. In [1], the same CIRS inserts as studied in our work yielded an RMSE of 0.68%.

Detailed information about the reference values, the mean and deviation of the estimated quantities, as well as the residuals, for the phantom inserts is listed in Table 3. The lung-mimicking inserts had relatively worse results, as also observed in previous studies [3, 5, 7]. Therefore, RMSEs for the Gammex phantom were calculated both with and without the two inserts, *LN-450 Lung* and *LN-300 Lung*.

5.3 Comparison Between Different Low- and High-Energy VMI Pairs

In the previous sections, results for one optimal pair of VMIs, 60 keV and 180 keV, were presented. In this section, we summarise the results of RED/EAN/RSP accuracy evaluation for different low- and high-energy VMI pairs used in our analysis.

Table 4 RMSEs of RED estimation for different VMI pairs. Results for the optimal VMI pair are shown in bold [4]

Phantom	Low energy [keV]	High energy [keV]	RMSE of RED
Gammex	**60**	**180**	**1.52%**
	160	180	1.65%
	160	190	1.65%
	170	190	1.91%
	Other cases		1.51%–1.54%
CIRS	**60**	**180**	**0.56%**
	160	180	0.58%
	160	190	0.58%
	170	190	0.71%
	Other cases		0.55%–0.56%

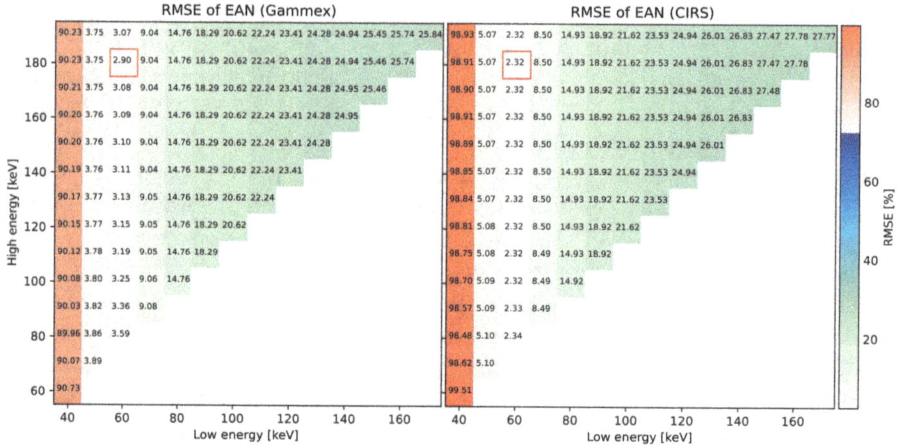

Fig. 5 RMSEs of EAN estimation for different VMI pairs [4]

For the RED estimation, the accuracy was not very sensitive to the choice of energy pairs. As listed in Table 4, the RMSEs for RED estimation mostly (102 out of 105 VMI pairs) ranged between 1.51% and 1.54% for the calibration phantom and 0.55% and 0.56% for the evaluation phantoms. The three exceptions are the VMI pairs with 160 keV or 170 keV as the low energy, which showed slightly worse RED accuracy.

For the EAN estimation, Fig. 5 shows the RMSEs obtained for different pairs of VMIs, on a background of different shades of colors corresponding to the RMSE value. For both phantoms, we can see that energy pairs sharing the same low energy had nearly the same accuracy, but still a best accuracy can be identified. The VMI pairs with 40 keV as the low energy had remarkably high errors for both phantoms, around 90% and 99%. This is potentially caused by dominant photoelectric absorption at lower energies. The HU-EAN calibration was consequently not so well fitted, resulting in worse EAN accuracy.

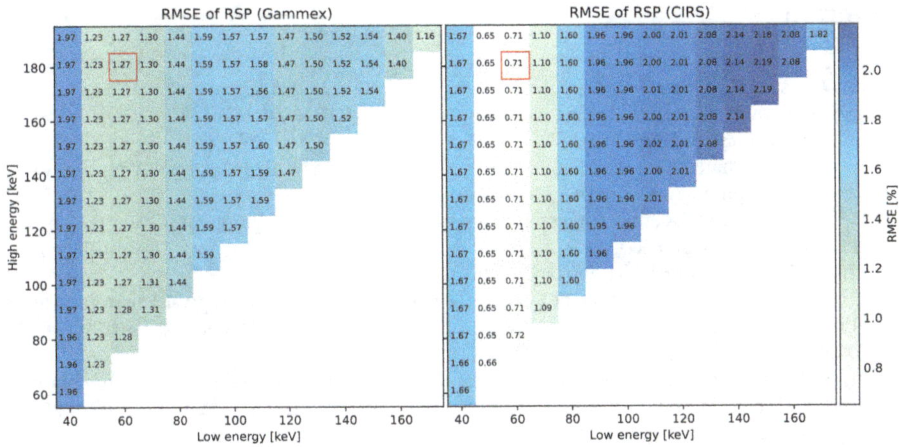

Fig. 6 RMSEs of RSP estimation for different VMI pairs [4]

For the RSP estimation, similar trends as observed for EAN can be seen in Fig. 6: using the same low-energy VMI, the RMSE barely changed despite the different choice of the high-energy VMI.

The optimal pair of VMIs, with 60 keV as the low energy and 180 keV as the high energy, which showed the overall best accuracy for EAN and RSP, is marked by red boxes in Figs. 5 and 6.

6 Conclusion and Outlook

This chapter summarises the results of a first study reporting tissue-equivalent material measurements at a first-generation clinical PCCT scanner, along with estimations of RED, EAN, and RSP of the materials using VMIs reconstructed at the scanner and evaluation of their accuracy. Using the optimal VMI pair with 60 keV as the low energy and 180 keV as the high energy, an accuracy of RSP of 1.27% (0.54% when excluding the porous lung-mimicking inserts) and 0.71% was obtained for the calibration and evaluation phantoms, respectively, which was found comparable to that reported in previous DECT studies [1, 5]. The results presented were obtained in the first year of operation of the scanner. Subsequent measurements are currently being analysed to further compare results of not only tissue-equivalent inserts but also biological tissue between PCCT, DECT, and proton CT in terms of RSP accuracy and spatial resolution [2].

Acknowledgments This work was funded by the German Research Foundation (DFG) within the Research Training Group GRK 2274 (grant number 299102935).

References

1. G. Dedes, et al., Experimental comparison of proton CT and dual energy x-ray CT for relative stopping power estimation in proton therapy. Phys. Med. Biol. **64**(16), 165002 (2019). https://doi.org/10.1088/1361-6560/ab2b72
2. E. Fogazzi, Characterizing the performance of a proton tomography system for x-ray CT crosscalibration, in *Third Ion Imaging Workshop 2022*, Munich, Germany (2022)
3. D.C. Hansen, et al., A simulation study on proton computed tomography (CT) stopping power accuracy using dual energy CT scans as benchmark. Acta Oncologica **54**(9), 1638–1642 (2015). https://doi.org/10.3109/0284186X.2015.1061212
4. G. Hu, et al., Assessment of quantitative information for radiation therapy at a first-generation clinical photon-counting computed tomography scanner. Front. Oncol. **12**, 970299 (2022). ISSN:2234–943X
5. N. Hudobivnik, et al., Comparison of proton therapy treatment planning for head tumors with a pencil beam algorithm on dual and single energy CT images. Med. Phys. **43**, 495 (2016). https://doi.org/10.1118/1.4939106
6. H.E. Johns, J.R. Cunningham, *The Physics of Radiology*. 3rd edn. (Thomas, Springfield, IL, 1978)
7. G. Landry, et al., Technical Note: Relative proton stopping power estimation from virtual monoenergetic images reconstructed from dual-layer computed tomography. Med. Phys. **46**(4), 1821–1828 (2019). https://aapm.onlinelibrary.wiley.com/doi/10.1002/mp.13404
8. C. Möhler, et al., Range prediction for tissue mixtures based on dual-energy CT. Phys. Med. Biol. **61**(11), N268–N275 (2016). https://doi.org/10.1088/0031-9155/61/11/n268
9. K.B. Niepel, et al., Animal tissue-based quantitative comparison of dual-energy CT to SPR conversion methods using high-resolution gel dosimetry. Phys. Med. Biol. **66**(7), 075009 (2021). https://doi.org/10.1088/1361-6560/abbd14
10. N. Peters, et al., Reduction of clinical safety margins in proton therapy enabled by the clinical implementation of dual-energy CT for direct stopping-power prediction. Radiotherapy Oncol. J. Eur. Soc. Therapeutic Radiol. Oncol. **166**, 71–78 (2022). https://doi.org/10.1016/j.radonc.2021.11.002
11. K. Rajendran, et al., First clinical photon-counting detector CT system: technical evaluation. Radiology **303**(1), 130–138 (2022). https://doi.org/10.1148/radiol.212579
12. M. Saito, Potential of dual-energy subtraction for converting CT numbers to electron density based on a single linear relationship. Med. Phys. **39**(4), 2021–2030 (2012). https://doi.org/10.1118/1.36941111
13. M. Saito, S. Sagara, A simple formulation for deriving effective atomic numbers via electron density from dual-energy CT data in the human body. Med. Phys. **44**(6), 2293–2303 (2017a). https://aapm.onlinelibrary.wiley.com/doi/10.1002/mp.12176
14. M. Saito, S. Sagara, Simplified derivation of stopping power ratio in the human body from dual-energy CT data. Med. Phys. **44**(8), 4179–4187 (2017b). https://aapm.onlinelibrary.wiley.com/doi/10.1002/mp.12386.
15. U. Schneider, E. Pedroni, A. Lomax, The calibration of CT Hounsfield units for radiotherapy treatment planning. Phys. Med. Biol. **41**(1), 111–124 (1996). https://doi.org/10.1088/0031-9155/41/1/009
16. S.M. Seltzer, et al., Report 90: key data for ionizing-radiation dosimetry: measurement standards and applications. J. ICRU **14**(1), 1–118 (2014)
17. S. Si-Mohamed, et al., Review of an initial experience with an experimental spectral photon-counting computed tomography system. Nucl. Instrum. Methods Phys. Res. A **873**, 27–35 (2017). https://doi.org/10.1016/j.nima.2017.04.014
18. P. Sigmund, A. Schinner, H. Paul, Errata and Addenda for ICRU Report 73, Stopping of ions heavier than helium. I C R U J. **5**(1), (2009). ISSN:1473–6691
19. A.R. Smith, Proton therapy. Phys. Med. Biol. **51**(13), R491–R504 (2006). https://doi.org/10.1088/0031%2D9155/51/13/r26

20. V.T. Taasti, et al., Validation of proton stopping power ratio estimation based on dual energy CT using fresh tissue samples. Phys. Med. Biol. **63**(1), 015012 (2017). https://doi.org/10.1088/1361-6560/aa952f
21. V.T. Taasti, et al., Comparison of single and dual energy CT for stopping power determination in proton therapy of head and neck cancer. Phys. Imag. Radiat. Oncol. **6**, 14–19 (2018a). https://www.sciencedirect.com/science/article/pii/S2405631617300842
22. V.T. Taasti, et al., Theoretical and experimental analysis of photon counting detector CT for proton stopping power prediction. Med. Phys. **45**(11), 5186–5196 (2018b). https://doi.org/10.1002/mp.13173
23. M.J. Willemink, et al., Photon-counting CT: technical principles and clinical prospects. Radiology **289**(2), 293–312 (2018). https://doi.org/10.1148/radiol.2018172656
24. P. Wohlfahrt, C. Richter, Status and innovations in pre-treatment CT imaging for proton therapy. Br. J. Radiol. **93**(1107), 20190590 (2020). https://doi.org/10.1259/bjr.20190590
25. P. Wohlfahrt, et al., Clinical implementation of dual-energy CT for proton treatment planning on pseudo-monoenergetic CT scans. Int. J. Radiat. Oncol. Biol. Phys. **97**(2), 427–434 (2017). https://www.sciencedirect.com/science/article/pii/S0360301616333429
26. M. Yang, et al., Theoretical variance analysis of single- and dual-energy computed tomography methods for calculating proton stopping power ratios of biological tissues. Phys. Med. Biol. **55**(5), 1343–1362 (2010). https://doi.org/10.1088/0031-9155/55/5/006
27. M. Yang, et al., Comprehensive analysis of proton range uncertainties related to patient stopping-power-ratio estimation using the stoichiometric calibration. Phys. Med. Biol. **57**(13), 4095–4115 (2012). https://doi.org/10.1088/0031-9155/57/13/4095

New Approaches for Metal Artifact Reduction in Computed Tomography

Julian Anhaus and Christian Hofmann

1 Introduction

Computed tomography (CT) has evolved into a pivotal tool for noninvasive image-based medical diagnosis since its development in the 1960s and 1970s. Today, there are nearly seven million CT examinations conducted on stationary patients in Germany, indicating a 126% increase over the last two decades and underscoring the significance of CT in modern diagnostics [1]. As old as CT itself, the challenge of metallic artifacts compromising image quality and, consequently, radiologists' diagnostic confidence persists. Recent advancements in implant biomaterials, coupled with heightened standards in routine clinical procedures, have led to a growing number of patients with metallic implants, including fixation systems for fractures, endoprostheses for joint replacement, cardiovascular support implants, and dental fillings or replacements.

Despite a nearly 50% decrease in the number of dental fillings over the past 40 years, Germany still witnesses over 47 million dental filling insertions annually [2]. While this statistic alone does not provide information on the prevalence of metallic inserts in CT examinations, a study has reported that almost 75% of all CT examinations in their emergency department involve artifacts induced by dental fillings [3]. Furthermore, the annual implantation of endoprostheses in the hips and extremities has steadily increased by 36% over the last two decades, reaching almost 440,000 implantations [1]. However, the presence of metal implants introduces severe metallic artifacts into the reconstructed image.

J. Anhaus (✉) · C. Hofmann
Siemens Healthineers, Forchheim, Germany
e-mail: julian.anhaus@siemens-healthineers.com

K. Iniewski, H. Gadey (eds.), *Emerging Radiation Detection*,
https://doi.org/10.1007/978-3-031-63897-8_8

In recent decades, various approaches addressing and reducing metal artifacts in CT have emerged in the literature, research, and clinical routine. Three main dominant groups have been identified:

Firstly, within the last decade, various deep learning-based (DL) methods have proven to be powerful tools for mitigating metal artifacts in CT. These methods can be categorized into subgroups, each offering unique advantages and trade-offs. Image-based DL methods focus on reconstructing the spatial domain by applying neural networks to segment metal-corrupted areas and learn patterns associated with metal-induced distortions [4–6]. This approach demonstrates the clear benefit of requiring only the network's inference without transformations to and from the sinogram space, which can be beneficial for the overall computational time. Sinogram-based methods benefit from the ability to identify metal projection-related artifacts and differentiate them from relevant details. An example within the sinogram involves correcting the metal trace to understand how the uncorrected sinogram would appear without metal, thereby introducing a reduced amount of artifact after image reconstruction [7]. Mixed-space methods exploit complementary information in both projection data and reconstructed images [8, 9].

Secondly, spectral metal artifact reduction techniques have emerged as a sophisticated approach to mitigating artifacts. The strength of metal artifacts within the image is mainly determined by the atomic number of an implant and the CT scans' spectrum. Conventional single-energy CT struggles to differentiate between photoelectric and Compton interactions of X-rays, but spectral CT scans, achieved through dedicated scan modes on energy-integrating CT systems (e.g., dual-source dual-energy, dual-spiral, or twin-beam acquisitions) or spectral CT detectors (including recent photon-counting detectors), provide spectral resolution. This resolution enables pixel-wise energetic interpolation and the computation of virtual monoenergetic images, proven to reduce metal artifacts in specific metal implants, such as spinal fixation systems or hip implants [10–12].

Thirdly, the most prominent correction method in clinical use involves replacing the metal trace in the sinogram to minimize the impact of metal projections in the reconstructed image. Those sinogram-inpainting techniques reduce metal artifacts but are often associated with a loss of detail since a single projection contributes to all image pixels along its projection path. In addition, modifications to areas within the sinogram, such as the metal trace, can lead to inconsistencies during image reconstruction due to inhomogeneous transitions between modified and unmodified projection traces. Various extensions have been introduced to address this issue, each with its own benefits and drawbacks. The first extension to interpolating linearly between the boundaries of the metal trace (LIMAR [13]) was described by the sinogram normalization toward the CT's main tissue classes (NMAR [14]) to perform the interpolation in a normalized sinogram and decrease the amount of interpolation-induced artifacts in the reconstructed image.

However, both LIMAR and NMAR introduce inconsistencies to the sinogram, violating assumptions for image reconstruction and introducing new artifacts. To address the loss of contrast introduced by both LIMAR and NMAR, an additional image-based frequency split was introduced (FSMAR [15]), which restores anatomical details but comes with an increase in image noise. Both the drawbacks of NMAR and FSMAR are discussed and assessed within this chapter.

2 The Cause of Metal Artifacts

Various factors contribute to metallic artifacts in CT images, arising from the X-ray attenuation composition, detector settings, and the reconstruction algorithm.

In general, the weighted filtered back projection (WFBP), as a prominent analytic CT image reconstruction method, generally involves a trade-off between the amount of Poisson noise and texture in the image, modulated by the reconstruction filter. However, this trade-off is disrupted when metal is present in the CT image, as the X-ray attenuation of metal significantly reduces the amount of valid photon information in respective projections, introducing a high Poisson error [16].

The basic photon-matter interactions (as illustrated in Fig. 1 for soft tissue, solid bone, gold (Au, which is commonly used in dental crowns), and titanium (Ti, which is frequently implemented in fixation systems in the spine or in implants in the extremities [10])) contribute to the overall X-ray attenuation. Two of these interactions, the photoelectric effect (or photo effect) and scattering, directly correlate to respective metal artifacts. The probability of the photoelectric effect occurring directly correlates with the atomic number of the attenuating tissue, with increased probability in metal (as shown in Fig. 1), contributing the most to the increased total attenuation compared to natural tissue within the body. This probability decreases with higher photon energies. Consequently, when metal is present, only a small, high energetic proportion of the X-ray spectrum penetrates the tissue without full attenuation, resulting in beam hardening. Beam hardening describes the divergence of the measured X-ray spectrum in comparison to the emitted X-ray spectrum toward higher energies. Metal-induced beam hardening introduces large, hypodense bands around the metal, most pronounced in projections with significant metal penetration lengths [16, 18]. As seen in Fig. 1 when comparing the photo effect's probability between titanium and gold, the materials' atomic number is the decisive factor if measuring with a harder X-ray spectrum leads to a decrease of image artifacts. In addition, the probability for Compton scattering does not directly correlate to the material's atomic number but depends on the material's electron and physical density, which contributes the most to the overall attenuation of water or soft tissue at high energies. For those natural tissue types, Rayleigh scattering is often neglected because it is only relevant for photon energies below 20 keV. For metallic implants with high atomic numbers, elastic Rayleigh scattering is strongly increased, with a higher probability at lower photon energies. Rayleigh scattered photons only have a small energy loss, and metal-induced scattering (both Compton

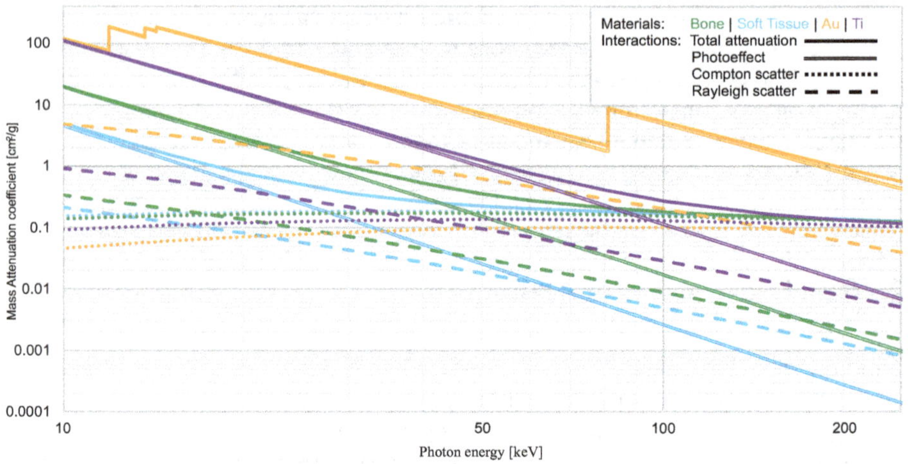

Fig. 1 Total mass attenuation coefficients and their composition within the diagnostic energy range for compact bone, soft tissue, gold (Au), and titanium (Ti). The material decompositions and attenuation coefficients are retrieved from the XCOM database [17]

and Rayleigh) introduces high-frequency streaks concentrically around the implant [18]. In addition, both scattering and the photo effect decrease the amount of valid photon information close to the metal, causing an overall increase in image noise in the tissue, referred to as photon starvation [16].

Furthermore, the CT image is susceptible to a final metal-related artifact introduced by the limited spatial resolution of the detector. In the computation of individual X-ray projections within the sinogram, the attenuation of various tissue types within the spatial coverage is subject to averaging [19]. This process involves calculating the arithmetic mean of both attenuation coefficients between a highly attenuating metal (or bone) and its surrounding tissue. This phenomenon, commonly referred to as partial volume averaging, results in the blurring of the metal implant's edge [20].

3 A New Approach for Normalized Sinogram Interpolation

Linear sinogram interpolation (LI) stands out as a highly effective approach in mitigating metal artifacts in clinical CT. Despite the limited introduction of new enhancements to this method within the last decade, which will be expounded upon subsequently, CT vendors continue to widely employ linear sinogram interpolation in their clinical routines for the correction of metal artifacts in CT acquisitions [21–24]. Moreover, a majority of published methods utilizing deep learning for metal artifact reduction (MAR) frequently refer to sinogram interpolation (either LIMAR or NMAR) as their gold standard reference [25–31].

The concept of linear sinogram interpolation involves correcting metal-corrupted projection data by replacing their values in the sinogram. This replacement is achieved by linearly interpolating between the boundaries of each projection angle, resulting in the computation of the LI sinogram P_{LI}. This approach considers all information within the metal sinogram as invalid. Other methodologies have proposed alternative metal trace replacement strategies beyond LI, including fitting higher-order interpolations [33], arithmetic approximations [34–36], or wavelets [37]. However, none of these alternatives have successfully established themselves in the metal artifact reduction routine. Despite the effectiveness and ease of implementation of linear interpolation metal artifact reduction (LIMAR) [13] (illustrated by the LIMAR image I_{LI} in Fig. 2), it is associated with drawbacks such as a loss of detail and the introduction of interpolation-related artifacts around the implant [15, 38]. These interpolation-related artifacts primarily introduce as a blurry impression around the metallic implant, leading to a deficiency in anatomic information and image contrast. This deficiency arises from the reduction of valid projection information surrounding the metal trace in the sinogram. The strength of the interpolation artifacts relies on defining the distance of the interpolation boundaries for each projection angle ϑ within the two-dimensional parallel-beam sinogram (following the notation from [19]). To mitigate the impact of the corrupted sinogram information on the reconstructed image, Meyer et al. introduced a tissue-based sinogram normalization before interpolation, termed normalized metal artifact reduction (NMAR). This normalization is based on the classification of water-equivalent soft tissue, bone, and air [14]. The tissue normalization flattens the sinogram, and the flatness before interpolation is strongly influenced by the precision of bone segmentation in the prior image. The precision of the bone directly adjacent to the metal is of high relevance for the interpolation. However, when severe metal artifacts disrupt anatomic information around the metal, image-based bone segmentation may still contain misclassifications. With an increasing amount of bone missing in the prior image, the normalized sinogram before linear interpolation becomes less flat. This can introduce steep edges to the sinogram, leading to bright, hyperdense, blooming-like artifacts in the reconstructed image. These artifacts result from the linear interpolation violating sinogram consistency and introducing inhomogeneous attenuation values between the projection angles ϑ_i. Based on the definition from [19], a single two-dimensional parallel projection is defined as the line integral through the object with the distance ξ as

$$p(\vartheta, \xi) = \int dxdy\ f(x, y)\, \delta(x \cos\vartheta + y \sin\vartheta - \xi)\,. \tag{1}$$

In theoretical terms, integrating both sides of Eq. (1) over ξ ends up at

$$\int p(\vartheta, \xi)\, d\xi = \int\int dxdy\ f(x, y)\, \delta(x \cos\vartheta + y \sin\vartheta - \xi)\, d\xi, \tag{2}$$

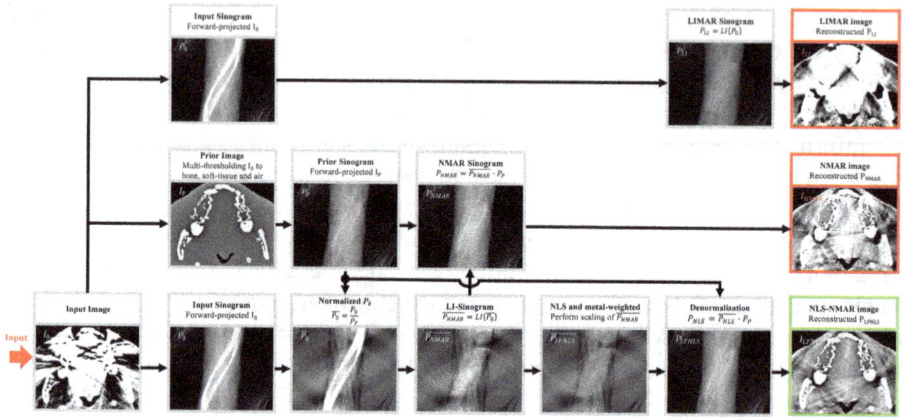

Fig. 2 Overview about a new low-frequency metal artifact correction method, which combines the established NMAR method (i.e., the linear tissue-normalized sinogram interpolation) with an intermediary scaling step between the linear interpolation and the subsequent tissue denormalization. This scaling is required to reduce residual artifacts after the reconstruction of the normalized and interpolated sinogram space. (Figure adapted, based on [32])

effectively removing the argument of the delta function. This ends up at the condition of the object function's continuity as

$$\int p\left(\vartheta, \xi\right) d\xi = \int dxdy\ f\left(x, y\right). \tag{3}$$

This condition must be satisfied for all rotation angles ϑ. However, angle-wise one-dimensional sinogram interpolation, relying on the definition of metal boundaries, manipulates this assumption, leading to its nonsatisfaction for some rotation angles ϑ. This implies that the interpolation causes the two-dimensional integral of the object function $f(x, y)$ to differ, depending from which angle the projection is generated. The issue of sinogram replacement-induced discontinuities has also been noted by [39]. To minimize image artifacts introduced by interpolation edges, regularization is necessary for all detector rows between all projection angles. A scaling step can be applied to the normalized and interpolated metal sinogram in each projection angle to reduce variance of the objects line integral between rotation angles, as introduced in [32]. Most steps are akin to conventional NMAR (refer to Fig. 2):

The initial step involves computing a prior image I_{P} through the image-based classification of the uncorrected image I_0; its prior sinogram P_{P} results from forward projecting I_{P}. Subsequently, linear sinogram interpolation $\mathrm{LI}\{P\}$ is performed in the prior-normalized sinogram domain, expressed as $\overline{P_{\mathrm{NMAR}}} = \mathrm{LI}\left\{P_0/P_{\mathrm{P}}\right\}$, as shown in Fig. 2. Depending on the precision of the prior image, this normalized sinogram domain can violate Eq. (3) for certain rotation angles. For this reason, a nonlinear scaling (NLS) function can be applied to the normalized projections close to the metal trace to reduce the impact of the sinogram inconsistencies, which are introduced by the interpolation edges. An arctan-based transfer function fulfils the requirement of flexible damping, dependent on the projection's height. This function is modified with two additional components, as introduced in [32, 40]: First, a threshold t prevents nonoutlier projections from being damped by the scaling function. Second, an adjustable parameter α defines the scaling strength of the transfer function. As most parameters within interpolation-based metal artifact reduction, this scaling parameter should be optimized toward different metal scenarios. Then, the scaling function f_s of an input projection P_i is defined as

$$f_s(P_i) = \begin{cases} \alpha\left(\tan^{-1}\left(\frac{P_i - t}{\alpha}\right)\right) + t & P_i > t \\ -\left(\alpha\left(\tan^{-1}\left(\frac{-P_i - t}{\alpha}\right) + t\right)\right) & P_i < -t \\ P_i & \text{otherwise} \end{cases} . \tag{4}$$

Various examples for transfer functions are depicted in Fig. 3. The projection scaling is then applied to the prior-normalized interpolated sinogram domain as

$$P_{\mathrm{LFNLS}} = P_{\mathrm{P}} \cdot \underbrace{\left(W \cdot f_s\left(\overline{P_{\mathrm{NMAR}}}\right) + (1 - W) \cdot \overline{P_{\mathrm{NMAR}}}\right)}_{\overline{P_{\mathrm{LFNLS}}}}, \tag{5}$$

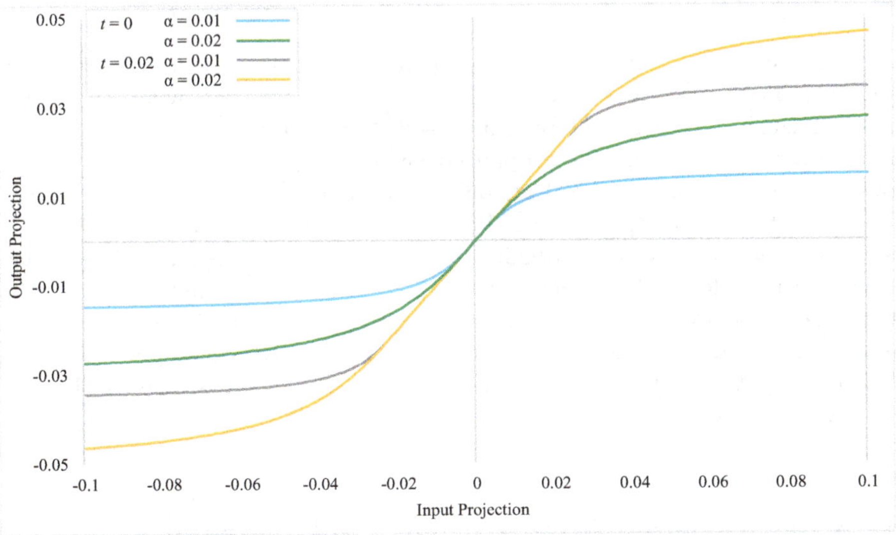

Fig. 3 Arctan-based scaling function to reduce interpolation-related sinogram inconsistencies in the normalized sinogram domain [32] and reduce directed high-frequency image noise to replace the image-based frequency split [40]

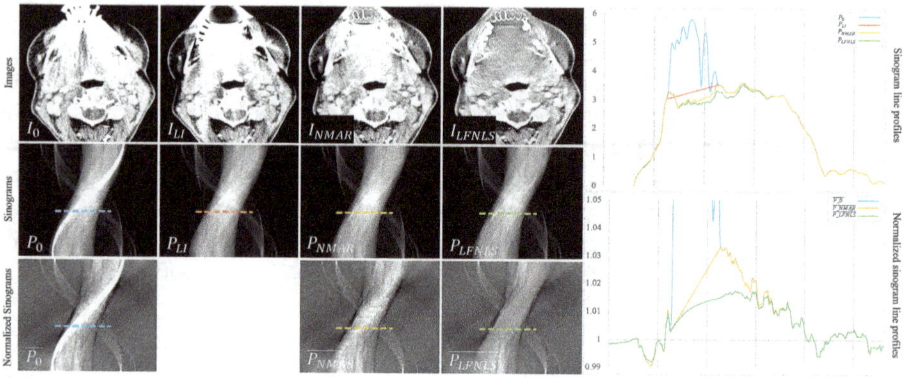

Fig. 4 Different CT images and their corresponding sinograms P, prior-normalized sinograms $\overline{P}$, and line-profiles through the sinograms. The unsteady transition of the interpolation edges of P_0 get smoothed by the sinogram normalization of P_{NMAR}. Still, the normalized sinogram domain shows a large interpolation edge of $\overline{P_{\mathrm{NMAR}}}$, introducing low-frequency artifacts to the image after reconstruction. A scaling of the sinogram inconsistencies, applied to all rotation angles, reduces respective image artifacts. Images: Center/Width = 40 HU/300 HU. Sinograms: Center/Width = 2/4 (unitless). Normalized sinograms: Center/Width = 1.010/0.060 (unitless)

where W represents a weighting mask designed to selectively scale metal projections and their proximity. A suitable mask computation is detailed in [40]. The multiplication with P_{P} describes the subsequent prior denormalization, similar to NMAR (Fig. 4).

Given the tissue normalization of P_0/P_P, it is essential to center the normalized sinogram around zero to ensure correct scaling. Therefore, removing the median of the normalized $\overline{P_{\mathrm{NMAR}}}$ before scaling with f_s and subsequently adding it back is necessary. In the centered normalized sinogram space, the threshold can be set to $t = 0$, with values $t \neq 0$ are only being necessary for the high-frequency scaling in the next section. The metal artifact-corrected images I_{NMAR} and I_{LFNLS} are then computed by reconstructing the denormalized sinograms P_{NMAR} and P_{LFNLS}. Similar to NMAR, denormalization results from anew multiplication of the normalized sinogram with P_{P}. Since the linear sinogram interpolation significantly reduces the image contrast, whether or not additional NLS is applied, an extra frequency split is required to restore anatomical detail, as introduced by [15] and detailed in the following section.

4 A New Sinogram-Based Frequency Split Technique

The preceding section provided an overview of various sinogram interpolation-based methods for metal artifact reduction. While these methods have demonstrated significant potential in reducing artifacts, they are accompanied by the drawback of tissue destruction around the metal implant. Sinogram normalization from NMAR partially mitigates this issue by providing more homogeneous transitions between metal and metal-free projections, assuming the inclusion of the bone within the prior image. However, capturing bone in the presence of metal artifacts can be challenging. To further restore tissue information and anatomical detail, an image-based frequency split technique, known as frequency split metal artifact reduction (FSMAR), has been introduced [15]. FSMAR combines NMAR with a Gaussian image-based high-pass filter and combines the low frequencies $I^{\mathrm{L}}_{\mathrm{NMAR}}$ with the high frequencies of the uncorrected image I_0^{H} as

$$I_{\mathrm{FSMAR}} = I^{\mathrm{L}}_{\mathrm{NMAR}} + \underbrace{W \cdot I_0^{\mathrm{H}} + (1 - W) \cdot I^{\mathrm{H}}_{\mathrm{NMAR}}}_{I^{\mathrm{H}}_{\mathrm{FSMAR}}}. \tag{6}$$

This is controlled by a weighting mask W, as depicted in the top row of Fig. 6. In this correction, the low-frequency space of the corrected NMAR image can be modularly substituted with the low frequencies $I^{\mathrm{L}}_{\mathrm{LFNLS}}$ from the previous section. Equation (6) provides two ways to control the insertion of uncorrected high frequencies.

First, the weighting mask W determines where to insert the uncorrected high frequencies. FSMAR recommends applying spatial weighting only in the vicinity of the metal to avoid unintended global increases in image noise. The kernel width of the image-based two-dimensional Gaussian low-pass filter, used to compute the weighting mask W, controls spatial weighting, as illustrated in the different columns of Fig. 5. The middle three columns exhibit various parameterizations of FSMAR, with permutations of the size of W, including a large, global mask on the left and a small, local mask on the right. Inserting high-frequency information as globally as possible is desirable to universally restore a high level of anatomical detail and the bone, often accompanied by an increase in image noise. The work from [40] did not incorporate different permutations of W to ensure optimal comparability of the different high-frequency spaces and address the impact of the dedicated sinogram scaling.

The incapability of an image-based two-dimensional Gaussian high-pass filter to distinguish high frequencies from metal projections and high frequencies from metal-free projections is the primary reason for the reintroduction of high image noise. This challenge can be overcome by performing the frequency split in the sinogram, as introduced in [40] and shown in Fig. 6.

The initial step involves extracting the high-frequency domain of the uncorrected sinogram P_0 through angle-wise one-dimensional Gaussian high-pass filtering to obtain P_0^{H}. All high-frequency projections within the metal trace exhibit higher absolute values than high frequencies from natural anatomic information, as evident in the line plots in Fig. 7. To mitigate the impact of these metal high frequencies on the reconstructed image, the nonlinear scaling function $f_{\mathrm{s}}(P_{\mathrm{i}})$ from Eq. (4) can be applied to the high-frequency metal projections. When reconstructing the images using weighted filtered back projection, this modification of a projection value

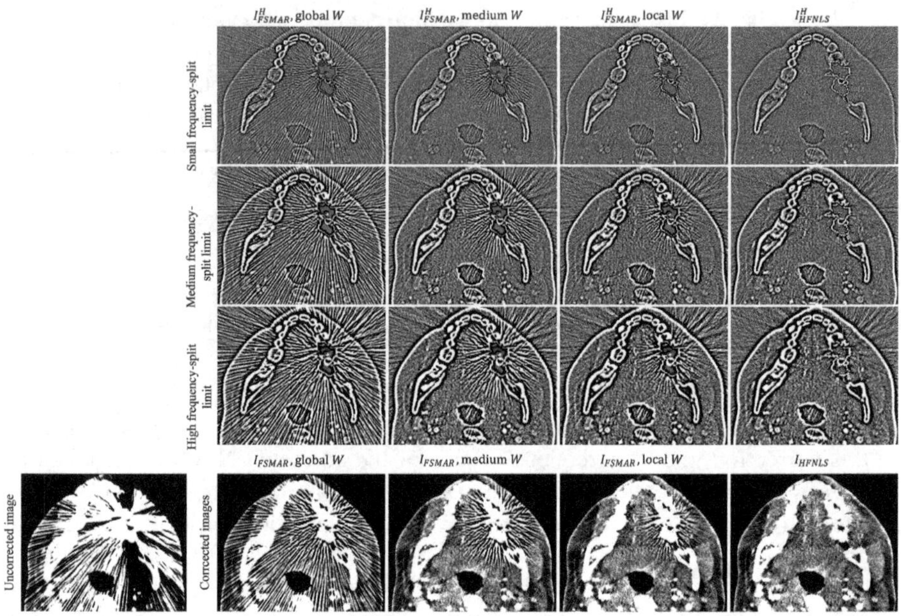

Fig. 5 Different parameterizations to compute a corrected high-frequency image. The different columns demonstrate permutations of W, while the rows increase the limit for the frequency split, both demonstrating the parameterization trade-off of conventional image-based FSMAR. The last column shows the corrected high frequencies with the new sinogram-based frequency split with the additional frequency scaling, also demonstrating its adaptability to different frequency split limits to extract the high frequencies. High-frequency images: Center/width = 0 HU /200 HU. Images: Center/width = 40 HU/300 HU

$P(\vartheta, \xi)$ would affect all image values of the image function $f(x, y)$ included on the line of $\xi = x\cos(\vartheta) + y\sin(\vartheta)$, introducing lowered bone contrast.

Given that the low-frequency correction of both I_{NMAR} and I_{LFNLS} requires computing the prior image, the bone included in the prior image can be removed before scaling to prevent a loss of bone detail. Since the prior image (and consequentially the prior sinogram) does not include high-frequency streak artifacts, its high frequencies $P_{\mathrm{P}}^{\mathrm{H}}$ (as seen in Fig. 7) can be removed before the application of the scaling function. Thus, a preliminary version for the corrected high-frequency domain $P_{\mathrm{NLS}}^{\mathrm{H}}$ result from

$$P_{\mathrm{NLS}}^{\mathrm{H}} = f_{\mathrm{s}}\left(\underbrace{P_0^{\mathrm{H}} - c \cdot P_{\mathrm{P}}^{\mathrm{H}}}_{P_{0,\mathrm{s}}^{\mathrm{H}}}\right). \tag{7}$$

Now, with the removal of the prior image's information, the scaling of $P_{0,\mathrm{s}}^{\mathrm{H}}$ mostly affects high-frequency noise. In this context, employing values with $t > 0$ and a

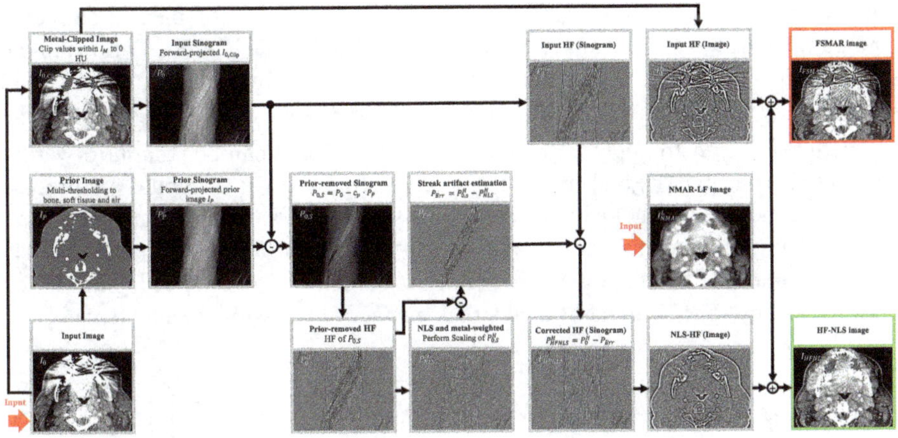

Fig. 6 Overview including all steps from a new method to compute a sinogram-based frequency split and a dedicated high-frequency correction. Conventional FSMAR combines the uncorrected high frequencies I_0^{H}, extracted by high-pass filtering of the uncorrected image, with the low frequencies of an NMAR-corrected image [15]. By performing the frequency split in the sinogram, an additional nonlinear scaling function can be applied to damp high frequencies, which originate from metal projections. To still preserve bone detail, the prior image, which is also applied within the normalization of NMAR, can be used to subtract bone before scaling [40]

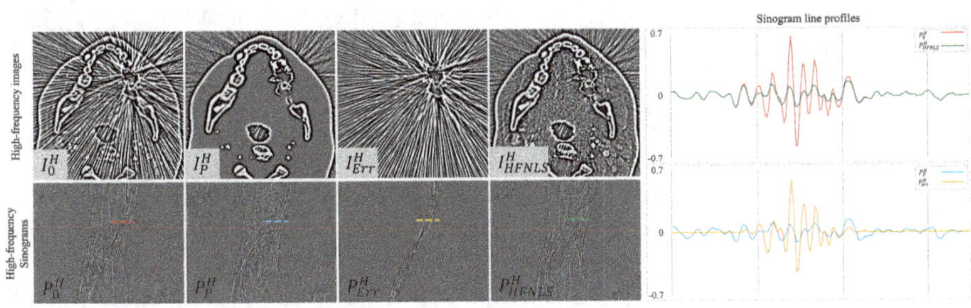

Fig. 7 Line profiles of different sinogram spaces of the new sinogram-based frequency split. The uncorrected I_0^{H} includes a high level of high-frequency noise, which can be reduced by performing an additional nonlinear scaling step to metal projections and their neighborhood. As such a scaling of the original high frequencies P_0^{H} would also introduce a loss of bone detail, the high frequencies of the prior image $P_{\mathrm{P}}^{\mathrm{H}}$ are removed before scaling to preserve bone from getting damped

too small α for the scaling function f_{s} would characterize as a bandpass-like filter. However, the arctan-based scaling of f_{s} does not include the rigid frequency cutoff typical of a bandpass, which could reintroduce new artifacts into the reconstructed high-frequency space. The subtraction strength of the prior sinogram is controlled by the variable c, which should be fine-tuned based on the scaling strength and the precision of the prior image. A straightforward threshold-based bone segmentation can be challenging in the presence of severe metal artifacts. It may misclassify hyperdense metal artifacts as the bone, potentially leading to new artifacts after reintroducing the prior sinogram's high frequencies to the nonlinearly corrected sinogram. Hence, a lower value for c is recommended in such cases. Following the subtraction of the prior sinogram, neither the scaled sinogram $P^{\mathrm{H}}_{\mathrm{NLS}}$ nor the unscaled sinogram $P^{\mathrm{H}}_{0,\mathrm{s}}$ includes bone information. Their direct comparison provides an estimation for high-frequency streak artifacts $P^{\mathrm{H}}_{\mathrm{Err}}$, which can be straightforwardly subtracted from the original uncorrected high frequencies to compute a corrected high-frequency space inclusive of dedicated streak artifact correction and preservation of the prior image's bone. Alternatively, the prior high frequencies $P^{\mathrm{H}}_{\mathrm{P}}$ can be simply added after scaling P^{H}_{0} to restore bone detail. Thus, the sinogram-based high-frequency split, with additional streak artifact suppression, results from

$$P^{\mathrm{H}}_{\mathrm{HFNLS}} = P^{\mathrm{H}}_{0} - \left(P^{\mathrm{H}}_{0,\mathrm{s}} - \left(\underbrace{W \cdot P^{\mathrm{H}}_{\mathrm{NLS}} + (1 - W) \cdot P^{\mathrm{H}}_{0}}_{P_{\mathrm{Err}}} \right) \right). \tag{8}$$

Similar to the weighting from Eq. (5), a sinogram-based metal weighting W is necessary to apply scaling only in metal projections and their vicinity. In all nonmetal projections, the uncorrected original high frequencies P^{H}_{0} can be applied because these cannot introduce high-frequency artifacts to the reconstructed image. In a final step, similar to the conventional FSMAR from eq. (6), the reconstructed high-frequency image space $I^{\mathrm{H}}_{\mathrm{HFNLS}}$ can be combined with the low-frequency space of a corrected images as

$$I_{\mathrm{HFNLS}} = I^{\mathrm{L}}_{\mathrm{X}} + I^{\mathrm{H}}_{\mathrm{HFNLS}}. \tag{9}$$

For the low-frequency corrected image $I^{\mathrm{L}}_{\mathrm{X}}$, either the low-frequency space of conventional NMAR can be used or the low-frequency space from NMAR with the additional dedicated regularization step from the previous chapter. In contrast to Eq. (6), no image-based weighting is required, as $I^{\mathrm{H}}_{\mathrm{HFNLS}}$ includes the original high frequencies outside the metal projections.

As the scaling function adjusts the scaling based on the amplitude of the input high frequency, it can be systematically applied to multiple reconstruction kernels. Figure 8 shows a sharp H60 kernel, where both the uncorrected image I_0 and its high frequencies I^{H}_{0}, computed through an image-based two-dimensional Gaussian high-pass filter, exhibit fine, radially directed image noise originating from the implants. This noise is reintroduced into the FSMAR image I_{FSMAR}. In contrast, the sinogram-based frequency split with additional frequency scaling in $I^{\mathrm{H}}_{\mathrm{HFNLS}}$ introduces a high level of bone detail.

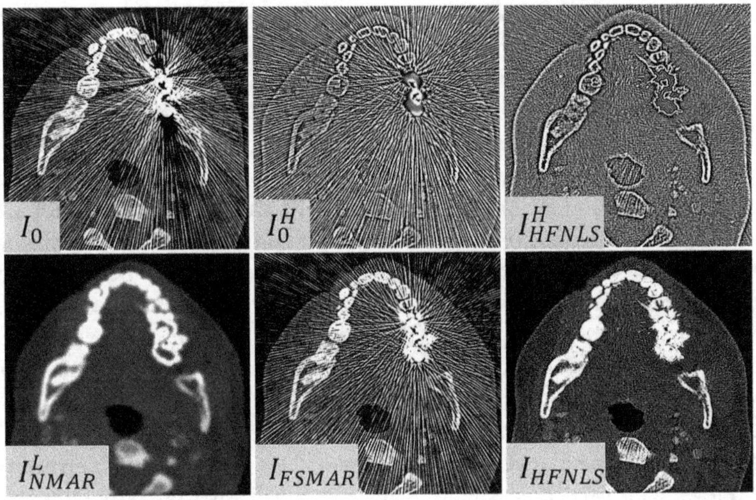

Fig. 8 Demonstration of the dedicated high-frequency split for reconstructions with a sharp reconstruction kernel (H60). The fine image noise structure from I_0 is carried over to the FSMAR image I_{FSMAR}. The dedicated high-frequency correction from [40] is capable of reducing the directed image noise, while preserving bone detail. Images: Center/width = 500 HU/2000 HU. High-frequency images: Center/width = 0 HU/1000 HU

5 Outlook and Discussion

Metallic objects have consistently posed a significant challenge in computed tomography, hindering a clear visualization of anatomical details and compromising image quality. Clinical CT system vendors commonly employ sinogram inpainting to address metal artifacts due to its simplicity and independence from prior knowledge or training data. This method, an old technique, has seen few modifications since its inception. One such modification is tissue-based sinogram normalization, which shifts linear interpolation from the standard uncorrected sinogram to a tissue-normalized sinogram to smooth the transition between the interpolated metal trace and surrounding projections. While this reduces interpolation-based artifacts, it introduces sinogram inconsistencies due to challenges in achieving a perfectly segmented prior image disrupted by metal artifacts. This overview, alongside the work by [32], introduces an additional scaling step applying a nonlinear scaling function to the interpolation edges introduced during NMAR. This regularization of metal traces across all rotation angles aims to minimize interpolation-related low-frequency artifacts in the reconstructed image. Post NMAR, the reinsertion of high frequencies is required to restore anatomical details such as soft tissue contrast or bone structure. FSMAR incorporates an image-based two-dimensional Gaussian high-pass filter to extract uncorrected high frequencies, which are then combined with the NMAR-corrected image. However, this introduces the challenge of increased image noise, dependent on the strength and location of the reinserted high frequencies. In contrast, by performing frequency extraction in the sinogram, the high frequencies of metal projections can be identified. An additional frequency damping can be applied by employing a nonlinear scaling function on all high-amplitude metal projections' high frequencies, reducing the strength of image noise, as introduced in this summary and the original work by [40].

Despite the promise demonstrated by these two extensions to the current state of the art, significant challenges persist. Metal artifact expressions exhibit a broad spectrum owing to variations in implant material composition, size, and surroundings. Consequently, each correction step should be finely tuned to individual implant settings, risking under- or overcorrection. Future advancements may leverage neural networks trained to optimize specific correction steps for individual metal scenarios within conventional inpainting metal artifact reduction. However, acquiring adequate training data remains a challenge, with phantom or simulated data being the primary alternatives. Phantom data lacks anatomical detail, especially near the metal, which is problematic, as those details can be of high clinical relevance for the assessment of the implant. However, fully simulated data tend to overfit in an end-to-end-trained network and produce oversmoothed images. The most viable ground truth may involve simulating metallic implants into real metal-free CT acquisitions, closely resembling real clinical data. Although such datasets may exhibit minor variations in metal surroundings, such as the density increase attributed to the hydroxyapatite coating of numerous implants being absent, they nonetheless offer the closest representation to authentic clinical data.

Furthermore, spectral capabilities hold potential for refining individual steps within interpolation-based metal artifact reduction. While current energy-integrating CT devices provide spectral information only in specific scan modes, the recently introduced photon-counting detectors offer potential advantages. Despite the applicability of the introduced low-frequency and high-frequency correction methods to CT scans from both energy-integrating and photon-counting detectors, they still operate nonspectrally, that is, they do not exchange information between low and high energetic images. For instance, spectral information can be employed for pixel-wise material decomposition, utilizing this knowledge to optimize the bone within NMAR's prior image and further diminish residual artifacts. However, the benefits of spectral information are specific to individual settings, as, similar to interpolation-based methods, spectral capabilities are strongly contingent on the implant. Dental fillings, characterized by dense and high-atomic number metals, often introduce more pronounced starvation and scattering artifacts, resulting in a complete loss of quantum statistics behind the implant. Conversely, fixation systems within the spine typically exhibit minor beam hardening, allowing the majority of the X-ray spectrum to remain detectable. Consequently, the greatest advantages of increased energy in virtual monoenergetic images are observed in metal scenarios with the lowest atomic numbers and shortest penetration lengths. As the implant's attenuation increases, less of the X-ray spectrum can be measured, further diminishing spectral capabilities.

References

1. Statistisches Bundesamt, Fallpauschalbezogene Krankenhausstatistik (DRG-Statistik). Operationen und Prozeduren der vollstationären Patientinnen und Patienten in Krankenhäusern (4-Steller), Statistisches Bundesamt (Destatis), 2005–2022.
2. Kassenärztliche Bundesvereinigung, Jahrbuch 2022 – Statistische Basisdaten zur vertragszahnärtzlichen versorgung, Köln: KZBV, 2022
3. J.T. Brandfass, A.C. Ulano, J.P. Nickerson, M.P. Bazylewicz, Dental caries on CT in the ER population: prevalence and reporting practices. Emerg. Radiol **26**, 263–267 (2019)
4. L. Zhu, Y. Han, L. Li, X. Xi, M. Zhu, B. Yan, Metal artifact reduction for x-ray computed tomography using U-net in image domain. IEEE Access **7**, 98743–98754 (2019)
5. L. Gjesteby, Q. Yang, Y. Xi, H. Shan, B. Claus, Y. Jin, B. De Man, G. Wang, Deep learning methods for CT image-domain metal artifact reduction, in *SPIE Optical Engineering + Applications*, San Diego, CA, USA, 2017
6. X. Huang, J. Wang, F. Tang, T. Zhong, Y. Zhang, Metal artifact reduction on cervical CT images by deep residual learning. Biomed. Eng. Online **17**, 1–15 (2018)
7. K. Liang, L. Zhang, H. Yang, Y. Yang, Z. Chen, Y. Xing, Metal artifact reduction for practical dental computed tomography by improving interpolation-based reconstruction with deep learning. Med. Phys. **46**(12), e823–e834 (2019)
8. Y. Zhang, H. Yu, Convolutional neural network based metal artifact reduction in x-ray computed tomography. IEEE Trans. Med. Imaging **37**(6), 1370–1381 (2018)
9. W.-A. Lin, H. Liao, C. Peng, X. Sun, J. Zhang, J. Luo, R. Chellappa, S.K. Zhou, Dudonet: dual domain network for ct metal artifact reduction, in *Proceedings of the IEEE/CVF Conference on Computer Vision and Pattern Recognition*, (2019), pp. 10512–10521
10. J.A. Anhaus, S. Schmidt, P. Killermann, A. Mahnken, C. Hofmann, Iterative metal artifact reduction on a clinical photon counting system—technical possibilities and reconstruction selection for optimal results dependent on the metal scenario. Phys. Med. Biol **67**(11), 115018 (2022)
11. M.N. Bongers, C. Schabel, C. Thomas, R. Raupach, M. Notohamiprodjo, K. Nikolaou, F. Bamberg, Comparison and combination of dual-energy-and iterative-based metal artefact reduction on hip prosthesis and dental implants. PLoS One **10**(11), e0143584 (2015)
12. T.R. Johnson, Dual-energy CT: general principles. Am. J. Roentgenol **5**, S3–S8 (2012)
13. W.A. Kalender, R. Hebel, J. Ebersberger, Reduction of CT artifacts caused by metallic implants. Radiology **164**(2), 576–577 (1987)
14. E. Meyer, R. Raupach, M. Lell, B. Schmidt, M. Kachelrieß, Normalized metal artifact reduction (NMAR) in computed tomography. Med. Phys. **37**(10), 5482–5493 (2010)
15. E. Meyer, R. Raupach, M. Lell, B. Schmidt, M. Kachelrieß, Frequency split metal artifact reduction (FSMAR) in computed tomography. Med. Phys. **39**(4), 1904–1916 (2012)
16. F.E. Boas, D. Fleischmann, CT artifacts: causes and reduction techniques. Imaging Med. **4**(2), 229–240 (2012)
17. M. Berger, J. Hubbell, S. Seltzer, J. Chang, J. Coursey, R. Sukumar, XCOM: Photon Cross Section Database (version 1.2), 1999. [Online]. [Accessed 2023]
18. T.M. Buzug, *Computed Tomography. From Photon Statistics to Modern Cone Beam CT* (Springer, Heidelberg, 2009)
19. W.A. Kalender, *Computertomographie - Grundlagen, Gerätetechnologie, Bildqualität, Anwendungen* (Publicis Corporate Publishing, Erlangen, 2006)
20. J. Hsieh, Nonlinear partial volume artifact correction in helical CT. IEEE Trans. Nucl. Sci. **46**(3), 743–747 (1999)
21. K.M. Andersson, P. Nowik, J. Persliden, P. Thunberg, E. Norrman, Metal artefact reduction in CT imaging of hip prostheses—an evaluation of commercial techniques provided by four vendors. Br. Inst. Radiol **88**(1052), 20140473 (2015)
22. H. Li, C. Noel, H. Chen, H. Harold Li, D. Low, K. Moore, P. Klahr, J. Michalski, H.A. Gay, W. Thorstad, S. Mutic, Clinical evaluation of a commercial orthopedic metal artifact reduction tool for CT simulations in radiation therapy. Med. Phys. **39**(12), 7507–7517 (2012)

23. F.W. Feldhaus, G. Böning, J. Kahn, U. Fehrenbach, M. Maurer, D. Renz, F. Streitparth, Improvement of image quality and diagnostic confidence using Smart MAR–a projection-based CT protocol in patients with orthopedic metallic implants in hip, spine, and shoulder. Acta Radiol. **61**(10), 1421–1430 (2020)
24. A. Kotsenas, G. Michalak, D. DeLone, F. Diehn, K. Grant, A. Halaweish, A. Krauss, R. Raupach, B. Schmidt, C.H. McCollough, J. Fletcher, CT metal artifact reduction in the spine: can an iterative reconstruction technique improve visualization? Am. J. Neuroradiol **36**(11), 2184–2190 (2015)
25. D.F. Bauer, C. Ulrich, T. Russ, A.-K. Golla, L.R. Schad, F.G. Zöllner, End-to-end deep learning CT image reconstruction for metal artifact reduction. Appl. Sci. **12**(1), 404 (2021)
26. L. Gjesteby, Q. Yang, Y. Xi, Y. Zhou, J. Zhang, G. Wang, Deep learning methods to guide CT image reconstruction and reduce metal artifacts, in *SPIE Medical Imaging*, Orlando, Florida, United States, 2017
27. B.E. Claus, Y. Jin, L.A. Gjesteby, G. Wang, B. De Man, Metal-artifact reduction using deep-learning based sinogram completion: initial results, in *Proceeding of the 14th International Meeting Fully Three-Dimensional Image Reconstruction Radiol. Nucl. Med*, 2017
28. M.U. Ghani, W.C. Karl, Deep learning based sinogram correction for metal artifact reduction. Electron. Imaging **15**, 472–1–472-8 (2018)
29. M.U. Ghani, W. Clem Karl, Fast enhanced CT metal artifact reduction using data domain deep learning. IEEE Trans. Comput. Imaging **6**, 181–193 (2019)
30. L. Gjesteby, H. Shan, Q. Yang, Y. Xi, Y. Jin, D. Giantsoudi, H. Paganetti, B. De Man, G. Wang, A dual-stream deep convolutional network for reducing metal streak artifacts in CT images. Phys. Med. Biol **64**(23), 1–11 (2019)
31. H.S. Park, S.M. Lee, H.P. Kim, J.K. Seo, Y.E. Chung, CT sinogram-consistency learning for metal-induced beam hardening correction. Med. Phys. **45**(12), 5376–5384 (2018)
32. J.A. Anhaus, P. Killermann, A.H. Mahnken, C. Hofmann, A nonlinear scaling-based normalized metal artifact reduction to reduce low-frequency artifacts in energy-integrating and photon-counting CT. Med. Phys. **50**(8), 4721–4733 (2023)
33. A.H. Mahnken, R. Raupach, J.E. Wildberger, B. Jung, N. Heussen, T.G. Flohr, R.W. Günther, S. Schaller, A new algorithm for metal artifact reduction in computed tomography: in vitro and in vivo evaluation after total hip replacement. Investig. Radiol. **38**(12), 769–775 (2003)
34. B.P. Medoff, W.R. Brody, M. Nassi, A. Macovski, Iterative convolution backprojection algorithms for image reconstruction from limited data. J. Opt. Soc. Am. **73**(11), 1493–1500 (1983)
35. H.K. Tuy, A post-processing algorithm to reduce metallic clip artifacts in CT images. Eur. Radiol **3**, 129–134 (1993)
36. R.L. Morin, D. Raeside, A pattern recognition method for the removal of streaking artifact in computed tomography. Radiology **141**(1), 229–233 (1981)
37. S. Zhao, D. Robeltson, G. Wang, B. Whiting, K.T. Bae, X-ray CT metal artifact reduction using wavelets: an application for imaging total hip prostheses. IEEE Trans. Med. Imaging **19**(12), 1238–1247 (2000)
38. J. Müller, T.M. Buzug, Spurious structures created by interpolation-based CT metal artifact reduction, in *Medical Imaging 2009: Physics of Medical Imaging*, (2009), pp. 636–643
39. C. Lemmens, D. Faul, J. Nuyts, Suppression of metal artifacts in CT using a reconstruction procedure that combines MAP and projection completion. IEEE Trans. Med. Imaging **28**(2), 250–260 (2008)
40. J.A. Anhaus, P. Killermann, M. Sedlmair, J. Winter, A.H. Mahnken, C. Hofmann, Nonlinearly scaled prior image-controlled frequency split for high-frequency metal artifact reduction in computed tomography. Med. Phys. **49**(9), 5870–5885 (2022)

Clinical Applications of Low-Dose Dental Cone-Beam Computed Tomography

Chang Min Hyun, Kiwan Jeon, Sung Ho Kang, Sang-Hwy Lee, Jin Keun Seo, and Hyoung Suk Park

1 Introduction

Recently, cone-beam computed tomography (CBCT) has been widely used in dental practice as an alternative to standard multi-detector CT (MDCT) [4, 9]. Dental CBCT is used to provide a sophisticated digital representation of patients' teeth, jaw bones, lips, and facial components, which facilitates advanced dental care such as prosthodontic and orthodontic treatments, cephalometric analysis, and maxillofacial surgery [15, 25, 29, 34, 35] (see Fig. 1). Along with recent advances in deep learning (DL) technologies, it has been evolving toward providing high-quality tomographic information while optimizing data collection for various clinical and industrial reasons such as minimized health risk and cost-effectiveness [9].

Dental CBCT imaging is to reconstruct a three-dimensional (3D) attenuation coefficient distribution of the human head from measured sinogram data. However,

C. M. Hyun
Department of Radiology, Perelman School of Medicine, University of Pennsylvania, Philadelphia, PA, USA
e-mail: chang.minhyun@pennmedicine.upenn.edu

K. Jeon · S. H. Kang · H. S. Park (✉)
National Institute for Mathematical Sciences, Daejeon, Republic of Korea
e-mail: jeonkiwan@nims.re.kr; runits@nims.re.kr; hspark@nims.re.kr

S.-H. Lee
Department of Oral and Maxillofacial Surgery, Oral Science Research Center, College of Dentistry, Yonsei University, Seoul, Republic of Korea
e-mail: sanghwy@yonsei.ac.kr

J. K. Seo
School of Mathematical and Computing (Computational Science and Engineering), Yonsei University, Seoul, Republic of Korea
e-mail: seoj@yonsei.ac.kr

K. Iniewski, H. Gadey (eds.), *Emerging Radiation Detection*,
https://doi.org/10.1007/978-3-031-63897-8_9

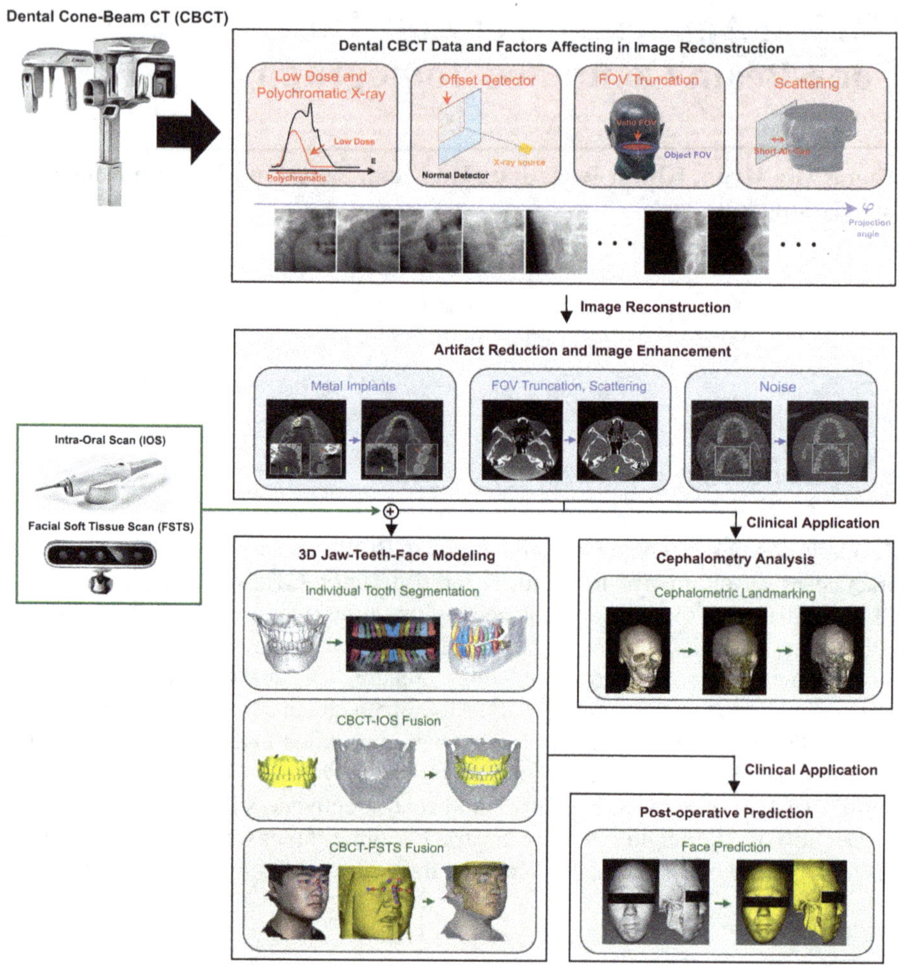

Fig. 1 Digital dentistry workflow using dental CBCT

the associated inverse problem is highly ill-posed [24], resulting in severe artifacts. It is related to the following imaging environment: In commercial dental CBCT devices, a small-size detector with offset array is employed for cost reduction while increasing the effective field-of-view (FOV) to the maximum extent [2, 9]. Low-dose X-ray radiation, circular source trajectory, and short air gaps (subject-to-detector distance) are used as well. Additionally, metal implants or fillings are common among dental patients and tend to occupy a significant area, inducing various nonlinear interactions among metals, bones, and tissues, such as beam hardening, photon scattering, and photon starvation [17, 20, 27]. Consequently, the quality of measured data is significantly reduced, attributing to serious mismatches between the reconstruction model and actual measurement [24].

For high-quality dental CBCT imaging, it is necessary to compensate for ill-posedness by taking advantage of strong prior knowledge of the desired solution. In this regard, DL has been shown to have a strong capability for learning prior information using numerous training samples [6, 7, 10]. Several studies [2, 8, 22–24] have attempted to use a learned image prior for dental CBCT imaging. Although the proposed methods have shown remarkable and promising performances, many limitations and challenges remain in practice. In particular, the handling of metal-related artifacts is far from satisfactory [11, 24]. The use of a supplementary image prior from an intraoral scan (IOS) or a facial soft-tissue scanner (FSTS) could be a potential solution [8, 11]. Hyun et al. [8] leveraged tooth surface information from an IOS as a concrete image prior and significantly improved a 3D segmented bone–teeth model damaged by metal-induced artifacts.

Recent advancements in dental imaging technologies, such as dental CBCT, IOS, and FSTS, have created a pressing need for the development of a smart digital platform known as the 3D jaw–teeth–face model in the field of digital dentistry [8, 13, 25]. This platform aims to integrate patient treatment plans into a cohesive digital anatomical model that encompasses the bone structure, teeth, gums, and face. The use of smart digital platforms offers a wide range of benefits, such as treatment planning, prediction of surgical outcomes during orthodontic treatments, and simulation of anticipated facial changes.

This chapter presents a comprehensive review of recent advances and challenges in low-dose dental CBCT with metal artifacts. Various DL techniques for artifact reduction and downstream applications are discussed along with their limitations and future research directions. We also investigate advanced digital dentistry applications of CBCT, including 3D bone–teeth–jaw model-based facial prediction and landmark detection for cephalometric analysis.

2 Dental CBCT Image Reconstruction

We begin with a mathematical model of CBCT with a circular trajectory and a flat-panel detector. Let $\mu(\mathbf{x})$ be a 3D CBCT image at position $(\mathbf{x}, z) = (x_1, x_2, z) \in \mathbb{R}^3$. CBCT projection data at a (scaled) detector position $(u, v) \in \mathbb{R}^2$ and projection angle $\varphi \in [0, 2\pi)$ can be modeled using the following cone-beam transform $\mathscr{C}$:

$$\mathscr{C}\mu(\varphi, u, v) = \int_0^\infty \mu(r(\Theta_\varphi^\perp, 0) + \tau\omega_{u,v})d\tau, \tag{1}$$

where $\Theta_\varphi^\perp = (\sin\varphi, -\cos\varphi)$ and $\omega_{u,v}$ is the cone-beam direction starting at $(r\Theta_\varphi^\perp, 0) = (r\sin\varphi, -r\cos\varphi, 0)$ and ending at detector position (u, v). Here, $(r\Theta_\varphi^\perp, 0)$ is related to how the source position moves around a circular trajectory of radius r.

According to Beer–Lambert law, measured sinogram data $P(\varphi, u, v)$ at the detector can be expressed as

$$P(\varphi, u, v) = -\ln \int_{\underline{E}}^{\overline{E}} \eta(E) \exp(-\mathscr{C}\mu_E(\varphi, u, v)) dE + noise, \tag{2}$$

where μ_E is an attenuation coefficient at energy E and $\eta(E)$ represents the fractional energy at photon energy $E \in [\underline{E}, \overline{E}]$ in the normalized spectrum of the X-ray beam with $\int_{\underline{E}}^{\overline{E}} \eta(E) dE = 1$. The noise originates from the properties of X-ray photons (quantum noise) and detector systems (electron noise).

Let us consider the ideal assumption of monochromatic X-rays (i.e., $\eta(E) = \delta(E - E_0)$ for some $E_0 > 0$) and negligible CT noise. The sinogram data (2) are then expressed as

$$P(\varphi, u, v) = \mathscr{C}\mu_{E_0}(\varphi, u, v), \tag{3}$$

where $\mu_{E_0}(\mathbf{x})$ denotes the distribution of attenuation coefficients at photon energy $E_0 \in [\underline{E}, \overline{E}]$. The following Feldkamp–Davis–Kress (FDK) reconstruction formula [5] is widely used:

$$\text{FDK}[P](\mathbf{x}, z) := \frac{1}{4\pi} \int_0^{2\pi} \frac{r^2}{U_{\varphi,\mathbf{x}}^2} \int P(\varphi, u, \frac{zr}{U_{\varphi,\mathbf{x}}}) \frac{r\, \hbar(r \frac{\mathbf{x}\cdot\Theta_\varphi}{U_{\varphi,\mathbf{x}}} - u)}{\sqrt{r^2 + u^2 + (\frac{zr}{U_{\varphi,\mathbf{x}}})^2}} du d\varphi, \tag{4}$$

where $\hbar(s) := \frac{1}{2\pi} \int_{\mathbb{R}} |\omega| e^{i\omega s} d\omega$ and $U_{\varphi,\mathbf{x}} := r + \mathbf{x} \cdot \Theta_\varphi^\perp$. Note that FDK is an approximate reconstruction algorithm (i.e., FDK$[P] \approx \mu_{E_0}$). A CBCT system with a circular trajectory violates Tuy's condition [30], which is the data sufficiency condition for exact CT reconstruction.

2.1 Ill-Posed Issues of Low-Dose Dental CBCT

Dental CBCT has been developed with the goal of minimizing radiation exposure while maintaining diagnostic image quality with minimal equipment costs. Most dental CBCT scans use a small-sized detector with an offset array to maximize the axial FOV while reducing the radiation dose and equipment cost [2, 9]. The FOV does not fully cover the entire region of the scanned object (see Fig. 2a).

Specifically, let P_{full} be measured data from a non-offset and sufficiently wide CBCT detector. We assume that P_{full} is compactly supported in $u \in [-\ell', \ell']$ for some $\ell' > 0$. The measured sinogram P can be represented as

$$P = \mathscr{T}_{\text{runc}}(P_{\text{full}}), \tag{5}$$

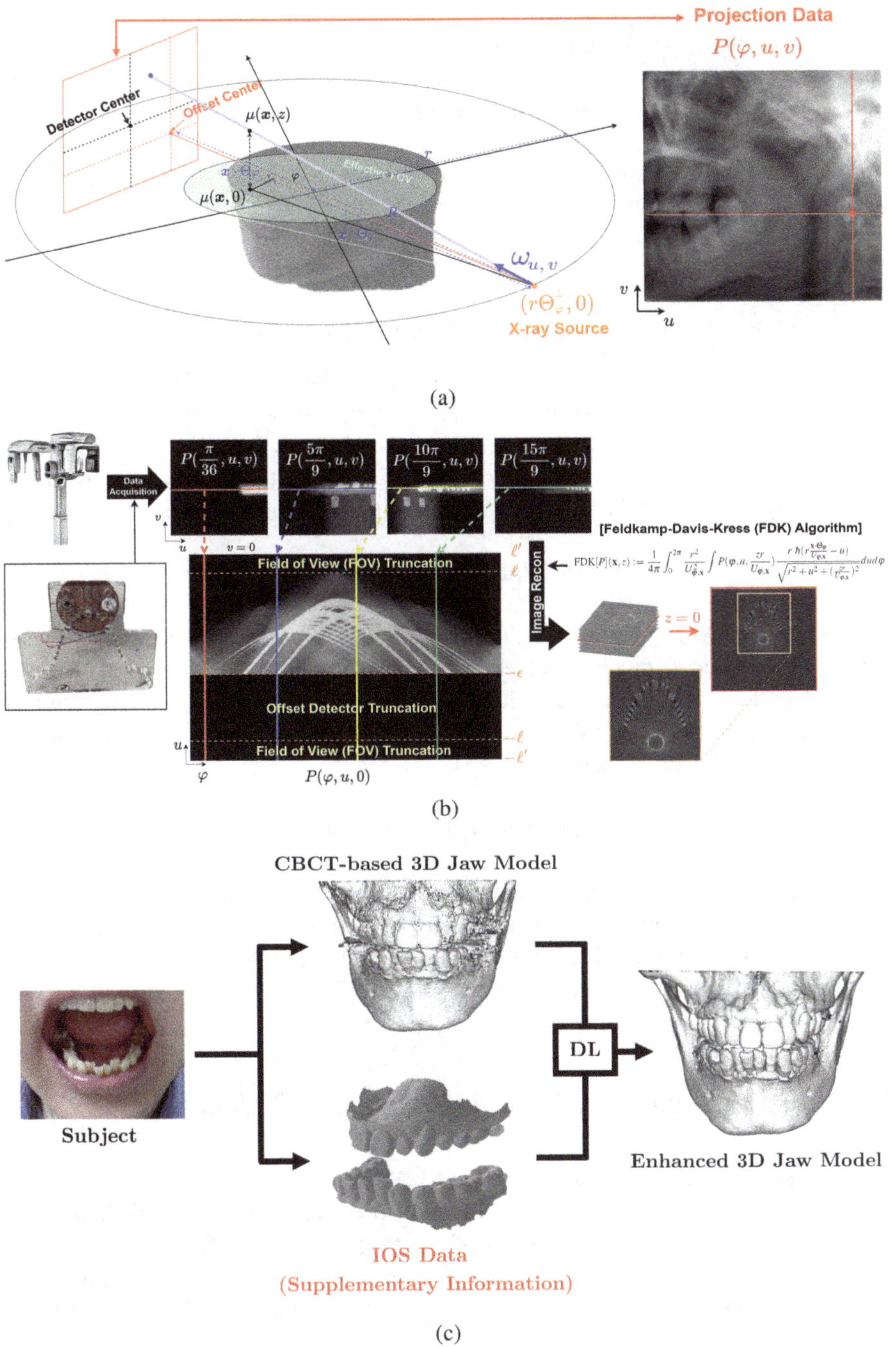

Fig. 2 (**a**) Geometry of dental CBCT system. (**b**) Data acquisition and standard FDK reconstruction in dental CBCT. (**c**) DL method [8] for metal artifact reduction in dental CBCT by using supplementary information from intraoral scan data

where $\mathscr{T}_{runc}$ is a truncation operation determined by the detector size and offset array. Figure 2b shows an example of $\mathscr{T}_{\text{runc}}$, where $[-\ell', -\ell) \cup (\ell, \ell']$ is a truncation region caused by the small effective FOV, and $[-\ell, -\epsilon)$ is caused by the offset array. Here, ϵ and ℓ are positive constants satisfying $\epsilon < \ell < \ell'$. Consequently, signoram P is supported in $u \in [-\epsilon, \ell] \subseteq [-\ell, \ell]$.

The dental CBCT imaging problem is to reconstruct a proper tomography image μ from P. The associated inverse problem is to solve

$$\mathscr{C}\mu = P, \tag{6}$$

which is highly ill-posed in the sense of Hadamard as follows [24]:

Nonexistence The standard CBCT reconstruction formula in (4) is based on the linear assumption of a monochromatic X-ray beam. However, owing to the polychromatic nature of the X-ray beam, the measured projection data P do not match the linear model. In particular, metallic implants, whose attenuation coefficients vary significantly with E, lead to serious model mismatches. In this case,

$$\nexists\ \mu \text{ such that } \mathscr{C}\mu = P \text{ for given } P. \tag{7}$$

Nonuniqueness Most dental CBCT machines use a small detector with a short subject-to-detector distance to achieve cost-effectiveness, resulting in data missing for P along the u- and v-axes. This leads to the need for solving an underdetermined problem, i.e., the number of equations is significantly lower than the number of unknowns (see Fig. 2b). Therefore,

$$\exists\ \tilde{\mu} \text{ such that } \mathscr{C}(\mu - \tilde{\mu}) = \mathbf{0} \text{ and } \tilde{\mu} \neq \mu. \tag{8}$$

Instability A local mismatch in P leads to global artifacts during the backprojection process [21]. In particular, a local singularity in the projection domain can generate global streaking artifacts in a CT image. It can be rigorously characterized using mathematical tools such as the wavefront set [20].

2.2 Reconstruction Methods for Dental CBCT Imaging

Over the past few decades, several efforts have been made to obtain high-quality CBCT images while overcoming the severe ill-posedness of dental CBCT imaging problem (i.e., reducing severe image artifacts as shown in Fig. 2b).

To mitigate artifacts induced by the sinogram truncation (5), the modified FDK or sinogram reflection algorithms can be applied. The modified FDK algorithm is defined as

$$\text{mFDK}[P](\mathbf{x}, z)$$
$$:= \frac{1}{4\pi}\int_0^{2\pi} \omega(u_{\varphi,\mathbf{x}})\frac{r^2}{U_{\varphi,\mathbf{x}}^2}\int \mathscr{P}_{\text{ad}}[P](\varphi, u, \frac{zr}{U_{\varphi,\mathbf{x}}})\frac{r\,\hbar(u_{\varphi,\mathbf{x}} - u)}{\sqrt{r^2 + u^2 + (\frac{zr}{U_{\varphi,\mathbf{x}}})^2}}du\,d\varphi, \tag{9}$$

where $u_{\varphi,\mathbf{x}} := r(\mathbf{x}\cdot\Theta_\varphi)/U_{\varphi,\mathbf{x}}$. Here, ω is a weighting function defined as

$$\omega(u) = \frac{1 - \cos(\pi(\epsilon - u)/(2\epsilon))}{2} \tag{10}$$

and $\mathscr{P}_{\text{ad}}$ is a constant padding defined as

$$\mathscr{P}_{\text{ad}}[P](\varphi, u, v) = \begin{cases} P(\varphi, u, v) & \text{if } u \in [-\epsilon, \ell] \\ P(\varphi, -\epsilon, v) & \text{if } u \in [-\epsilon', -\ell') \\ P(\varphi, \ell, v) & \text{if } u \in (\ell, \ell'] \end{cases} \tag{11}$$

The sinogram reflection is based on the following approximation of missing values in $[-\ell, -\epsilon]$:

$$P_{\text{full}}(\varphi, u, v) \approx P_{\text{full}}(\varphi + \pi + 2\tan^{-1}(-\frac{u}{R}), -u, v),\ \forall\, u \in [0, \ell]. \tag{12}$$

See [2, 10] for more details.

Now, we consider a reconstruction model for the polychromatic X-ray beams. Assuming that μ_E is differentiable with respect to E, the mismatch between P and $\mathscr{C}\mu_{E_0}$, which is called the *beam hardening effect*, can be expressed as follows [20]:

$$[P - \mathscr{C}\mu_{E_0}] \approx \int_{\underline{E}}^{\overline{E}} \eta(E)\int_{E_0}^{E} \exp(-\mathscr{C}[\mu_{E'} - \mu_{E_0}])\mathscr{C}\left[\frac{\partial\mu_{E'}}{\partial E'}\right]dE'dE. \tag{13}$$

High-attenuating materials such as dental crowns and braces considerably vary with E (i.e., $|\partial\mu_E/\partial E| \gg 0$) and hence result in the severe mismatch $P - \mathscr{C}\mu_{E_0}$. It generates global image artifacts that appear as streaks and shadows.

Bayaraa et al. [2] proposed a two-stage approach for reducing metal-related beam hardening artifacts. The first stage uses a sinogram intensity adjustment procedure to generate a CBCT image with reduced beam hardening artifacts. The second stage consecutively applies a DL network that suppresses the remaining artifacts such as noise. Park et al. [22] adopted an iterative DL framework to correct metal artifacts while enforcing data fidelity in the projection domain. To be precise, the corrected image is obtained by the following iterative procedure: For $k = 0, 1, \cdots, K$,

$$\mathbf{u}^{(k+1/2)} = \mathbf{u}^{(k)} - f_{\text{MA}}^{(k)}(\mathbf{u}^{(k)}) \tag{14}$$

$$\mathbf{u}^{(k+1)} = g(\mathbf{u}^{(k+1/2)}, \boldsymbol{P}), \tag{15}$$

where $f_{\text{MA}}^{(k)} : \mathbf{u}^{(k)} \rightarrow \xi^{(k)} (= \mathbf{u}^{(k)} - \mathbf{u}^*)$ is a DL-based artifact extractor and g is a function defined by

$$g(\mathbf{u}, \mathbf{P}) := \underset{\tilde{\mathbf{u}}}{\text{argmin}} \, \|(\boldsymbol{I} - \boldsymbol{M})(\boldsymbol{A}\tilde{\mathbf{u}} - \mathbf{P})\|_2^2 + \gamma \|\tilde{\mathbf{u}} - \mathbf{u}\|_2^2. \tag{16}$$

Here, $\boldsymbol{I}$ is the identity matrix, $\boldsymbol{M}$ is the metal trace mask, γ is a regularization parameter, $\mathbf{u}$ and $\mathbf{P}$ denote the discrete version of μ and P, and $\mathbf{u}^*$ is the artifact-free image corresponding to $\mathbf{u}$. The DL network $f_{\text{MA}}^{(k)}$ in (14) is trained to satisfy $f_{\text{MA}}^{(k)}(\mathbf{u}^{(k)}) \approx \mathbf{u}^*$.

Obtaining a paired dataset of metal-free and metal-affected CBCT scans is challenging in clinical practice. To circumvent this issue, realistic simulation datasets can be used to train the DL network [8, 21]. Refer to [9] for realistic metal artifact simulations in dental CBCT.

Although DL-based approaches provide promising and excellent outcomes, there remain challenges in practical applications. In dental CBCT imaging, metal-related artifacts are considerably more complex due to other physical effects (e.g., photon starvation, photon scattering, and partial volume effect) and their nonlinear interactions. To address these problems, an effective DL strategy that is optimized to real-world dental CBCT imaging environments should be designed. The use of a supplementary image prior from an IOS or an FSTS can be a potential solution [8, 11]. Hyun et al. [8] attempted to utilize IOS data to obtain a high-quality maxillofacial image in the presence of serious metal artifacts, as shown in Fig. 2c.

3 3D Jaw–Teeth–Face Model for Digital Dentistry

Recent advances in dental imaging technologies, including dental CBCT, IOS, and FSTS, have created a pressing demand for the development of a smart digital platform that can generate an integrated patient treatment plan by incorporating a single digital anatomical model of the bone, teeth, gingiva, and face [4, 29]. This 3D jaw–teeth–face model enables a wide range of dental applications such as digital implant (positioning) simulation, maxillofacial surgery simulation, and orthodontic treatment planning.

However, the automatic integration of dental CBCT, IOS, and FSTS data poses significant challenges because of the inherent dissimilarities in the data acquired from these devices. The registration process typically involves determining the rigid transformation of partially overlapping areas. Jang et al. [13] proposed an integration method by aligning individually segmented teeth from dental CBCT and IOS data, while Park et al. [25] utilized facial surface landmarks with minimal geometric variation to integrate dental CBCT and FSTS data (see Fig. 3).

CBCT-IOS Registration Let $\mathbf{X}$ be a tooth point cloud from the CBCT consisting of individual tooth point clouds as follows:

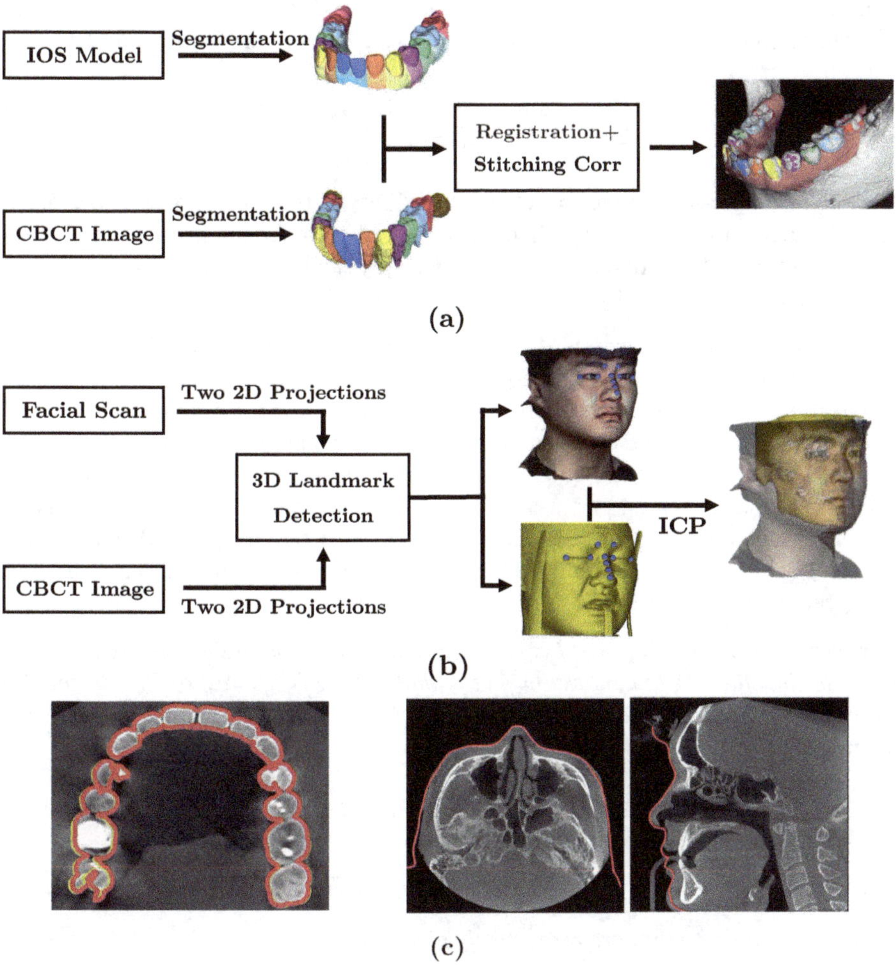

Fig. 3 Overall flow diagram of registration method of CBCT with (**a**) IOS and (**b**) FSTS. (**c**) Registration results for a test subject (2D CT slice view). Here, red lines represent the registered IOS and FSTS data

$$\mathbf{X} = \mathbf{X}_{t_1} \cup \mathbf{X}_{t_2} \cdots \cup \mathbf{X}_{t_J}, \tag{17}$$

where $\mathbf{X}_{t_j}$ represents the t_j-tooth for $j = 1, \ldots, J$. Here, J denotes the number of teeth. An individual tooth can be segmented and identified using the DL-based method [12]. Similarly, let $\mathbf{Y}$ be a 3D point set of the IOS scan given by

$$\mathbf{Y} = \mathbf{Y}_{t_1} \cup \mathbf{Y}_{t_2} \cdots \cup \mathbf{Y}_{t_J} \cup \mathbf{Y}_{gingiva}, \tag{18}$$

where $\mathbf{Y}_{t_j}$ represents a tooth with code t_j in $\mathbf{Y}$ and $\mathbf{Y}_{gingiva}$ denotes the rest including the gingiva in $\mathbf{Y}$. Here, segmentation can be achieved using the DL technique [31]. Given two 3D point sets of $\mathbf{X}$ and $\mathbf{Y}$, the optimal transformation $\mathscr{T}^*$ can be found by solving the following least squares minimization problem:

$$\mathscr{T}^* = \underset{\mathscr{T}}{\text{argmin}} \sum_{(\mathbf{x},\mathbf{y})\in Corr} \|\mathbf{y} - \mathscr{T}(\mathbf{x})\|_2^2, \tag{19}$$

where $Corr$ is a correspondence set given by $Corr = \{(\mathbf{x}, \mathbf{y}) \in \mathbf{X} \times \mathbf{Y}\}$. To achieve both global and local alignment effectively, the fast point feature histograms (FPFH) [26] and improved iterative closest point (ICP) [3] can be applied, respectively. The registration result is visually shown in the left part of Fig. 3c.

CBCT-Facial Scan Registration Let $\mathbf{X}$ represent a 3D point cloud of facial surfaces obtained from CBCT data. Let $\mathbf{X}_{\text{ldmk}} = \{\mathbf{x}_j = (x_j, y_j, z_j) : j = 1, \cdots, J\}$ represent the landmark points in $\mathbf{X}$. Here, J is the number of landmarks. Let $\mathbf{X}^{\varphi}$ be a rotated facial surface of $\mathbf{X}$ around the z-axis by an angle of φ. For each $\mathbf{X}^{\varphi}$, the 2D projection image $\mathbf{I}^{\varphi}$ can be generated according to Formula (4) in [25]. From $\mathbf{I}^{\varphi}$, we obtain a set $\{(x_j^{\varphi}, z_j) : j = 1, \cdots, J\}$, which represents the 2D landmarks corresponding to the 3D landmarks in $\mathbf{X}_{\text{ldmk}}$. For two different sets of 2D landmarks, $\{(x_j^{\varphi_1}, z_j) : j = 1, \cdots, J\}$ and $\{(x_j^{\varphi_2}, z_j) : j = 1, \cdots, J\}$, the 3D landmarks in $\mathbf{X}_{\text{ldmk}}$ can be detected using the following theorem.

Theorem 1 (Park et al. [25]) *Let two different rotation angles φ_1 and φ_2 be given. Suppose that the 2D landmarks $(x_j^{\varphi_1}, z_j)$ and $(x_j^{\varphi_2}, z_j)$, corresponding to the 3D landmark $\mathbf{x}_j$, are obtained from the two projection images $\mathbf{I}^{\varphi_1}$ and $\mathbf{I}^{\varphi_2}$, respectively. Let $\mathbf{x}_j$ be expressed by*

$$\mathbf{x}_j = (x_j, y_j, z_j) = (-L \sin\theta, L\cos\theta, z_j), \tag{20}$$

where $L = \sqrt{x_j^2 + y_j^2}$ and $\theta = \tan^{-1}(\frac{y_j}{x_j}) - \frac{\pi}{2}$. Then, L and θ are functions of $\varphi^1, \varphi^2, x_j^{\varphi_1}$, and $x_j^{\varphi_2}$ that are respectively given by

$$L = L(\varphi^1, \varphi^2, x_j^{\varphi_1}, x_j^{\varphi_2}) = \sqrt{\frac{(x_j^{\varphi_2} - x_j^{\varphi_1}\cos(\varphi_1 - \varphi_2))^2}{\sin^2(\varphi_1 - \varphi_2)} + (x_j^{\varphi_1})^2}$$

and

$$\theta = \theta(\varphi^1, \varphi^2, x_j^{\varphi_1}, x_j^{\varphi_2}) = \sin^{-1}\left(-\frac{x_j^{\varphi_1}}{L}\right) - \varphi_1. \tag{21}$$

Denoting $\mathbf{Z}$ by a 3D point cloud of the facial surface obtained from FSTS data, the corresponding 3D landmarks $\mathbf{Z}_{\text{ldmk}}$ can be detected in the same manner. In the field of computer vision, various methods have been proposed for 2D facial landmark detection [32, 33]. The open-source library Dlib [16] provides an effective pretrained model for detecting 68 facial feature points in 2D images. Some of the points with minimal positional changes on the surfaces of CBCT and FSTS can be selected as facial landmarks. Given a pair of 3D landmarks from CBCT and FSTS, the optimal rigid transformation $\mathscr{T}$ can be obtained by solving the least squares minimization problem (19). Finally, the ICP method can be applied to accurately estimate $\mathscr{T}$ using the subsurfaces $\mathbf{X}_{\text{sub}}$ and $\mathbf{Z}_{\text{sub}}$ that contain the selected 3D landmarks $\mathbf{X}_{\text{ldmk}}$ and $\mathbf{Z}_{\text{ldmk}}$, respectively. The registration result is shown in the right part of Fig. 3c.

Nevertheless, the aforementioned methods focus on the primary task of aligning the maxillary region but do not provide seamless integration of the mandibular region and soft tissues. Hence, future research should explore the robust integration of CBCT, IOS, and FSTS, enabling the comprehensive and accurate creation of a 3D jaw–teeth–face model that incorporates occlusion and articulation. This can be accomplished by developing an advanced digital platform.

3.1 Postsurgical Facial Prediction

Recent years have witnessed a growing interest in virtual modeling techniques to enhance the predictability of postoperative facial appearance following orthognathic surgery [28]. With the rapid development of dental CBCT and 3D FSTS with high resolution and precision, surgical simulation has become possible in virtual environments. Additionally, owing to the competitive developments in mathematical modeling, 3D computer graphics, and high-performance computing, postoperative facial prediction has become accessible.

Finite-element-based approaches (FEAs) [19] can simulate the mechanical behavior of tissues and bones under various surgical interventions based on the linear elastic equation, allowing the evaluation of changes in facial morphology and biomechanics. However, FEAs suffer from a limitation with respect to individual patient postoperative facial prediction in terms of computational complexity and associated costs.

Utilizing the linear elastic equation, we attempt to determine a displacement, denoted by $\mathbf{f}$, as follows:

$$\mathbf{f} = \underset{\mathbf{f}}{\operatorname{argmin}} \sum_{i=1}^{N} |\mathbf{f}(\mathbf{x}_i) - \mathbf{y}_i|^2 + \lambda \int_{\mathbb{R}^3} |\nabla^2 \mathbf{f}(\mathbf{x})|^2 d\mathbf{x}, \tag{22}$$

where $\mathbf{x}_i$ and $\mathbf{y}_i$ represent the paired points obtained before and after virtual surgery on the 3D skull model, respectively. The second term in (22) involves the fundamental solution $\phi(|\mathbf{x}|)$ of the biharmonic equation in $\mathbb{R}^3$. By leveraging this, we can express $\mathbf{f}$ as

$$\mathbf{f}(\mathbf{x}) = \sum_{i=1}^{N} \mathbf{w}_i, \phi(|\mathbf{x} - \mathbf{x}_i|) + \mathbf{v} \cdot \begin{bmatrix} 1 \\ \mathbf{x} \end{bmatrix}, \tag{23}$$

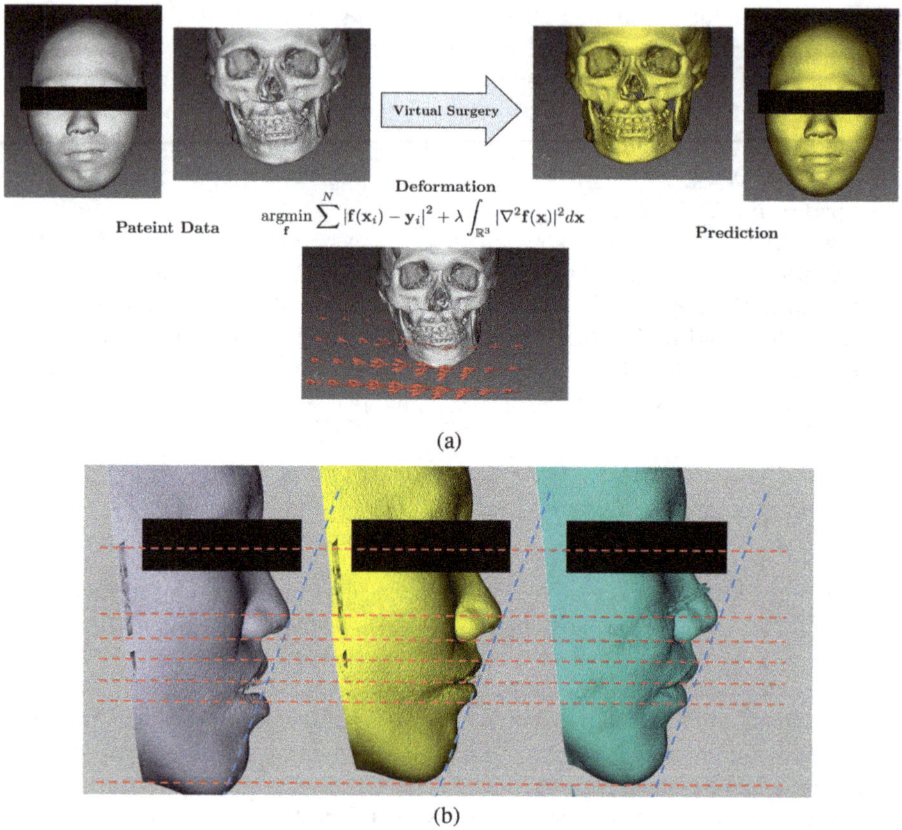

Fig. 4 (**a**) Postoperative prediction method. (**b**) Test results using clinical data: (purple) before surgery, (yellow) prediction, and (mint) after surgery—one year

where $\{\mathbf{w}_i \in \mathbb{R}^3\}_{i=1}^N$ and $\mathbf{v} \in \mathbb{R}^{3\times 4}$ are unknowns to be determined. By putting (23) into (22), the minimization problem in (22) is converted into a linear quadratic minimization problem. Hence, the minimizer (i.e., optimal $\{\mathbf{w}_i\}_{i=1}^N$ and $\mathbf{v}$) can be obtained by solving the subsequent symmetric and positive definite system.

Figure 4a and b shows the overall postoperative prediction process and corresponding results using clinical data. The predicted face is fairly close to the actual postoperative face; however, there are still discrepancies associated with nonlinear dynamics between soft tissue and bony structures.

Recently, DL techniques have been employed in the field of postoperative facial prediction [1]. DL methods can learn and capture highly nonlinear facial deformation relationships. Combining DL techniques with (22) is expected to provide a more practical and effective solution than existing approaches.

4 Landmark Detection for Cephalometric Analysis

Traditionally, 2D cephalometry using plain radiographs has been the standard practice for over four decades. However, recent advancements in imaging technologies and computer-assisted surgery have paved the way for the transition from 2D to 3D cephalometry, offering a multitude of advantages, including the precise identification of anatomical structures and comprehensive evaluation of complex cranial structures.

Previously, 3D cephalometric landmarking primarily relied on the manual identification of points representing craniofacial morphological characteristics. This process demands a high level of expertise, significant time, and considerable efforts even for experienced professionals. Consequently, there is a growing demand for automatic landmarking systems to streamline processes and improve workflow efficiency [35].

Recently, significant progress has been made in the automatic annotation of 3D cephalometric landmarks in CT scans thanks to the rapid advancements in DL technology [15, 18, 34]. Kang et al. [15] developed an advanced automatic 3D cephalometric annotation system based on multistage deep reinforcement learning (RL) and volume-rendered imaging. This innovative system considers the geometric characteristics of landmarks and simulates the sequential decision-making process employed by human experts during landmarking. It involves constructing appropriate two-dimensional cutaways or 3D model views, followed by implementing either single-stage RL with gradient-based boundary estimation or multistage RL to accurately determine the 3D coordinates of target landmarks.

Specifically, the RL agent learns the optimal trajectory toward a labeled target position through a sequential decision-making process. It can be formulated as a Markov decision process defined as $M = (S, A, P, R)$, where S represents a set of states, A is a set of actions, P represents the state transition probabilities, and R is the reward function. An environment (E) consists of images obtained through the volume rendering of 3D CBCT images, with reference landmark positions. The actions of the agent ($a \in A$) are defined by the movements (right, left, up, and down) on the 2D image plane along the orthogonal axis within E. A state ($s \in S$) is defined as the ROI within E where the agent is located. The reward function (R_t) is given by

$$R_t = \|AP_{t-1} - TP\|_2^2 - \|AP_t - TP\|_2^2, \tag{24}$$

where AP_t represents the predicted position at step t and TP is the target reference position. At each t, the agent receives a reward from the environment upon the execution of a valid action. Accordingly, a state–action function $Q(s, a)$ is defined by the expectation of the cumulative future reward with a discount factor γ as follows:

$$Q(s, a) = \mathbb{E}[R_{t+1} + \gamma \max_{a'} Q(s_{t+1}, a')|s, a]. \tag{25}$$

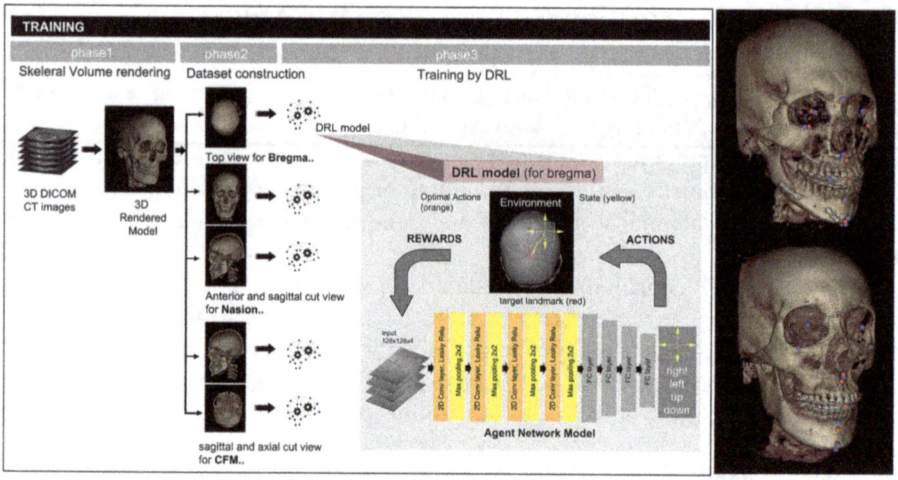

(a) (b)

Fig. 5 (**a**) Schematic diagram of the 3D cephalometric landmark detection framework using deep reinforcement learning [15]. (**b**) Visualization of prediction landmark points (blue dots) and reference points (red dots)

Q-learning finds an optimal action-selection policy by solving equation (25) iteratively. Figure 5a and b shows the overall process and prediction results, respectively.

The integration of CBCT imaging, DL techniques, and automatic landmarking systems has a great potential for transforming the field of cephalometry and facilitating more accurate diagnoses, treatment planning, and evaluation of craniofacial conditions with reduced reliance on manual intervention.

5 Discussion

Dental CBCT has been widely adopted in dentistry due to its exceptional benefits, notably a significantly lower radiation dose than traditional imaging methods, as well as its relative affordability in terms of both initial purchase and ongoing maintenance costs. In digital dentistry, CBCT is set to play a pivotal role in harnessing recent advancements in AI tools, IOS, and FSTS. The combination of these cutting-edge innovations with established techniques, such as 3D CAD/CAM and 3D printing, is driving a transformative wave across the dental industry, expanding their influence into various domains, including implants/prosthodontics, oral and maxillofacial surgery, and orthodontic treatments.

With the increasing number of elderly patients opting for metal implants, the challenge of metal-induced artifacts undermining the diagnostic performance of CBCT has become increasingly significant. Nevertheless, there is an expectation that recent rapid advancements in MAR technology, along with the integration of

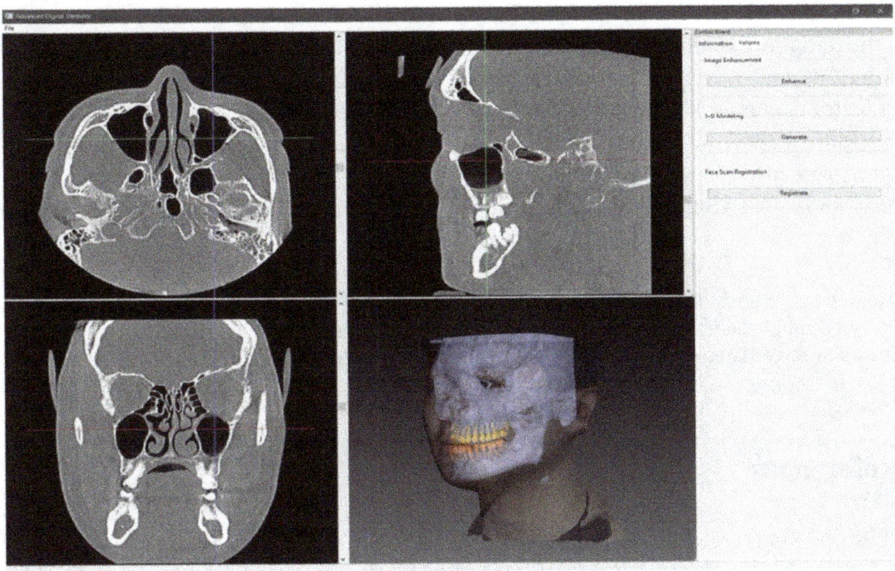

Fig. 6 Digital dentistry software [14] providing an integrated 3D jaw–teeth–face model from dental CBCT, IOS, and FSTS data

IOS and AI-based image enhancement methods, will offer a promising solution for effectively alleviating metal-induced artifacts in the dental CBCT environment. This combined approach has great potential for enhancing the quality and accuracy of CBCT images, thereby enabling dentists to make informed diagnoses and treatment decisions.

The rapid advancement of AI is expected to revolutionize the amalgamation of CBCT, oral scanners, and facial scanners, offering groundbreaking tools for managing dental care and optimizing oral health, with profound benefits for both patients and dentists. Through the seamless integration of CBCT and IOS technologies, the precision of digital impressions can be significantly improved, effectively surpassing the limitations of traditional impression-making methods. Moreover, factors that impede the accuracy of conventional approaches, such as patient movement, impression tearing, deformation, and soft-tissue contraction, can be successfully addressed through this innovative fusion. Consequently, this transformative integration not only eliminates the cumbersome procedures associated with traditional impressions but also streamlines the workflow for both dentists and patients, leading to significantly reduced treatment times and enhanced overall patient experience.

Digital dentistry should empower dentists with a seamless access to a wide range of software, even in the absence of specialized software expertise. Although remarkable progress has been made in individual technologies, there is an urgent need for developing a comprehensive platform that enables dentists without software skills to utilize diverse digital tools freely and efficiently, as shown in Fig. 6. To address the evolving needs of dentists and patients, it is crucial for digital dentistry to prioritize enhancing satisfaction levels and revenue generation. By embracing innovative solutions and optimizing the integration of cutting-edge technologies, the dental industry can unlock a host of opportunities to enhance patient experience, streamline workflows, and boost profitability. Embracing these advancements will lead to improved efficiency, precise treatment planning, and enhanced communication, ultimately fostering a positive impact on the overall success and growth of dental practice.

Acknowledgments This work was supported by a grant of the Korea Health Technology R&D Project through the Korea Health Industry Development Institute (KHIDI), funded by the Ministry of Health & Welfare, Republic of Korea (grant number : HI20C0127).

References

1. R. Ali, R. Lei, H. Shi, J. Xu, Cranio-maxillofacial post-operative face prediction by deep spatial multiband VGG-NET CNN. Am. J. Transl. Res. **14**(4), 2527–2539 (2022)
2. T. Bayaraa, C.M. Hyun, T.J. Jang, S.M. Lee, J.K. Seo, A two-stage approach for beam hardening artifact reduction in low-dose dental CBCT. IEEE Access **8**, 225981–225994 (2020)
3. P.J. Besl, N.D. McKay, Method for registration of 3-d shapes, in sensor fusion IV: control paradigms and data structures. Int. Soc. Opt. Photon. **1611**, 586–606 (1992)
4. N.H. Elnagar, S. Aronovich, B. Kusnoto, Digital workflow for combined orthodontics and orthognathic surgery. Oral Maxillofacial Surg. Clin. **32**(1), 1–14 (2020)
5. L.A. Feldkamp, L.C. Davis, J.W. Kress, Practical cone-beam algorithm. JOSA A **1**(6), 612–619 (1984)
6. C.M. Hyun, K.C. Kim, H.C. Cho, J.K. Choi, J.K. Seo, Framelet pooling aided deep learning network: the method to process high dimensional medical data. Mach. Learn. Sci. Technol. **1**(1), 015009 (2020)
7. C.M. Hyun, S.H. Baek, M. Lee, S.M. Lee, J.K. Seo, Deep learning-based solvability of underdetermined inverse problems in medical imaging. Med. Image Anal. **69**, 101967 (2021)
8. C.M. Hyun, T. Bayaraa, H.S. Yun, T.J. Jang, H.S. Park, J.K. Seo, Deep learning method for reducing metal artifacts in dental cone-beam CT using supplementary information from intra-oral scan. Phys. Med. Biol. **69**, 101967 (2022)
9. C.M. Hyun, T. Bayaraa, S.M. Lee, H. Jung, J.K. Seo, *Deep Learning for Dental Cone-Beam Computed Tomography*. Deep Learning and Medical Applications (Springer Nature, 2023), pp. 101–175
10. C.M. Hyun, J.K. Seo, *Deep Learning for Ill Posed Inverse Problems in Medical Imaging*. Deep Learning and Medical Applications (Springer Nature, 2023), pp. 319–339
11. C.M. Hyun, K. Jeon, H.S. Park, Data-driven approach for metal artifact reduction in dental cone-beam CT with an extra-condition of intra-oral scan data, in *Proceedings of the 17th International Meeting on Fully 3D Image Reconstruction in Radiology and Nuclear Medicine* (2023). arXiv:2310.16846
12. T.J. Jang, K.C. Kim, H.C. Cho, J.K. Seo, A fully automated method for 3D individual tooth identification and segmentation in dental CBCT. IEEE Trans. Pattern Anal. Mach. Intell. (2021). https://doi.org/10.48550/arXiv.2102.06060
13. T.J. Jang, H.S. Yun, C.M. Hyun, J.E. Kim, S.-H. Lee, J.K. Seo, Fully automatic integration of dental CBCT images and full-arch intraoral impressions with stitching error correction via individual tooth segmentation and identification. Preprint (2021). arXiv:2112.01784

14. K. Jeon, H.S. Park, Software of integrated module for digital dentistry. South Korea Software Copyright, C-2023-025645 (2023)
15. S.H. Kang, K. Jeon, S.H. Kang, S.H. Lee, 3D cephalometric landmark detection by multiple stage deep reinforcement learning. Sci. Rep. **11**(1), 1–13 (2021)
16. D.E. King. Dlib-ml: A machine learning toolkit. J. Mach. Learn. Res. **10**, 1755–1758 (2009)
17. S.M. Lee, T. Bayaraa, H. Jeong, C.M. Hyun, J.K. Seo, A direct sinogram correction method to reduce metal-related beam-hardening in computed tomography. IEEE Access **7**, 128828–128836 (2019)
18. S.M. Lee, H.P. Kim, K. Jeon, S.-H. Lee, J.K. Seo, Automatic 3D cephalometric annotation system using shadowed 2D image-based machine learning. Phys. Med. Biol. **64**(5), 055002 (2019)
19. W. Mollemans, F. Schutyser, N. Nadjmi, F. Maes, P. Suetens, Predicting soft tissue deformations for a maxillofacial surgery planning system: From computational strategies to a complete clinical validation. Med. Image Anal. **11**, 282–301 (2007)
20. H.S. Park, J.K. Choi, J.K. Seo, Characterization of metal artifacts in X-ray computed tomography. Commun. Pure Appl. Math. **70**(11), 2191–2217 (2017)
21. H.S. Park, S.M. Lee, H.P. Kim, J.K. Seo, Y.E. Chung, CT sinogram-consistency learning for metal-induced beam hardening correction. Med. Phys. **45**, 5376–5384 (2018)
22. H.S. Park, J.K. Seo, C.M. Hyun, S.M. Lee, K. Jeon, A fidelity-embedded learning for metal artifact reduction in dental CBCT. Med. Phys. **49**(8), 5195–5205 (2022)
23. H.S. Park, K. Jeon, S.-H. Lee, J.K. Seo, Unpaired-paired learning for shading correction in cone-beam computed tomography. IEEE Access **10**, 26140–26148 (2022)
24. H.S. Park, C.M. Hyun, J.K. Seo, Nonlinear ill-posed inverse problems in dental cone-beam computed tomography. IMA J. Appl. Math. (2023). https://doi.org/10.48550/arXiv.2303.01678
25. H.S. Park, C.M. Hyun, S.-H. Lee, J.K. Seo, K. Jeon, Automatic 3D registration of dental CBCT and face scan data using 2D projection images. Preprint (2023). arXiv:2305.10132
26. R.B. Rusu, N. Blodow, M. Beetz, Fast point feature histograms (FPFH) for 3D registration, in *Proceeding of 2009 IEEE International Conference on Robotics and Automation*, pp. 3212–3217 (2009)
27. R. Schulze, U. Heil, D. Grob, D. Bruellmann, E. Dranischnikow, U. Schwanecke, E. Schoemer, Artifacts in CBCT: a review. Dentomaxillofacial Radiol. **40**, 265–273 (2011)
28. P. Scolozzi, S. Schendel, Soft-tissue changes and predictions of orthognathic surgery. *Peterson's Principles of Oral and Maxillofacial Surgery* (pp. 2019–2038) (2011)
29. S. Shujaat, M.M. Bornstein, J.B. Price, R. Jacobs, Integration of imaging modalities in digital dental workflows-possibilities, limitations, and potential future developments. Dentomaxillofacial Radiol. **50**, 20210268 (2021)
30. H.K. Tuy, An inversion formula for cone-beam reconstruction. SIAM J. Appl. Math. **43**(3), 546–552 (1983)
31. Y. Wang, Y. Sun, Z. Liu, S.E. Sarma, M.M. Bronstein, J.M. Solomon, Dynamic graph CNN for learning on point clouds. arXiv (2018)
32. N. Wang, X. Gao, D. Tao, H. Yang, X. Li, Facial feature point detection: A comprehensive survey. Neurocomputing **275**, 50–65 (2018)
33. Y. Wu, Q. Ji, Facial landmark detection: A literature survey. Int. J. Comput. Vis. **127**, 115–142 (2019)
34. H.S. Yun, C.M. Hyun, S.H. Baek, S.-H. Lee, J.K. Seo, A semi-supervised learning approach for automated 3D cephalometric landmark identification using computed tomography. Plos One **17**(9), e0275114 (2022)
35. H.S. Yun, A. Jargal, C.M. Hyun, J.K. Seo, *Nonlinear Representation and Dimensionality Reduction*. Deep Learning and Medical Applications (Springer Nature, 2023), pp. 1–49

High-Efficient Compton Scattering Computed Tomography Based on a Thomson Scattering X-Ray Source

Zhijun Chi, Yingchao Du, Wenhui Huang, and Chuanxiang Tang

1 Introduction

Compared with conventional transmission computed tomography (CT), Compton scattering computed tomography (CSCT) has many advantageous properties. Firstly, the one-sided scanning scheme makes it possible to examine the internal structure of extended objects (e.g., installed aircraft sections, explosive charge in shells, and filled storage tanks) for which a transmission measurement is infeasible. Secondly, a higher contrast sensitivity to small density variation in low-density objects can be obtained in CSCT [1]. Thirdly, CSCT can provide similar contrast for both low-Z and high-Z materials [2], which can resolve the worse contrast problem of low-Z samples (e.g., body soft tissue) encountered in transmission imaging. Finally, the electron density, rather than the linear attenuation coefficient, is reconstructed in CSCT, which extends the ability of material discrimination. Therefore, CSCT has many important applications in medical imaging and nondestructive testing (NDT) [3–8].

For CSCT, the commonly used light sources are conventional X-ray tubes and radioisotopes (e.g., ^{109}Cd, ^{241}Am, ^{137}Cs, and ^{60}Co). Due to the broad spectrum characteristics of X-ray tubes, the CSCT modality using an X-ray tube is often used in a direct 3D spatial definition geometry without considering the energy spectrum of Compton scattered X-rays [9], because of which a quantitative analysis of the imaging results will be compromised. The monochromatic X-rays produced

Z. Chi (✉)
Key Laboratory of Beam Technology of Ministry of Education, School of Physics and Astronomy, Beijing Normal University, Beijing, China
e-mail: chizj18@bnu.edu.cn

Y. Du · W. Huang · C. Tang
Key Laboratory of Particle and Radiation Imaging of Ministry of Education, Department of Engineering Physics, Tsinghua University, Beijing, China

K. Iniewski, H. Gadey (eds.), *Emerging Radiation Detection*,
https://doi.org/10.1007/978-3-031-63897-8_10

by radioisotopes make it possible to realize energy-coded imaging [10, 11], and several novel reconstruction theories for CSCT have been developed by using monochromatic X-ray sources [12–18]. Since the photon flux of radionuclide sources is relatively low, the data acquirement of CSCT using this type of light source is proven to be time-consuming. The high-flux and monochromatic X-rays produced by synchrotron radiation facilities are very suitable for CSCT [5, 19]. However, the large-scale footprint and the expense of such a facility cannot meet the flexibility requirement of clinical imaging and industrial NDT.

As an alternative, a Thomson scattering (also known as inverse Compton scattering at high-energy region) X-ray source provides valuable prospects for CSCT, since it can produce quasi-monochromatic, continuously energy-tunable, precisely polarization-controllable, and high-brightness X-rays. The X-ray energy of this type of light source, typically ranging from tens of keV to several MeVs, can cover all CSCT applications in medical imaging and industrial NDT. Meanwhile, its high beam qualities will help to improve the reconstruction precision of CSCT. Furthermore, the implementation of compactness and mobility of such a light source, e.g., the very compact inverse Compton scattering gamma-ray source (VIGAS) facility under construction at Tsinghua university [20], makes it very suitable for hospital- and industrial-scale applications. In this chapter, we will propose a high-efficient imaging modality of CSCT by using linearly polarized X-rays and investigate the feasibility of linearly polarized CSCT based on a Thomson scattering X-ray source.

2 Theoretical Basis of High-Efficient CSCT: Compton Scattering Improvement Effect

For linearly polarized and unpolarized X-rays, the differential cross sections of Compton scattering are given by the Klein–Nishina formulas:

$$\frac{\mathrm{d}\sigma_{\mathrm{KN,LP}}}{\mathrm{d}\Omega} = \frac{1}{2} r_{\mathrm{e}}^2 \varepsilon^2 (\varepsilon + \varepsilon^{-1} - 2\sin^2\theta\cos^2\phi), \tag{1a}$$

and

$$\frac{\mathrm{d}\sigma_{\mathrm{KN,UP}}}{\mathrm{d}\Omega} = \frac{1}{2} r_{\mathrm{e}}^2 \varepsilon^2 (\varepsilon + \varepsilon^{-1} - \sin^2\theta), \tag{1b}$$

respectively, where θ and ϕ are polar angle and azimuthal angle, respectively, and $\varepsilon = \frac{1}{1+\frac{E_{\mathrm{i}}}{m_{\mathrm{e}}c^2}(1-\cos\theta)}$, with E_{i} and $m_{\mathrm{e}}c^2$ being the incident X-ray energy and the rest energy of an electron, respectively. At the direction of $\phi = 90°/270°$, the differential cross section in the linear polarization case can reach a value higher than the counterpart in the unpolarization case at any polar angles except for $\theta = 0°/180°$, as shown in Fig. 1. The maximum difference of the differential cross

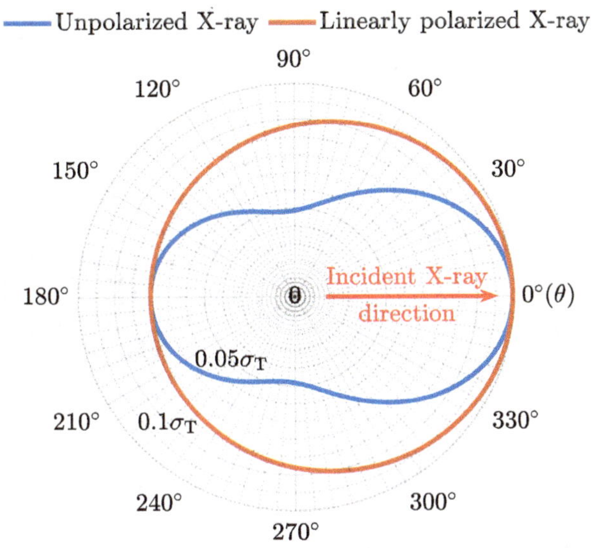

Fig. 1 Differential Compton scattering cross sections for linear polarization and unpolarization cases at X-ray energy of 60 keV ($\phi = 90°/270°$). Data are depicted in angular coordinate

section between the two polarization cases occurs at the direction of $\theta = 90°/270°$. Therefore, the single Compton scattering signal in CSCT, compared with the unpolarization case (e.g., X-ray tube or radionuclide sources-based CSCT), will be greatly improved by using linearly polarized X-rays and placing the X-ray Compton scattering detector at the direction of $\theta = 90°/270°$ and $\phi = 90°/270°$.

3 Monte Carlo Simulation

In order to demonstrate the feasibility of CSCT based on a Thomson scattering X-ray source, a Monte Carlo (MC) simulation was performed using the Geant4 toolkit (version 10.05) [21].

3.1 Imaging System and Phantom

The imaging layout of CSCT simulation is shown in Fig. 2, which is similar to the one previously used in X-ray fluorescence computed tomography (XFCT) simulation. Details of the imaging system can be found elsewhere [22, 23]. In the MC simulation, X-rays with vertical (y-axis) polarization were used. The rms source spot size was 10.0 μm, and the incident X-ray energy was $E_0 = 60$ keV with an rms bandwidth of 1.0%.

According to the Compton scattering improvement theory in Sect. 2, a photon counting detector (PCD) used to measure the Compton scattered X-ray photons

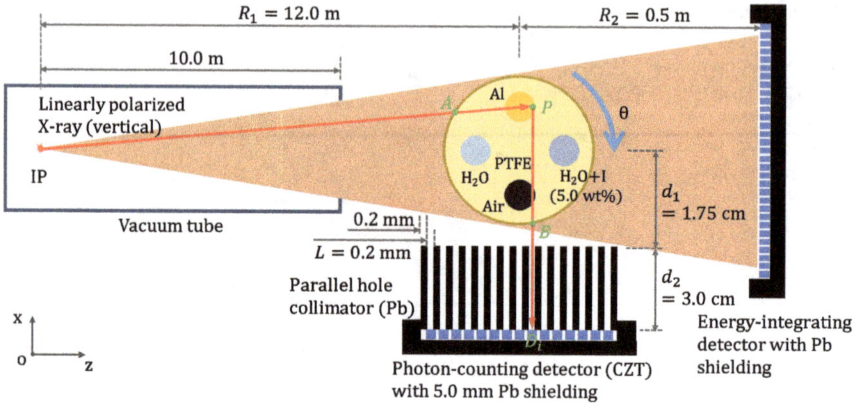

Fig. 2 Schematic for the CSCT based on a Thomson scattering X-ray source (not to scale)

was placed perpendicular to the x-axis in the x–z plane for the highest Compton scattering signal. The PCD was made of cadmium zinc telluride (CZT) and modelled with a pixel size of 200 μm and a sensitive energy range of 20–100 keV. These PCD parameters can be easily available at present [24]. For simplicity, other parameters (e.g., detection efficiency, energy resolution, and noise) of the PCD were assumed to be ideal. The distance between the PCD and the sample was 4.75 cm. In front of the PCD, there was a parallel hole collimator, made of lead (Pb), with a thickness of 0.2 mm, a length of 3.0 cm, and a hole width of $L = 0.2$ mm. For CSCT reconstruction, the attenuation data of the sample are necessary (see Sect. 3.2). Hence, an ideal energy integrating detector with pixel size 100 μm was placed $R_2 = 0.5$ m downstream of the sample to acquire the attenuation data of the sample.

The phantom was a cylindrical polytetrafluoroethylene (PTFE) of diameter 2.5 cm, inside which there were four cylinders of diameter 5.0 mm. Three of them were made of aluminum (Al), water (H_2O), and iodine (I)-loaded water solution with I concentration of 5.0 wt%, respectively, and the last one was left for empty (Air). For a CSCT scan, 360 projections were acquired at a sampling interval of 1°. The incident photon number used in each projection was 2.0×10^9.

3.2 *Image Reconstruction*

According to the imaging geometry in Fig. 2, the Compton scattered X-ray intensity $I_{\text{det}}(P, D_i)$ coming from point P and detected at the ith pixel bin D_i of the PCD can be calculated as

$$
\begin{aligned}
I_{\text{det}}(P, D_i) = & I_0 \exp\left[-\int_A^P \mu(E_0, \vec{r})\text{d}s\right] \rho_\text{e}(P) \frac{\text{d}\sigma_{\text{KN,LP}}}{\text{d}\Omega} \\
& \times \Delta\Omega_\text{c} \exp\left[-\int_P^B \mu(E_{\text{CS}}, \vec{r})\text{d}s\right],
\end{aligned} \tag{2}
$$

where I_0 is the incident X-ray intensity, $\rho_\text{e}(P)$ is the local electron density at P, $\Delta\Omega_\text{c}$ is the collecting angle of the parallel hole collimator corresponding to D_i, μ is the linear attenuation coefficient of the sample, E_{CS} is the photon energy of Compton scattered X-rays and its value is 53.70 keV calculated based on the detection geometry in Fig. 2, and $\vec{r}$ denotes a vector in Euclidean space. Thus, the total Compton scattered X-ray intensity detected by the pixel bin D_i is a volume integral of $I_{\text{det}}(P, D_i)$,

$$
I_{\text{det},i} = \iiint_{V_{P\to D_i}} I_{\text{det}}(P, D_i)\text{d}V_P, \tag{3}
$$

with the domain of integration $V_{P\to D_i}$ being the pixel region in the sample subtended by D_i toward the parallel hole collimator. The relation between the detected Compton scattered X-ray intensity $\mathbf{I}_{\text{det}}$ of size $M \times 1$ and the unknown electron density $\boldsymbol{\rho}_\text{e}$ of size $N \times 1$, after sample discretization, can be simplified as

$$
\mathbf{I}_{\text{det}} = \mathbf{A}\boldsymbol{\rho}_\text{e}, \tag{4}
$$

with $\mathbf{A} = [a_{ij}]$ being the system matrix of size $M \times N$. Then, the electron density distribution $\boldsymbol{\rho}_\text{e}$ can be reconstructed using the maximum-likelihood expectation maximization (MLEM) algorithm [25].

However, the attenuation terms $\mu(E_0, \vec{r})$ and $\mu(E_{\text{CS}}, \vec{r})$ in Eq. (2) are usually unknown and cannot be neglected, which makes the CSCT reconstruction of ρ_e nonlinear. In order to convert this nonlinear problem into a linear one, attenuation CT scan of the phantom becomes necessary. For a Thomson scattering X-ray source, the attenuation CT scan of the sample at arbitrary X-ray energies can be easily achieved since the energy of this type of light source is continuously tunable. To determine the attenuation of the phantom for Compton scattered X-rays, another attenuation CT scan was simulated just by reducing the X-ray energy to 53.70 keV. In this MC simulation, 180 projections were acquired at a sampling interval of 1° and other parameters were unchanged. The attenuation CT of the phantom at both 60 keV and 53.70 keV were reconstructed using the filter back projection (FBP) algorithm where 180 projections at a sampling interval of 1° were used.

4 Results and Discussions

In order to examine the efficiency improvement of Compton scattering by using linearly polarized X-rays, three typical X-ray polarizations, i.e., vertical polariza-

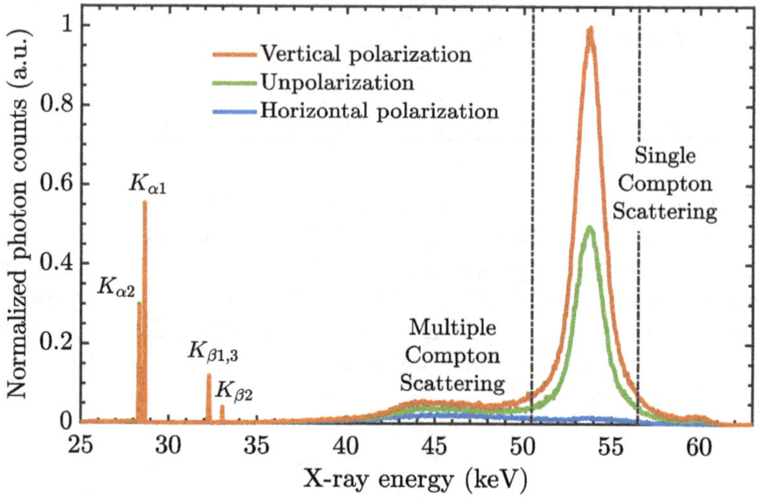

Fig. 3 Comparison of the Compton scattered X-ray spectra detected by the PCD in three typical polarizations

tion (y-axis), horizontal polarization (x-axis), and unpolarization, were used for comparison in the MC simulation. The Compton scattered X-ray spectra detected by the PCD with an ideal energy resolution in the three polarizations are shown in Fig. 3. The spectrum data were obtained by summing all pixels of the PCD after a full CSCT scan. The intensities of K lines of I are the same in the three spectra since the polarization of incident X-rays does not affect the X-ray fluorescence process. Compared with the unpolarization case, the Compton scattering signal in the vertical polarization case is improved approximately 2 times, which agrees with the theoretical prediction in Sect. 2 (see Fig. 3). In the case of horizontal polarization, the Compton scattering signal is completely compressed, which has an important application in the XFCT [22].

Due to the Doppler broadening of Compton scattering caused by the momenta of bound electrons and the influence of multiple Compton scattering, the measured single Compton scattering spectra will be broadened, as shown in Fig. 3. To reduce the influence of the spectrum broadening of single Compton scattering on the CSCT reconstruction, energy thresholds (50.5 keV and 56.5 keV for the low and high energy thresholds, respectively, black dash-dotted lines in Fig. 3) in the spectrum are selected, within which the area of the spectrum is calculated as the effective Compton scattering signal. Using the effective Compton scattering signals detected by the PCD in the vertical polarization case, the electron density distribution of the phantom was reconstructed. The reconstruction results with and without attenuation correction of the phantom are shown in Fig. 4. Also shown in Fig. 4 are the attenuation CT results of the phantom at 60 keV and 53.70 keV. The attenuation terms in Eq. (2) and the system matrix $\mathbf{A}$ in Eq. (4) are calculated using the data in Figs. 4c and d. Compared with the reconstruction result in Fig. 4a, the image

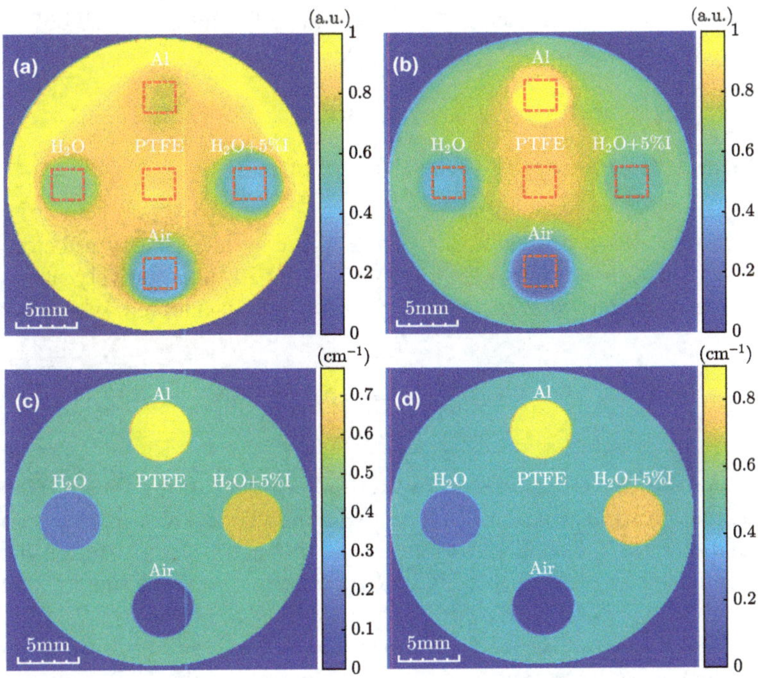

Fig. 4 Electron density maps reconstructed (**a**) without and (**b**) with attenuation correction and the attenuation CT of the phantom at (**c**) 60 keV and (**d**) 53.70 keV. The red dash-dotted squares in (**a**) and (**b**) are ROIs chosen for the quantitative analysis of the reconstruction results

quality of CSCT in Fig. 4b reconstructed with attenuation correction is significantly improved. In Fig. 4a, all of the four inner cylinders of the phantom, compared with the PTFE, exhibit a negative contrast, which is not consistent with the theoretical prediction since the electron density of Al is higher than that of PTFE. After attenuation correction, the Al, as predicted, exhibits a positive contrast compared with the PTFE, as shown in Fig. 4b. In the attenuation CT results, both the Al and the 5% I solution show a positive contrast compared with the PTFE, while the two other inner cylinders of the phantom exhibit a negative contrast compared with the PTFE, as shown in Figs. 4c and d.

For quantitative analysis of the reconstructed electron density distribution, five regions of interest (ROIs) highlighted by red dash-dotted squares in Figs. 4a and b are taken into consideration. The H_2O is used to calibrate the reconstructed pixel values of the CSCT image, so that the relative value of the electron density in Figs. 4a and b can be expressed by its absolute value. The theoretical electron density is defined as $\rho(Z/A)N_A$, with ρ, Z, A, and N_A representing the mass density, atomic number, atomic weight, and Avogadro's constant, respectively. A comparison between the reconstructed electron density of the phantom after attenuation correction and the theoretical result calculated based on the data from the National Institute of Standards and Technology (NIST) [26] shows that the relative deviations for the reconstructed 5% I solution, PTFE, and Al are 18.06%, 3.57%, and 6.34%, respectively, leading to a mean value of 9.32%.

4.1 The Influence of Phantom Attenuation

The CSCT reconstruction result of the phantom can be seriously affected by the phantom attenuation, as shown in Figs. 4a and b. To quantitatively analyze this influence, the parameter contrast-to-noise ratio (CNR) is adopted, which is defined as

$$\mathrm{CNR} = \frac{S - S_{\mathrm{BG}}}{\sigma_{\mathrm{BG}}}, \tag{5}$$

where S and S_{BG} are the means of the ROI pixel values in the inner cylinder and PTFE regions, respectively, and σ_{BG} is the standard deviation in the same ROI region of the PTFE. The CNR results of the phantom are illustrated in Fig. 5. After attenuation correction, the CNR of Al becomes positive and its magnitude is improved from 7.67 to 13.10. For H_2O and air, the CNR values with attenuation correction are improved 2.1 and 1.6 times than that of the corresponding results without attenuation correction, respectively. Meanwhile, the CNR value of 5% I solution is reduced 1.3 times after attenuation correction.

Due to the influence of phantom attenuation, projections at opposing angles in a quasi-parallel imaging geometry will provide additional information of the phantom. To demonstrate this conclusion, a CSCT of the phantom is reconstructed using 180° angular sampling while keeping the same photon statistics as the case of 360° angular sampling. The reconstruction result shows that the artifacts are so strong that the Al cylinder cannot be accurately reconstructed. Therefore, a 360° sampling becomes necessary when the phantom attenuation cannot be neglected.

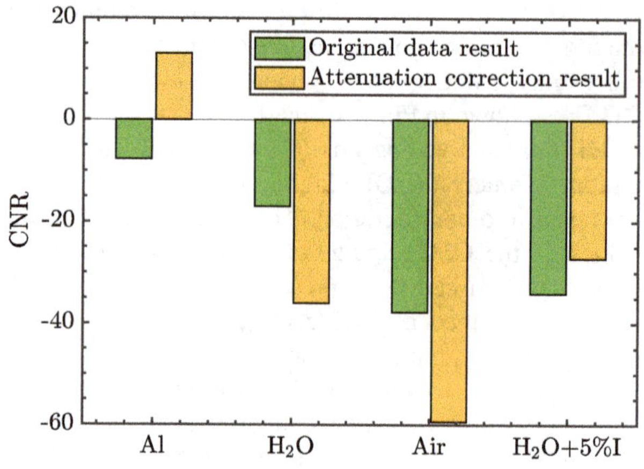

Fig. 5 Comparison of the CNRs in CSCT reconstructed with and without attenuation correction

Fig. 6 Relation between the relative error R_e and the sampling interval at different PCD detector samplings

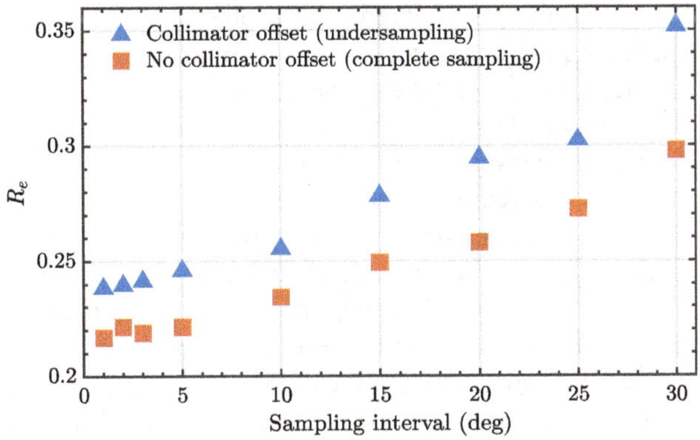

4.2 The Influence of Sampling Interval

The sampling interval has a significant influence on the reconstruction quality of CSCT and the imaging time. To demonstrate this influence, the phantom is reconstructed at different sampling intervals over 360°. In the CSCT reconstruction, the attenuation data of the phantom in Figs. 4c and d are used here and below, neglecting the errors introduced by the attenuation CT reconstruction due to the sparse sampling. The reconstruction result shows that the sampling interval plays a significant role in the accurate CSCT reconstruction. With the increase of the sampling interval, the reconstruction quality of CSCT is degraded by the artifacts, especially at large sampling intervals.

To quantitatively analyze the influence of sampling interval on the reconstruction quality of CSCT, the relative error R_e is calculated, which is defined as $R_e = \sqrt{\frac{\sum_{j=1}^{N}(\rho_{e,j}-\rho^*_{e,j})^2}{N}}$ with $\rho_{e,j}$, $\rho^*_{e,j}$ and N representing the reconstructed electron density of the jth pixel, its corresponding theoretical value, and the number of pixels in the reconstruction region, respectively. The relation between R_e and the sampling interval is illustrated in Fig. 6 (blue data). Note that R_e is a monotonically increasing function of the sampling interval. When the sampling interval is less than 5°, the increase of R_e is slow and the influence of the sampling interval on the reconstruction quality is trivial. However, if the sampling interval exceeds 10°, the increase of R_e becomes rapid and the reconstruction quality gets worse seriously. In order to reduce the imaging time without significantly losing the reconstruction quality, a 5° sampling interval is a better choice.

4.3 The Influence of Detector Undersampling

In the imaging geometry Fig. 2, half of the PCD is shielded by the parallel hole collimator and cannot detect the Compton scattered X-rays, which will lead to the undersampling problem of the phantom. To evaluate the influence of detector undersampling on the reconstruction quality of CSCT, a CSCT scan of the phantom was simulated using a complete detector sampling at the same geometry parameters. The complete detector sampling is realized by moving the parallel hole collimator to a 0.2 mm position offset along the z-axis in each projection. Combining the CSCT projections at two position offsets of the collimator, CSCT of the phantom with a complete detector sampling can be reconstructed. In order to keep the same photon statistics as the case of detector undersampling, the incident X-ray photon number in each collimator offset per projection is reduced to 1.0×10^9. In this case, the dependence of the relative error R_e on the sampling interval is depicted in Fig. 6 (red data). Compared with the case of detector undersampling, the relative error R_e is reduced in the case of complete detector sampling. Therefore, a complete detector sampling helps to improve the reconstruction quality of CSCT at the expense of a movement of the collimator. However, the movement of the collimator will cause other problems, such as the misregistration of projections and the increase of imaging time. This can be resolved by arranging another detection system in the opposite direction of the phantom with a 0.2 mm position offset of the collimator.

4.4 The Minimum X-ray Photons Requirement per Projection

In order to reconstruct the CSCT of the phantom accurately, enough incident X-ray photons are needed since the Compton scattering cross section described by Eq. (1) is relatively low (less than the total Thomson scattering cross section σ_T). However, more incident X-ray photons mean a longer imaging time and a higher radiation dose. To optimize the imaging time and radiation dose, it is necessary to calculate the requirement of minimum X-ray photons per projection. Simulations using different incident X-ray photons per projection were carried out with the sampling strategy unchanged (i.e., 360° sampling at 1° intervals, detector undersampling). The result shows that the reconstruction quality of CSCT deteriorates with the reduction of the incident X-ray photons. To quantitatively analyze this influence, the relative error R_e is also evaluated. When the incident X-ray photon number per projection is more than 8×10^8, the value of R_e varies slowly with the incident X-ray photon number and the four inner cylinders can be identified in the reconstructed CSCT image. However, when the incident X-ray photon number per projection is less than 8×10^8, R_e increases rapidly with the reduction of the incident X-ray photon number and the 5% I solution cylinder cannot be discriminated from the background. Therefore, 8×10^8 is the minimum incident X-ray photon number required per projection to accurately reconstruct the CSCT of the phantom.

4.5 Estimation of Imaging Time

Based on the analysis in Sect. 4.4, 8×10^8 incident X-ray photons per projection are required for the accurate CSCT reconstruction of the phantom in our imaging geometry. Using the similar imaging geometry, beam parameters, and sampling strategy, we have already calculated that the total photon yield requirement for a whole CT scan, when the incident X-ray photon number per projection is 5×10^8, is 2.0×10^{15} in our previous work [22]. Hence, the total photon yield requirement for a whole CSCT scan in the present work is 3.2×10^{15} [i.e., $(8\times10^8)/(5\times10^8)\times(2.0\times10^{15})$].

For a typical low-repetition Thomson scattering light source, e.g., the Tsinghua Thomson scattering X-ray source (TTX), its photon yield is 2×10^8 phs/s (2×10^7 phs/pulse, 10 Hz) [27, 28]. To finish a whole CSCT scan using the TTX in our imaging geometry, it would take $\sim$ 185 days. Even though an increased sampling interval of 5° can reduce the imaging time to $\sim$ 37 days without significantly affecting the reconstruction quality according to the result in Sect. 4.2, such a long imaging time cannot be acceptable in practical CSCT applications. For a typical high-repetition Thomson scattering light source, e.g., the Munich Compact Light Source (MuCLS), its photon yield is 10^{10} phs/s [29]. Using this type of light source and the optimized sampling strategy, the data acquisition time of CSCT can be reduced to $\sim$ 18 hours, which is acceptable in a practical CSCT application. Meanwhile, many high-flux Thomson scattering light sources, based on the high-repetition design technology, have been proposed and are under development at present, aiming to reach a photon yield as high as 10^{13} phs/s [30, 31]. Using these high-flux light sources, the imaging time of a whole CSCT scan can be further reduced to the minute level, which is very attractive for in vivo imaging.

4.6 Analysis of the Potential Pulse Pile-Up Effect

For a Thomson scattering light source, the typical X-ray pulse length (fs-ps) is far below the time resolution of a PCD. Hence, it is necessary to analyze the potential pulse pile-up problem in our imaging geometry. To avoid the pulse pile-up, the photon number of Compton scattered X-rays detected per pixel on the PCD cannot exceed 1 in one incident X-ray pulse, which requires that the photon yield Y per pulse should meet the following condition:

$$\begin{aligned}\frac{YF}{\pi\left[R_1\tan(\frac{\theta_c}{2})\right]^2}\frac{d\sigma_{KN,LP}}{d\Omega}|_{\theta=90^\circ,\phi=90^\circ}\\ \times\rho_e(P)V_c\frac{L^2}{4\pi d_p^2}\leq 1,\end{aligned} \tag{6}$$

where $F = 0.023$, calculated in our previous work [22], is the fraction of X-ray photons confined in the cone of an angle $\theta_c/2$ ($\theta_c = 2.56$ mrad, confined by the collimator at the end of the vacuum tube in our imaging geometry [22]), $\frac{d\sigma_{KN,LP}}{d\Omega}|_{\theta=90^\circ,\phi=90^\circ} = 6.4 \times 10^{-24}$ mm^2 is the differential cross section of Compton scattering when $\theta = 90^\circ$ and $\phi = 90^\circ$, $V_c = 1.0$ mm^3 (i.e., $L^2 \times 25$ mm, for conservative estimate) is the phantom volume confined by a parallel hole of the collimator, and $d_p \simeq (d_1 + d_2) = 47.5$ mm is the distance between the point P and the PCD end of the collimator. For conservative estimate, the attenuation of the phantom in Eq. (6) is neglected, and the phantom is assumed to be fully filled with Al, i.e., $\rho_e(P) = \rho_{e,Al} = 7.8 \times 10^{20}$ mm^{-3} calculated based on the NIST data [26]. Other parameters in Eq. (6) are the same as before. Taking these parameters into Eq. (6), we can obtain the photon yield $Y \leq 4.6 \times 10^{12}$ phs/pulse.

For a low-repetition Thomson scattering light source, e.g., the TTX, the pulse photon yield is below the above limitation and the pulse pile-up cannot occur for Compton scattered X-ray photons caused by adjacent incident X-ray pulses due to its low repetition; hence, the pulse pile-up problem can be neglected in this case. For a high-repetition Thomson scattering light source, the pulse pile-up for Compton scattered X-ray photons caused by adjacent incident X-ray pulses may occur due to its high repetition [usually in continuously wave (CW) mode]. To evaluate this effect, the photon flux of 10^{13} phs/s is taken as an example. Based on the above analysis, the photon flux of Compton scattered X-rays on the PCD is ~2.2 phs/pixel/s [i.e., $10^{13}/(4.6\times10^{12})$]. At present, the count rate for a practical PCD can reach 0.01–10 Mcounts/pixel/s [24], which is far above the photon flux of Compton scattered X-rays. Therefore, the pulse pile-up problem can also be neglected for a high-repetition Thomson scattering X-ray source with photon flux of 10^{13} phs/s.

4.7 Evaluation of Radiation Dose

For CSCT imaging of small animals or human beings, the radiation dose is a key factor that needs to be taken seriously. In this section, we will estimate the typical radiation dose in our imaging parameters. Considering a water cylindrical phantom with a radius of $r = 1.25$ cm and radiated by a parallel monochromatic X-ray beam that is perpendicular to the axis of the phantom and covers the phantom, the theoretical radiation absorbed dose D of the phantom can be calculated as [32]

$$D = \frac{2}{\pi r \rho_w}(1 - g)E_0\phi_0 T \mu_w(E_0)/\mu_{w,en}(E_0), \tag{7}$$

where ρ_w is the density of water, ϕ_0 is the fluence rate of the incident X-ray beam in front of the phantom, T is the exposure time, $\mu_w(E_0)$ and $\mu_{w,en}(E_0)$ are the linear attenuation coefficient and the linear energy absorption coefficient of water at X-ray energy $E_0 = 60$ keV, respectively, and the factor $g =$

$\int_0^{\frac{\pi}{2}} \exp\left[-2r\mu_{w,en}(E_0)\cos\theta\right]\cos\theta d\theta$. In our imaging geometry, the incident X-ray fluence $\phi_0 T$ can be calculated using the optimized sampling strategy (i.e., 360° at 5° intervals) and incident X-ray photon number (8×10^8 per projection) and its value is 1.96×10^{16} phs/m^2. The values of $\mu_w(E_0)$ and $\mu_{w,en}(E_0)$ can be obtained from the NIST data, and their values are 20.59 m^{-1} and 3.19 m^{-1} [26], respectively. Based on these parameters, the factor g can be numerically calculated and the result is 0.9394. Thus, the radiation absorbed dose D is 3.7 Gy, which is about half of LD_{50} (50% lethal dose, ~7 Gy [33]) for mice.

5 Conclusion

The excellent X-ray beam qualities of Thomson scattering light sources provide a promising prospect for CSCT. In this chapter, a high-efficient CSCT based on a Thomson scattering light source is proposed by using linearly polarized X-rays, and its feasibility has been studied by Monte Carlo simulations. Compared with CSCT using unpolarized X-rays, almost 2 times efficiency improvement has been verified by using X-rays with vertical polarization. The influence of phantom attenuation, sampling strategy, and detector undersampling on the reconstruction quality of CSCT has been discussed. The practical problems for CSCT based on this type of light source, concerning the imaging time, potential pulse pile-up, and radiation dose have been analyzed. Our results show that a CSCT scan can be realized at the minute level without causing any pulse pile-up problem by using a high-flux Thomson scattering X-ray source (e.g., 10^{13} phs/s), and the radiation dose is acceptable for small animals (e.g., mice), which is very promising for *in vivo* imaging.

Acknowledgments This work was supported by the National Natural Science Foundation of China (Grant Nos. 11905011 and 12375157).

References

1. G. Harding, E. Harding, Appl. Radiat. Isotop. **68**, 993 (2010)
2. A. Lange, M.P. Hentschel, A. Kupsch, G.-R. Jaenisch, N. Kardjilov, C. Tötzke, H. Markötter, A. Hilger, I. Manke, in *11th European Conference on Non-Destructive Testing (ECNDT 2014)* (2014) pp. 6–10
3. R. Guzzardi, G. Licitra, Crit. Rev. Biomed. Eng. **15**, 237 (1987)
4. B.L. Evans, J. Martin, L. Burggraf, M. Roggemann, IEEE Trans. Nucl. Sci. **45**, 950 (1998)
5. P. Duvauchelle, G. Peix, D. Babot, NDT E Int. **33**, 23 (2000)
6. A. Brunetti, R. Cesareo, B. Golosio, P. Luciano, A. Ruggero, Nucl. Instrum. Methods Phys. Res. Sect. B Beam Interact. Mater. Atoms **196**, 161 (2002)
7. V. Gorshkov, M. Kroening, Y. Anosov, O. Dorjgochoo, Nondestruct. Test. Evaluat. **20**, 147 (2005)
8. O. Adejumo, F. Balogun, G. Egbedokun, J. Sustain. Dev. Environ. Protect. **1**, 73 (2011)
9. G. Harding, H. Strecker, R. Tischler, Philips Tech. Rev. **41**, 46 (1983)

10. F. Farmer M.P. Collins, Phys. Med. Biol. **19**, 808 (1974)
11. B.L. Evans, J.B. Martin, L.W. Burggraf, M. Roggemann, T. Hangartner, Nucl. Instrum. Methods Phys. Res. Sect. A Accelerat. Spectromet. Detect. Assoc. Equip. **480**, 797 (2002)
12. S.J. Norton, J. Appl. Phys. **76**, 2007 (1994)
13. M. Nguyen, T. Truong, Inverse Probl. **26**, 065005 (2010)
14. T. Truong, M. Nguyen, Inverse Probl. **27**, 125001 (2011)
15. T. Truong, M. Nguyen, Inverse Probl. **35**, 054005 (2019)
16. J.W. Webber, W.R. Lionheart, Inverse Probl. **34**, 084001 (2018)
17. J. Webber, E.L. Miller, Inverse Probl. **36**, 025007 (2020)
18. S.J. Norton, Inverse Probl. Sci. Eng. **28**, 894 (2020)
19. B. Golosio, A. Simionovici, A. Somogyi, L. Lemelle, M. Chukalina, A. Brunetti, J. Appl. Phys. **94**, 145 (2003)
20. Y. Du, H. Chen, H. Zhang, Q. Gao, Q. Tian, Z. Chi, Z. Zhang, H. Zha, J. Shi, L. Yan, et al., High Power Laser Particle Beams **34**, 104010 (2022)
21. S. Agostinelli, J. Allison, K.a. Amako, J. Apostolakis, H. Araujo, P. Arce, M. Asai, D. Axen, S. Banerjee, G. Barrand, et al., Nucl. Instrum. Methods Phys. Res. Sect. A Accelerat. Spectromet. Detect. Assoc. Equip. **506**, 250 (2003)
22. Z. Chi, Y. Du, W. Huang, C. Tang, J. Synchrotron Radiat. **27**, 737 (2020)
23. Z. Chi, in *Advanced X-ray Detector Technologies* (Springer, 2022) pp. 195–214
24. K. Taguchi, J.S. Iwanczyk, Med. Phys. **40** (2013)
25. L.A. Shepp, Y. Vardi, IEEE Trans. Med. Imag. **1**, 113 (1982)
26. See https://www.nist.gov/pml/x-ray-mass-attenuation-coefficients for the basic information of different materials (last accessed July 10, 2023)
27. Z. Chi, L. Yan, Y. Du, Z. Zhang, W. Huang, H. Chen, C. Tang, Nucl. Instrum. Methods Phys. Res. Sect. B Beam Interact. Mater. Atoms **402**, 364 (2017)
28. Z. Chi, Y. Du, L. Yan, D. Wang, H. Zhang, W. Huang, C. Tang, J. Synchrotron Radiat. **25** (2018)
29. E. Eggl, M. Dierolf, K. Achterhold, C. Jud, B. Günther, E. Braig, B. Gleich, F. Pfeiffer, J. Synchrotron Radiat. **23**, 1137 (2016)
30. M. Jacquet, Nucl. Instrum. Methods Phys. Res. Sect. B Beam Interact. Mater. Atoms **331**, 1 (2014)
31. K. Deitrick, G. Krafft, B. Terzić, J. Delayen, Phys. Rev. Accelerat. Beams **21**, 080703 (2018)
32. T. Yuasa, E. Hashimoto, A. Maksimenko, H. Sugiyama, Y. Arai, D. Shimao, S. Ichihara, M. Ando, Nucl. Instrum. Methods Phys. Res. Sect. A Accelerat. Spectromet. Detect. Assoc. Equip. **591**, 546 (2008)
33. E.J. Hall, A.J. Giaccia, et al., *Radiobiology for the Radiologist*, vol. 6 (Lippincott Williams & Wilkins, 2006)

Machine Learning for Contaminant Detection in X-Ray Spectral Photon Counting

Laszlo Zalavari, Jericho O'Connell, Devon Richtsmeier, Joanna Nguyen, Krzysztof Iniewski, and Magdalena Bazalova-Carter

1 Motivation

Detection of contaminants in food quality assurance is one of the most important tasks of the food industry [1, 2]. Among the existing methods, X-ray systems are most commonly used because these systems provide superior resolution and enable nondestructive testing. The ability of X-rays to traverse through matter and reveal hidden contaminants has also led to their extensive use in baggage scanning. However, X-ray inspection systems are known to have difficulties in differentiating low-density, often organic, materials. This is due to the fact that existing X-ray inspection systems utilize conventional energy-integrating detectors (EIDs) that are not capable of recording the energy of the incoming X-ray. This additional information is available with state-of-the-art multicolor detector technology and

L. Zalavari
Physics & Astronomy Department, University of Victoria, Victoria, BC, Canada

Stanford Health Care, Stanford, CA, USA

J. O'Connell
Physics & Astronomy Department, University of Victoria, Victoria, BC, Canada

Dana-Fraber Cancer Instituteand and Harvard Medical School, Boston, MA, USA

D. Richtsmeier · M. Bazalova-Carter
Physics & Astronomy Department, University of Victoria, Victoria, BC, Canada

J. Nguyen
Physics & Astronomy Department, University of Victoria, Victoria, BC, Canada

GE HealthCare, Waukesha, WI, USA

K. Iniewski (✉)
Redlen Technologies, Saanichton, BC, Canada

K. Iniewski, H. Gadey (eds.), *Emerging Radiation Detection*,
https://doi.org/10.1007/978-3-031-63897-8_11

could help with identification of low-density materials, which is a crucial aspect for food inspection [3].

The topic of this chapter is to present food inspection software solutions [4] that include image segmentation and machine learning that can be used to enhance detection performance. The X-ray spectral photon-counting hardware detection technology and test results obtained using the bench top test system at UViC XCITE laboratory (http://web.uvic.ca/~bazalova/) have been presented in the literature [5–9].

2 Deep Learning for X-Ray Testing

2.1 Artificial Intelligence

From movie recommendations on Netflix to conversations with ChatGPT, we hear about artificial intelligence (AI) on many occasions. Movie selection is not a big deal, but ChatGPT might at some point in the future become an incredible expert medical system, providing rapid diagnostic and therapeutic information to physicians at the point of care. It can warn about interactions and side effects, remind physicians of best practices, and generally improve the care delivered.

Big tech companies like Google and Facebook invest billions of dollars into AI developments, and countries like the USA and China vie for worldwide AI supremacy. Computer science is pushing forward the ability of computers to handle complex calculations needed to help discover new medical drugs, identify trends and risks in financial modeling, handle large database searches, and perform massive calculations needed in many fields of science. From writing new books to fixing computer programs, AI techniques are expected to proliferate in all aspects of our lives eventually. How did we get here?

Let us start our little AI journey with chess. Chess is a well-known board game between two players. The current form of the game emerged in Spain and the rest of Southern Europe during the second half of the fifteenth century after evolving from chaturanga, a similar but much older game of Indian origin. Unlike bridge or poker, chess involves no hidden information. It is played on a chessboard with 64 squares arranged in an eight-by-eight grid with 16 pieces by each player; see Fig. 1. The number of possible moves in chess is large, but not astronomically large, on average every player might have 30 reasonable moves to choose from and the average game lasts 40 moves. The complexity of the game is high, with millions of people playing it and thousands of books written about it.

It was only natural that one of the goals of early computer scientists was to create a chess-playing machine. In 1997, IBM Deep Blue became the first computer to beat the reigning World Champion Garry Kasparov. The game was close, two wins for Deep Blue, one for Kasparov, and three draws. The match lasted several days and received massive media coverage around the world. It was the classic plot line of man vs. machine. Once machine won that game, the game was over; machines will always be better than humans forever. Today's chess engines are significantly

Fig. 1 Chess (left) and Go (right) board games

stronger than the best human players and have deeply influenced the development of chess theory.

The win for IBM was impressive but the Deep Blue program was quite limited in its capability; it could just play chess, nothing else. And its "intelligence" was quite narrow, critics said at the time as the machine was simply calculating all possible move combinations. Deep Blue was supported by 480 special VLSI (very large scale of integration) chips that could see 12 moves ahead while Kasparov arguably could see about 10 moves ahead. Deep Blue could search and rank possible outcomes of 12 moves ahead, an advantage that piled on and led to the victory. That strategy was fine for chess but would never work for the game of Go.

Go is an abstract strategy board game for two players in which the aim is to surround more territory than the opponent. The game was invented in China more than 2500 years ago and is believed to be the oldest board game continuously played to the present day. Two players, using either white or black stones, take turns placing their stones on a board; see Fig. 1. The goal is to surround and capture their opponent's stones or strategically create spaces of territory. Once all possible moves have been played, both the stones on the board and the empty points are tallied. The highest number wins.

Despite decades of work, the strongest Go computer programs could only play at the level of human amateurs. Standard AI methods, which test all possible moves and positions using a search tree, could not handle the sheer number of possible Go moves or evaluate the strength of each possible board position. As simple as the rules may seem, Go is profoundly complex. There are an astonishing 10 to the power of 170 possible board configurations—more than the number of atoms in the known universe. This makes the game of Go a googol (10^{10}) times more complex than chess. Go is known as the most challenging classical game for artificial intelligence because of its complexity.

AlphaGo was the first computer program to defeat a Go world champion; Lee Sedol, arguably the strongest Go player in history, was defeated in 2016. AlphaGo is a computer program that combines advanced search trees with deep neural networks. These neural networks take a description of the Go board as an input and process it through a number of different network layers containing millions of neuron-like connections. These layers are hidden deep in the network; hence the term deep learning is used for such software. One neural network, the "policy network," selects the next move to play. The other neural network, the "value network," predicts the winner of the game. At the beginning, AlphaGo was introduced to numerous amateur games to help it develop an understanding of reasonable human play. Then the developers had it played against different versions of itself thousands of times, each time learning from its mistakes. Over time, AlphaGo improved and became increasingly stronger and better at learning and decision-making. This process is known as reinforcement learning. Once AlphaGo conquered the game, it was expected to move onto something else more productive and beneficial to society at large. Which it did and we can now observe the enormous progress that has been made in this field since 2016. It is anticipated that within 5 years AI will completely change the world we live in.

2.2 Machine Learning

As the above short chess/Go story illustrates, artificial intelligence has changed enormously over the last 25 years since Deep Blue victory in 1997. Various terms like AI, deep learning, machine learning, or unsupervised training are being used in literature and popular science, so it is important to explain these terms in more detail.

Machine learning approaches use training samples to discover patterns and ultimately achieve an accurate prediction of future data or trends. With the development of machine learning approaches and the increasing scale and difficulty of data analysis, traditional machine learning methods have often become unsuitable. As a result, deep learning methods have been developed, because they have more complex architectures and improved data analysis capacity. The relationship between artificial intelligence, machine learning, and deep learning can be understood as three circles, deep learning being the smallest and fully contained within machine learning; this is in turn contained within the artificial intelligence circle; see Fig. 2. In the following subsections, the algorithms and applications based on machine learning and deep learning methods will be introduced.

Traditional machine learning involves manually extracting features on small sample sets, to balance the validity of the learning results with the interpretability of the learning model and to provide a framework for solving the learning problem when limited samples are available. Traditional machine learning can be subdivided into supervised learning, unsupervised learning, and reinforcement learning, based on the training method, depending on whether the training data have labels or

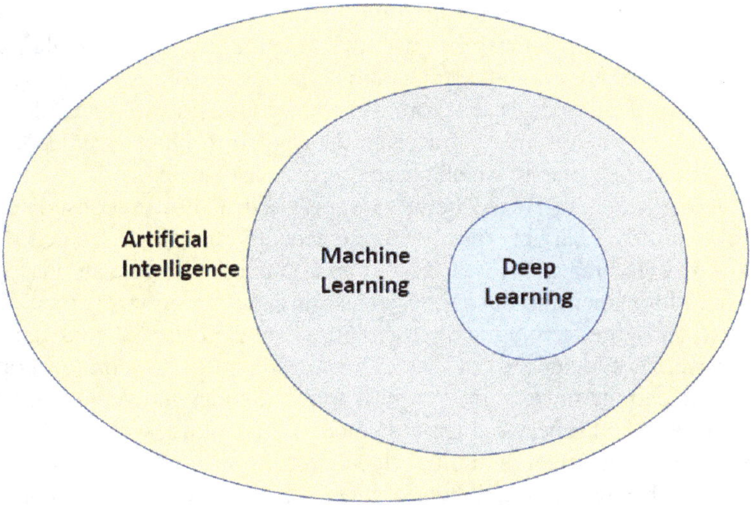

Fig. 2 Schematic illustration of the relationship between deep learning, machine learning, and artificial intelligence

not. One of the important theoretical foundations of traditional machine learning methods is statistics, and the main analysis includes pattern classification, regression analysis, probability density estimation, or other analyses.

2.3 *Supervised vs. Unsupervised Learning*

Let us briefly discuss the difference between supervised learning and unsupervised learning approaches. Supervised learning is a machine learning approach that is defined by its use of labeled datasets. These datasets are designed to train or "supervise" algorithms into classifying data or predicting outcomes accurately. Using labeled inputs and outputs, the model can measure its own accuracy and learn over time. Supervised learning can be separated into two types of problems when data mining: classification and regression. Classification problems use an algorithm to accurately assign test data into specific categories, such as separating apples from oranges. Regression is another type of supervised learning method that uses an algorithm to understand the relationship between dependent and independent variables. A simple example of regression algorithms is linear regression.

Unsupervised learning uses machine learning algorithms to analyze and cluster unlabeled datasets. These algorithms discover hidden patterns in data without the need for human intervention and hence their name. Unsupervised learning models are used for two main tasks, clustering and association. Clustering is a data mining technique for grouping unlabeled data based on their similarities or differences.

For example, K-means clustering algorithms assign similar data points into groups, where the K-value represents the size of the grouping and granularity. This technique is helpful for market segmentation and image compression. Association is another type of unsupervised learning method that uses different rules to find relationships between variables in a given dataset. An example would be Netflix suggesting a certain movie to watch based on previous viewing.

What is the main difference between supervised and unsupervised learning? The main distinction between the two approaches is the use of labeled datasets, and supervised learning uses labeled input and output data, while an unsupervised learning algorithm does not. In supervised learning, the algorithm "learns" from the training dataset by iteratively making predictions on the data and adjusting for the correct answer. While supervised learning models tend to be more accurate than unsupervised learning models, they require upfront human intervention to label the data appropriately. Unsupervised learning models, in contrast, work on their own to discover the inherent structure of unlabeled data.

There are other differences between supervised and unsupervised approaches. In supervised learning, the goal is to predict outcomes for new data. You know up front the type of results to expect. With an unsupervised learning algorithm, the goal is to get insights from large volumes of new data. Machine learning itself determines what is different or interesting from the dataset. Supervised learning models are ideal for processes like spam detection or inflation predictions. In contrast, unsupervised learning is a great fit for anomaly detection and medical imaging, so it is a more natural choice for food inspection. Finally, supervised learning is a relatively simple method for machine learning, typically calculated through the use of standard programs written in the Python programming language. In unsupervised learning, you need powerful tools for working with large amounts of unclassified data. Unsupervised learning models are inherently computationally complex because they need a large training set to produce intended outcomes.

There are serious risks and drawbacks using both approaches. Supervised learning models can be time-consuming to train, and the labels for input and output variables require expertise. Meanwhile, unsupervised learning methods can have wildly inaccurate results unless you have human intervention to validate the output variables.

Classifying large datasets can be a challenge in supervised learning, but the results are highly accurate and trustworthy. In contrast, unsupervised learning can handle large volumes of data in real time. But there is a lack of transparency into how data is clustered and a higher risk of inaccurate results. The only reliable way to select supervised vs. unsupervised training programs is to try to use both. Examples of supervised and unsupervised training experiments for NDT food inspection applications are presented in the next sections.

3 X-Ray Detectors for NDT Applications

3.1 Food Inspection Challenges

X-ray imaging is a widely used technique for nondestructive, inline food inspection. These products are typically processed on a conveyor belt on the factory floor. One of the important applications of X-rays is the automatic detection of foreign object inclusions that might appear in food products. Examples of such objects are bone fragments in a chicken fillet, plastic debris, and bones in fish. The standard technique is to use energy-integrated detectors (EIDs), but there are several limitations imposed by this technology.

The reasons that in-line foreign object detection in food samples on a conveyor belt possesses some major challenges are as follows. First, high-throughput X-ray acquisition leads to a significant noise level that greatly impacts the detection process. Noise reduction methods have to be applied to reduce this effect. Second, typical foreign objects in the food industry have similar X-ray attenuation properties to that of food samples, resulting in low contrast between foreign object and main object. Contrast-enhancing methods have to be applied to mitigate this effect. Third, a variation in the thickness of the main object causes ambiguities when detecting a foreign inclusion in a sample. For example, a thin sample of a chicken fillet with a bone fragment might have similar attenuation to a thicker sample of a fillet object without a bone.

To resolve the challenges related to NDT food inspection, better detector technology is required. The next section describes a spectral photon-counting technology that is expected to dramatically improve performance of X-ray detection systems.

3.2 Photon-Counting Detectors

Photon-counting detectors (PCDs) are direct converters where the X-ray photon energy is directly translated into electronic signal by the semiconductor sensor. The height of this signal is proportional to the photon energy. If the signal duration is short enough, the PCD is able to detect almost every photon impinging on the sensor signal. In order to be able to absorb X-ray photons effectively, High-Z materials, such as CdTe and CdZnTe (CZT), are used for sensor fabrication.

Photon-counting detectors (PCDs) offer several advantages over traditional energy-integrated detectors (EIDs) that utilize indirect conversion process of X-ray conversion to visible light by a scintillating material follow by light to electronic signal conversion that uses standard photodiodes. First, PCDs are not subject to detector intrinsic noise, also called electronic noise since their lowest-energy threshold is set above the electronic noise floor. This means that every photon counts, unlike in EID where electronic noise adds to the measured signal and thus becomes relevant in low flux detection. For that reasons, PCDs are frequently referred to as "noise-free" although that is highly inaccurate as PCD like every electronic devise suffers from noise effects; it is just that the level of noise in PCDs is much lower than in EIDs.

The second advantage of PCDs is their ability to resolve photon energy. For example, by binning the photons into two or more energy bins, their signals are energy-selective, and thus, PCD intrinsically offers a spectral imaging capability. Whether to use the spectral information can be determined retrospectively after the NDT scan and the readout systems can be configured appropriately depending on the final application.

The third advantage of PCDs is their spatial resolution as they are built using small sensor pixels. The small pixel size also helps avoiding or minimizing pulse pileup effect, that is, counting of two photons that arrive very close in time as one single photon. In contrast in EID, smaller pixel size requires cutting the sensor into smaller pixels and thus creating more dead space around the pixels. PCDs do not require such cutting because the pixel dimensions are determined by the pixel metal electrodes which are defined in lithography process during sensor fabrication.

3.3 PCCT330 Detector and Experimental Setup

The X-ray test results presented in this chapter have been collected using the X-ray detection system installed at the XCITE laboratory at University of Victoria, Victoria, BC, Canada (http://web.uvic.ca/~bazalova/). The detection system uses a PCCT330 detector from Redlen Technologies (www.redlen.com) that is being used to build commercial spectral photon-counting computed tomography (PCCT) scanners. The PCD detector uses a 2-mm-thick CZT sensor that is pixelated into 330 um pixels. The sensor is connected to the high-speed application-specific integrated circuit (ASIC) that counts photons up to 62 Mcps offering PCD counting capability up to 600 Mcps/mm^2 which is more than sufficient for these studies. In fact, in NDT tests described here, the flux is low enough to avoid pileup conditions.

The PCD detector and X-ray source (XRS-160, Comet Technologies, Bern, Switzerland) were both mounted on vertical and horizontal linear motion stages (Newport technologies, Irvine, CA). These stages were oriented perpendicular to each other to allow for easy navigation while imaging the phantom, which was placed on a stage in between the X-ray source and the CZT detector. The X-ray source used was a module XRS-160 from the Comet Technologies. The whole system was controlled by a personal computer (PC) and is shown schematically in Fig. 3.

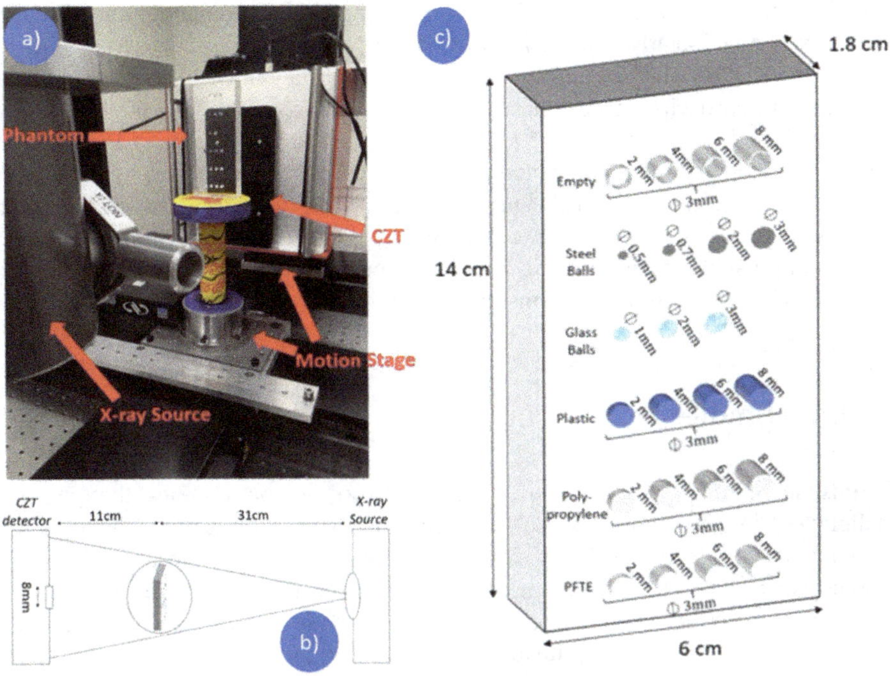

Fig. 3 Experimental XCITE laboratory setup used in the experiments described in this chapter

3.4 Material Discrimination Techniques

The ability of spectral energy photon counting to decompose a material into its constituent elements relies on the energy and element-dependent nature of X-ray attenuation. In the typical energy range of 20–200 keV, the attenuation of X-rays by matter is primarily due to the photoelectric effect (PE) and Compton scattering (CS) processes. This dependence is a monotonic and smoothly varying function of energy for elements without K-edges within the energy range. By modeling the dependence of the photoelectric and Compton interaction processes on material mass density and atomic number, a mass density-atomic number map can be generated and material-specific information obtained. As the X-ray attenuation depends primarily on these two interaction mechanisms, any material's attenuation coefficient can be expressed as a linear combination of the attenuation coefficients of two underlying basis materials.

For these reasons, it was proposed by Alvarez [3] that there is little improvement to spectral analysis when using more than two energies: since the photoelectric component (PE) and the Compton scattering (CS) component of the photon interactions nearly completely describe all attenuation, two components can span the space of all images generated by attenuation. Therefore, two images at different energies can solve for the CS and PE components of the attenuation, and further images will give linearly dependent information.

To exploit availability of multiple energy bin information, we have explored a set of different approaches. In one case study, we have explored using the possibility of fitting based on the intensity [5]. Multienergy images contain more information about the underlying energy dependence of attenuation. Thus, we have hypothesized that a fitting based on more than two energy bins could better describe the underlying physics processes rather than just relying on the CS-PE components. In another approach, we have explored using properties of count maps without relying on underlying physical mechanisms that determine X-ray attenuation properties.

4 Case Studies

In this section, we are presenting the results of various studies that use AI techniques, in particular the use of deep learning networks, to enhance food contamination detection using spectral photon-counting X-ray testing. The case studies mirror hardware development research which is described elsewhere. The first section describes using a deep learning network for supervised learning while the second case study is using unsupervised learning for classification. As a result, the second section is looking at approaches that do not require a user to label the data. Finally, the third section described object detection based on edge data analysis using AI classification methods. The first case study [5] relied purely on simulations, so the ground truth is known, and accuracy of developed techniques is therefore clear. The second [6] and the third case studies use experimental data collected using spectral photon-counting detector system developed by the Redlen Technologies described earlier.

4.1 Case Study: Supervised Image Segmentation [5]

This study presents a novel supervised learning algorithm designed to detect cartilage in projection images of chicken meat, a task considered unachievable through conventional projection radiography. Leveraging the potential of spectral projection imaging, the proposed algorithm aims to identify cartilage by exploiting the higher calcium content, leading to increased absorption of low-energy radiation via the photoelectric effect in comparison to the surrounding background meat. To achieve accurate cartilage detection, this research focuses on optimizing photon-counting energy bins and further proposes the development of a machine learning-based cartilage segmentation workflow, enabling automated cartilage detection in the images. Further, the study aims to determine the benefit of spectral imaging as compared to the simpler method of dual-energy imaging. The combination of spectral projection imaging and advanced machine learning techniques shows

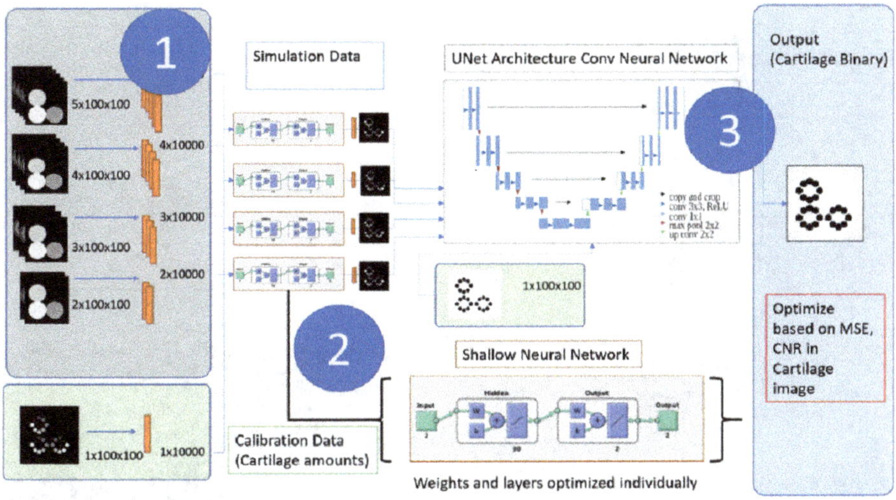

Fig. 4 The complete workflow composed of the three steps: image simulation, multienergy subtraction, and image segmentation/classification

Table 1 Atomic composition of the materials used. Compositions are given as percentage weights

Tissue	H	C	N	O	Ca	P	Na	S	Cl
Cartilage	9.6	9.9	2.2	74.4	0	2.2	0.5	0.9	0.3
Water	8.09	67.22	2.40	19.84	2.32	0	2.32	0	0.13

Cartilage density is 1.1 g/cm^3 compared to 1.05 g/cm^3 for water

promising potential for noninvasive, efficient, and precise cartilage detection in chicken meat projection images, with potential implications for the broader field of food quality assessment.

An overview of the total workflow for cartilage segmentation is outlined in Fig. 4. The first step consists of image simulation while the second step, multienergy subtraction, is used to optimize the acquisition parameters of the photon-counting detector. Sitting on top of this optimization step is the third and final step, that is, image segmentation/classification, which demonstrates the limitations of the higher-level feature classification described below in detail in the original publication [5].

The simulated phantoms in this study were composed of multiple thicknesses of cartilage embedded in different thicknesses of water (Table 1) in this way mimicking the real-life situation of finding cartilage in chicken breast.

The simulation phantoms (Fig. 5) were imaged with a 60 kVp photon beam generated from a tungsten anode filtered by 0.8 mm beryllium window and 0.1 mm copper. The spectrum was generated using SpekCalc an analytical X-ray spectrum generator (http://spekcalc.weebly.com). At the detector location, energy bins for all of the simulation were calculated so as to have equal fluence in each energy bin. This was done for optimal statistics as noise increases quadratically with

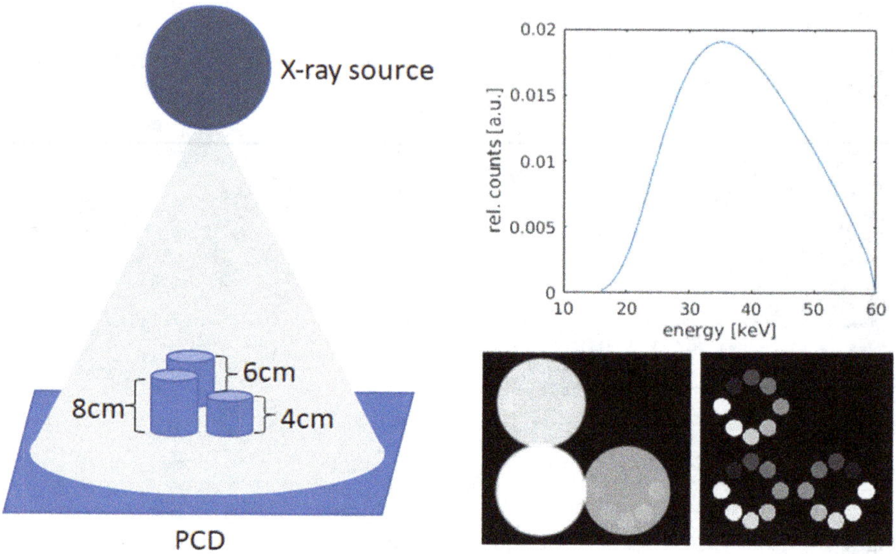

Fig. 5 Simulation phantoms of planar X-ray imaging with PCDs used in this work, shown on the left. Typical X-ray spectrum (kVpp = 60 keV) is shown on the upper right while calibration phantoms are shown on the bottom right

decreasing signal. Image analysis of the simulation data was conducted using MATLAB (The Mathworks, Natick, MA) to map the multidimensional photon-counting detector (PCD) data to estimated thicknesses of cartilage. To expand the well-established method of dual-energy subtraction to multienergy subtraction, two different techniques were employed: a linear method using linear regression and a machine learning method using a multilayer perceptron. The image intensities were extracted from predefined regions of interests (ROIs) encompassing the circular inserts of the calibration phantom, as well as from the body of the phantom and the surrounding air. These intensity values were then compared to the ground truth cartilage depth obtained from the simulation, serving as a reference for accuracy validation.

Material decomposition equations for the nonlinear method were formulated by fitting a linear combination of the PCD bins using a feed-forward neural network [5]. The Levenberg-Marquardt fitting function was utilized with the default set of hyperparameters. To determine the optimal number of hidden layers, we performed 10 trials for each combination of signal-to-noise ratio (SNR), ranging from 5 to 100 layers in intervals of 5. The validation data consisted of cartilage inserts in the validation phantom. The ideal number of layers was selected based on the result with the lowest mean squared error (MSE) derived from the validation data.

In both the linear and nonlinear cases, the same data from the calibration phantom was used. For the linear case, data fitting was accomplished using a least squares method. Conversely, in the nonlinear case, the data was fitted with a feed-forward neural network, introducing the difference between the fitting algorithms.

The linear and nonlinear regressions were evaluated by means of CNR and MSE as a function of depth of cartilage and SNR. The CNR and SNR were defined relative to the background consisting of 4 cm of water. In the subsequent stage of the method, a convolutional neural network (CNN) was employed and trained on the cartilage images' output to enhance the robustness of the detection process. The decision to use CNN stemmed from its advantageous capability to detect geometric structures within images, making it well suited for identifying the characteristic features of cartilage in chicken breast. Unlike simpler multilayer perceptrons, CNNs excel at recognizing intricate patterns and shapes, which is particularly beneficial in this context. The input images for the CNN were generated using randomly generated SW polygons as the background. These polygons were assigned image values derived from the cone-beam simulation images. Poisson noise was added to each region to achieve a desired SNR of 100. The foreground was composed of circular regions of differing depth of combined SW and cartilage, also derived from the cone-beam calibration dataset. Thirty of these simulated images as well as segmented binary images denoting the cartilage were provided to the U-Net for training. A data augmentation algorithm was used to generate training samples from 30 images.

When given the PCD images, the trained U-Net outputs two arrays with values corresponding to each pixel of the image. One array represented the probability of that pixel being cartilage, and another array represented the probability of it being water. The error was investigated using the Keras's categorical error function. The training of the U-Net can be accessed through an online environment on an Internet browser [5].

The accuracy of automated cartilage tissue segmentation for planar X-ray imaging with a PCD was investigated and reported in the original publication [5]. Across all studied SNRs, four energy bins resulted in the lowest MSE in cartilage depth calculation for investigated imaging geometries. Compared to the dual-energy case, the MSE was 19% and 4% lower for the cone-beam and parallel-beam cases, respectively. This implies that there are potential gains in soft tissue segmentation when using a PCD with more than two energy bins. Likewise, a nonlinear regression method should be used instead of least squares fit as this resulted in lower MSE in all cases. Overall, four energy bins and the nonlinear method would result in the highest accuracy in lesion detection for a given imaging dose.

The reduction in MSE for cartilage depth calculation was SNR dependent with the largest changes in MSE at SNR of 50–100 in the cone-beam case. The average MSE was reduced to 29% in this region compared to the two-energy bin case. This shows potential for this method to be used in low-dose imaging. However, in the parallel-beam Monte Carlo simulation, which included scatter, the error was relatively constant over all SNR. Thus, an optimal X-ray imaging method with PCDs should include scatter reduction techniques or should be performed with a fan-beam geometry to limit scatter. It is expected that the simulation results are dependent on the imaging beam energy. Note that in this work we used a 60 kVp beam,

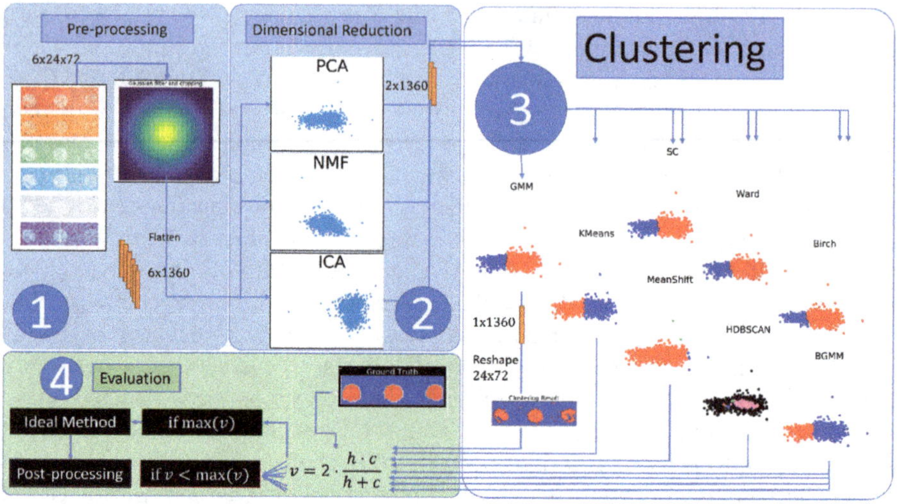

Fig. 6 Algorithm flow for the unsupervised learning technique

and for imaging thicker objects, a higher energy beam would be more appropriate and should be investigated for spectral projection imaging. Overall, we note that PCDs coupled with the machine learning methods outlined above show promise over methods such as dual energy radiography in the detection of materials such as cartilage.

4.2 *Case Study: Unsupervised Image Segmentation [6]*

In this section, we report on the unsupervised learning method investigation for material segmentation in projection X-ray imaging with a spectral detector. A phantom containing two hard materials (glass, steel) and three soft materials (PVC, polypropylene, and PFTE) all embedded in PMMA were imaged with a five-energy bin spectral photon-counting detector from the Redlen Technologies. The projection images were utilized to test nine unsupervised learning algorithms for automated material segmentation. The proposed algorithm testing workflow is shown in Fig. 6 while the experimental setup was shown earlier in Fig. 3. Details can be found in the original publication [6].

Each algorithm was investigated using single-energy, dual-energy, and multi-energy images. Clustering results were scored based on homogeneity and completeness of the clusters, which were combined into the Rosenberg and Hirshberg's V-measure. Principal component analysis (PCA), independent component analysis (ICA), and nonnegative matrix factorization (NMF) were tested as dimensional reduction methods. Spectral, dual-energy, and single-energy material segmentation was performed using five-, two-, and single-energy images, respectively. Spectral energy had the highest V-measure on the soft materials using PCA and a novel interpolating Bayesian gaussian mixture model (BGMM) clustering with a V-measure of 0.71. This number was 3.5% better than dual energy and 20.3% better than single energy. Conversely, single-energy imaging was most capable of hard tissue segmentation using the standard BGMM, with a V-measures of 0.84. This was 6.3% better than dual energy and 5.0% better than spectral energy. This work demonstrated that spectral X-ray imaging might be superior in segmenting soft tissues compared to conventional single-energy X-ray imaging.

Various signal processes were explored to reduce the dimensionality of the data and improve unsupervised learning algorithm performance as described in detail in [6]. The best results were obtained using principal component analysis (PCA). PCA decomposes the covariance matrix of the feature space into eigenvalues and eigenvectors; eigenvectors are sorted in terms of the magnitude of their eigenvalue to find the directions of highest variance in the data. The data was then projected into a lower-dimensional orthogonal space defined by the eigenvectors with the largest eigenvalues. This method results in a loss of information; however, this loss of information is usually relatively small and ideally the discarded dimensions in the data amount to noise.

For clustering, a number of methods were used; K-means clustering and a diverse range of additional algorithms were such as mean shift, spectral clustering, ward hierarchical clustering, HBDSCAN, and both Bayesian and non-Bayesian Gaussian mixture models (GMM). A gaussian mixture model (GMM) is used as a clustering method that implicitly accounts for the Gaussian nature of a dataset. A GMM was expected to perform well due to the Poisson statistics present in X-ray imaging. First, we will discuss a general mixture model, and then we will see how it is changed by assuming a Gaussian distribution.

General mixture models contain N data points, each assumed to be a mixture of K components, with all of these components assumed to be from the same family of distributions, in this case a Gaussian distribution. However, these distributions are allowed to have different parameters, such as different means and variances. The model also includes the identity of the mixture component to which each data point belongs. A weighting of each of the K components and sets of parameters for each of the distributions, in this case means and variances for the Gaussians, are necessary.

In addition to the previous methods, a new clustering method was also proposed. A demonstration of this method can be seen in Fig. 7. A BGMM was trained in the GMM section above. Within each cluster, the Mahalanobis distance from all the points to the center was calculated. Points at a distance greater than 90% of the maximum distance were then identified as potential outliers. Each potential outlier was then interpolated spatially using biharmonic interpolation. The BGMM was then retrained using a starting point defined by the end state of the previous iteration. The process was then repeated until convergence was indicated by an increasing AIC, typically 5–10 iterations.

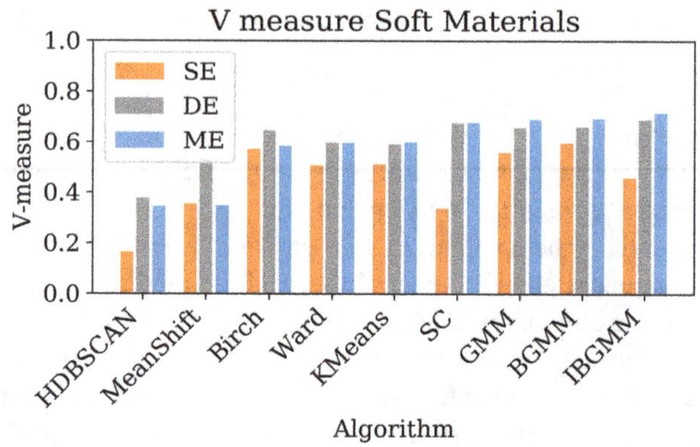

Fig. 7 Comparison of spectral V-measure to dual-, single-, and multienergy (DE, SE, ME) imaging with results averaged over all of the soft tissues for all of the algorithms, respectively. SC denotes spectral clustering, GMM is a gaussian mixture model, BGMM is a Bayesian GGM, and IBGMM is an iterative Bayesian GMM

Imaging modalities are constantly improving. The equipment of 5–10 years ago often has different capabilities than the state of the art today. Thus, having a machine learning system that is dependent on experimental data for training goes contrary to the development cycle of imaging modalities. Large experimental datasets will not be acquired until the modality has been adopted by a large number of providers. This begs the question; if we are to implement new imaging modalities to leverage deep learning, how do we validate their utility without large experimental datasets?

The work reported in [6] offers an alternative approach. Although deep learning segmentation would likely result in superior segmentation, a clinical dataset for spectral imaging does not yet exist. This could change, if spectral imaging is introduced into clinical practice in coming years. Thus, these methods could exist in the interim, between when spectral imaging is implemented, and before large clinical datasets become available. Further, having an automated segmentation method would be of particular importance in clinical spectral imaging since manual segmentation is not as effective: clinicians are disadvantaged compared to a computer when visualizing five-dimensional images as compared to one-dimensional images.

It was shown that PCA is the optimal dimensional reduction method for spectral imaging in this task. Although ICA shows promise in hyperspectral imaging, it did not show improvement over the unreduced data in this implementation. In the hyperspectral implementation domain, knowledge was used to calculate the exact spectral response of the materials to be segmented. Currently, we were not able to model the spectral response of the materials and thus could not apply this domain knowledge. With an analytical model of the CZT detector response and a ray tracer, one could calculate the mean attenuation for each energy bin. These theoretical mean values could then be compared to the mean value of a GMM's cluster to identify a material.

For single-energy imaging, the BGMM still showed the best results in this application. This approach could be used as an effective automatic segmentation technique superior to windowing for soft tissue segmentation. For spectral and dual energy, the IBGMM approach could yield superior segmentations and could see applications in cartilage detection. IBGMM could serve as an automated second look, working in tandem with professionals in NDT. Further work could also investigate a semiautomated workflow, working with a trained person to cluster using user-selected regions to be used as the mean of the GMM, letting the user define the number of clusters and select materials they would like the clusters to include.

In summary, segmentation of spectral, dual-energy, and single-energy data has been presented as well as an improved method for image segmentation of soft materials. It was seen that spectral imaging had the highest V-measure on the soft materials using PCA and a novel IBGMM method with a V-measure of 0.71. This method was 3.5% better than dual energy and 20.3% better than single energy. Conversely, single-energy imaging is most capable of hard tissue segmentation using a BGMM having the highest V-measures of 0.84. This method was 5.0% better than spectral and 6.3% better than dual energy on the same task.

4.3 Case Study: Machine Learning Using Edge Detection

In this case study, we have investigated the utility of machine learning models as applied to the food contaminant detection problem using spectral photon-counting X-ray data. At the most basic level, this entails correctly identifying scans of food products that contain contaminants, based on a previous set of scans that were accurately labelled as contaminated or not contaminated. The need for machine learning arises because a simple thresholding algorithm rejecting food above some cutoff in some characteristic of the X-ray images would not be able to deal with the instantaneously changing environment of food production lines, which may involve the influx of steam and other factors.

As an example of many possible food products, we have decided to use chicken breast fillets as our material vessel for contaminants and purchased several pieces from the local grocery store. On food production lines, these pieces would be scanned in the anterior-posterior direction and moving at speeds close to 100 mm/s. This is inconsistent with what was available to us in the lab because the anode-cathode direction of our x-ray tube was vertical and parallel with the detector surface, while a motion stage was set to move in between these two suspended pieces of equipment. Simply placing a sample on this stage would have led to lateral scans of the chicken breasts, an incorrect orientation for our studies. Therefore, we first had to devise a mechanical system that could suspend the chicken in a way that its anterior side would be perpendicular to the direction of the X-rays. This system can be seen in Fig. 8 with the chicken attached and wrapped in cling film to avoid any damage to and soiling of underlying equipment.

For contaminants, we have chosen materials with a range of different densities, atomic numbers, and thicknesses: 1 mm aluminum ($Z = 13$), 1 mm copper ($Z = 29$), 1–3 mm shattered glass ($Z \approx 10$), and 3–5-mm-thick rocks collected from the

Fig. 8 Skewer system attached to the horizontal motion stage in front of the CZT detector. Chicken breasts are pierced by two skewer rods, holding them in the right orientation with respect to the X-rays and the CZT detector to mimic the scanning setup of food production lines

beach ($Z \approx 6$). The equivalent square sizes of the contaminant pieces varied, with aluminum pieces of approximately 5 mm, copper pieces ranging from 2 mm to 15 mm, glass shards ranging from 5 mm to 15 mm, and rocks ranging from 10 mm to 20 mm. The photon-counting CZT detector supplied by the Redlen Technologies had pixel dimensions of 24 x 576 with a pixel pitch of about 0.3 mm and was set to acquire 1 ms images for 5 s during a scan. For PCD operation, the five energy readout channels were set to 16–35, 35–50, 50–70, 70–90, and 90–120 keV with an additional sum bin. Lastly, the X-ray tube was operated at 120 kVp, with a tube current of 2 mA, without any filtration, and using a focal spot.

During data acquisition, air and dark scans were recorded in order to be able to apply a flat-field correction to the scans and counter gain variations and dark currents. Three air and two dark images were acquired with an acquisition time of 5 s. Next, a piece of thawed (from frozen) chicken breast was set up with the motion stage as shown in Fig. 8. The contaminants were arranged into groups according to their type and enclosed in tape, with an additional group of a mix of contaminants containing aluminum, copper, and glass. Each of these contaminant groups were taped to the center of the wrapped chicken breast and scanned with the motion stage moving at 10-, 20-, and 30 mm/s between the X-ray tube and the detector.

For a control set with no contaminants the plain, wrapped chicken breast was also scanned on its own at the same speeds and equipment settings.

The resulting image sets were saved in a NumPy format for later analysis. After an initial manipulation of the data involving the removal of images with too much noise, correcting for defective pixels, calculating attenuation maps with flat-field corrections, and using a time-delay integration (TDI) procedure to smooth out the scans, the mean and variance of each image across all energy bins were collected, and the presence of contaminants was manually noted. This was followed by segmenting the regions in the images that could potentially contain contaminants, calculating the ROI mean, contrast, and CNR for these segments and then manually noting both the presence and type of contaminant contained within.

For our purposes of contaminant detection, we have used a batch size of 24 (i.e., summed 24 images using TDI), which was the smaller pixel dimension of the detector used. As mentioned above, after performing the TDI, the mean and variance of all images across all channels were collected and paired with a label describing whether the image had contained contaminants of any kind. The first two variables would serve as the predictors for some of the more simple machine learning models predicting contaminant presence, that is, the label. This was the last time the images were considered and manipulated as a whole, and this last set will be henceforth referred to as processed images.

For the more involved analysis of identifying the type of contaminants present in the chicken breast fillets, it was necessary to first locate the potentially harmful structures in the scans. For this purpose, we have used an edge detection algorithm (canny) inherent to our software of choice (Python) that utilized large gradients in the data (i.e., sharp changes in the pixel intensity values) to discover the outlines of embedded contaminants. It should be noted that across all energy channels, this algorithm performed best on the 35–50 keV data, and so this combination was used for further analysis of contaminant localization. Figure 9 shows a processed image from the 35–50 keV bin (top) and the edges detected by our algorithm in the image (bottom). It can be seen that the rock contaminants were accurately outlined as was the end of the chicken breast.

With the edges established, the next step in contaminant localization was finding where these edges were. Although this is a trivial task for a human observer, it is considerably more challenging for a computer software. This was most easily done by splitting the image into horizontal segments by vertical lines. To do so, we summed the columns of the edge array and identified transitions between zero and nonzero sums as the beginning or end of an outline and noted the column number

Fig. 9 Performance of the edge-detection algorithm on a processed image of chicken breast with rock contaminants

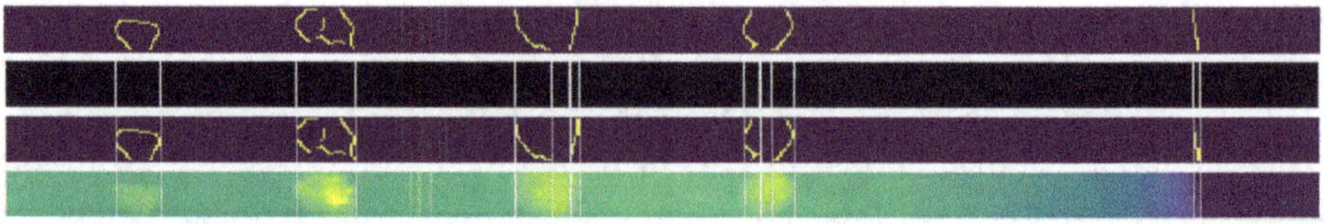

Fig. 10 The image segmentation process. In order from top to bottom, the first image is the result of our edge-detection protocol shown in Fig. 9, the second image shows where the outlines begin and end using our segmentation algorithm, the third strip shows how these segments are related to the original edge array, and the bottom picture depicts the segments' relationship to the actual processed image

where the transition had occurred. The parts of the image between these columns were what we had considered to be the segments of the image.

This process is depicted in Fig. 10, where the first strip shows the edge array calculated using the edge-detection method described earlier and shown in Fig. 9. The second image is what would result from applying our segmentation algorithm to the first strip; and the third and fourth rows depict how the established segments relate to the edge array and the original processed scan.

Once the images had been segmented this way, we tried to classify what material was sitting inside each segment the options being contaminant vs. no contaminant on the one hand and plain chicken, aluminum, copper, glass, or rock on the other. To do so, we established a threshold value inherent to each segment that has split the data inside the segment into two. The pixels above this threshold were what we called our ROIs and those below our background. Using this thresholding technique, we had collected the ROI mean, the contrast (i.e., the absolute difference between the ROI and background means), and the CNR for every segment of every image across every energy channel. To each such data point, we had also manually appended a value representing the true contaminant presence and type.

Recall that the data we have collected from each image so far were the mean and variance of the entire processed frame in each energy bin, along with whether the frame had a contaminant in it or not and the ROI mean, contrast, and CNR for ROI in every energy channel, in every segment, combined with the presence and type of contaminant (if any) present in the segment. A representative plot of the data points resulting from this process is shown in Fig. 11.

The striking thing about this graph is that even the simplest of metrics like the mean pixel value and the standard deviation of the pixel values in the entire image indicate that the images with no contaminants (0) and some contaminants (1–4) clearly separate. This feature shows amazing sensitivity of the scan to food contaminants. Bins 1–4, and Bin 7 (representing total counts) clearly show some contaminant clustering with Bin 1 showing the greatest promise for contaminant selection. However, Bin 5 which has high-energy photon counts and more surprisingly even Bin 6 which only contains very few pileup events show some clustering.

As for the differentiation between the different contaminants, there is unfortunately not much conclusion that can be drawn here. Although the clustering in Fig.

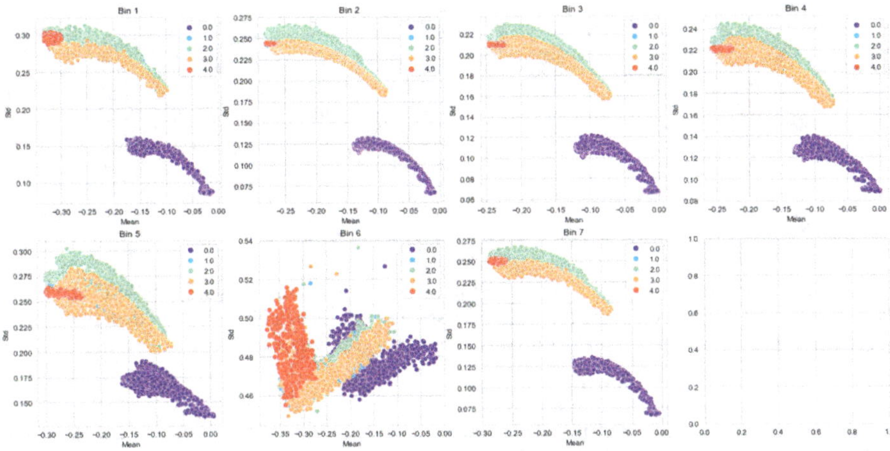

Fig. 11 Mean and variance of the entire processed frame recorded in each energy bin. Indices 0–5 have been assigned as 0, phantom only; 1, glass; 2, steel; 3, polypropylene; and 4, polytetrafluoroethylene (PTFE)

11 seems to imply that contaminant identification is possible, what is not shown is that some of the data points are laid over each other. For instance, in Bin 1, there are green data points "behind" the red ones, and this would lead to misidentifying frames containing steel as having PTFE as contaminants using these very simple metrics. This subtlety is why the more sophisticated segmentation tools already described in detail above had to be sought out. That process allowed for using local rather than global metrics that are more characteristics of the contaminant types. This is where machine learning and associated contaminant training helped the most as discussed below.

The collected data (mean and variance of the entire processed frame in each energy bin, along with whether the frame had a contaminant in it or not and the ROI mean, contrast, and CNR for ROI in every energy channel) allowed us to train models to detect contaminants in two ways: either by looking at the statistics of the entire frame or by looking at the segments separately. In the latter case, if all segments were identified to be contaminant-free, then the frame itself would also be identified as containing no contaminants. The information gathered from the segments was also used to train models to classify the contaminants located therein.

On top of this distinction in purpose, models could also be trained with balanced or unbalanced data. In the case of balanced datasets used for detection, equal numbers of contaminated and noncontaminated frames/segments were involved in the training (and the contaminated frames/segments were equally sampled among the different contaminant types), whereas during training for classification, equal numbers of segments for each contaminant were used. In unbalanced datasets, no such normalization among the different classes was performed.

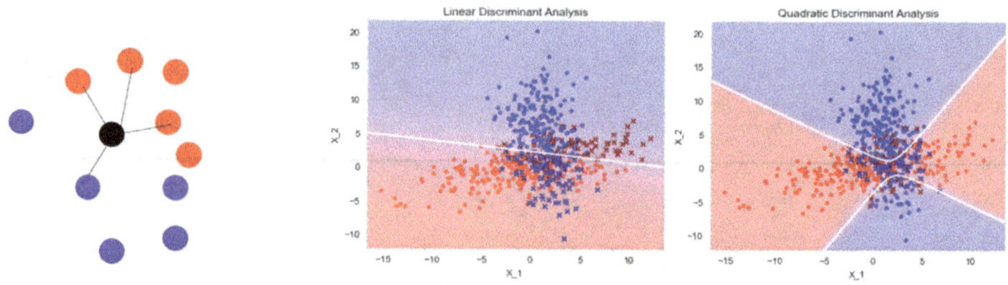

Fig. 12 The working of the KNN (K, nearest neighbors) algorithm shown on the left. The black data point becomes assigned to the red class because more of its neighbors are red than blue. KNN outperforms linear and quadratic discriminant algorithms, shown on the right

Simultaneously, the motion stage speed also played a factor in the data used for training. Models were trained either on 10 mm/s data, 10 and 20 mm/s data, or 10, 20, and 30 mm/s data. If multiple speeds were involved, the sample sizes from each speed were maintained equal (so that speed would always be a "balanced" parameter).

Our model of choice for our machine learning purposes was the K-nearest neighbor (KNN) model that has been previously evaluated as superior to logistic regression and quadratic discriminant analysis in a similar context considering contaminant detection in an acrylic phantom. According to this algorithm, each scan is represented by a data point in the multidimensional space of the prechosen variables (ROI mean, contrast, and CNR across all energy channels) and labelled appropriately for the given problem (either "contaminated" or "not contaminated" for detection or as the contaminant type for classification). Then, scans that the model is used to make predictions on appear as additional data points in this same space, and the most prevalent class among their K closest neighbors is the class the model assigns to the new data point as shown in Fig. 12.

With the above variations, we have trained models separately for each energy bin, the PCD sum bin, and the information from the five bins combined (i.e., all inputs from all channels were used as variables in the model). To protect against sampling bias, we have used fivefold cross-validation, and to ensure finding the best models, we have implemented a grid search algorithm for our KNN models. It should also be noted that the models were primed to maximize accuracy and so rewarded true positive outcomes the most.

Detection models were scored based on their accuracy producing true negatives (TN), that is, frames without contaminants predicted not to contain contaminants and true positives (TP), that is, frames with contaminants predicted to contain contaminants. Accuracy was defined as correctly identified instances over all identified instances in each class, and the accuracies of the "no-contaminant" and "contaminant" classes were added together for a combined score, with 200 being the maximum (100% TN + 100% TP). Classification models were evaluated in a similar manner with combined TP scores across all contaminant classes (not including plain chicken as a class because the detection model already identifies this feature and so this information becomes superfluous).

It was found that the contaminant detection model of the 16–35 keV energy bin—trained on a balanced set of frames across all speeds on frame statistics—was 100% successful at predicting no contaminant presence when there had been none and was 70% accurate at predicting contaminant presence when a contaminant had been present in the frame. Alternatively, a 16–35 keV energy bin model trained to classify contaminants in the segments of frames across all speeds based on an unbalanced dataset had been found to be 81% accurate in predicting TN and 95% accurate in predicting TP.

For contaminant classification, we had found that the most successful model that was able to categorize the various contaminant types was the model trained on a balanced set of features coming from the 10 and 20 mm/s data in the 50–70 keV energy bin with average accuracies of 95%, 97%, 91%, and 93% on the classes of aluminum, copper, glass, and rock, respectively.

In conclusion, adding a machine learning analysis could be a valuable asset to the food contaminant detection problem. It is possible to train models that on the one hand do not waste any uncontaminated chicken and on the other hand capture 70% of the contaminated pieces. Furthermore, it is possible to improve the amount of contaminated chicken pieces identified to 95%, although this comes at the expense of also discarding 19% of uncontaminated chicken breasts. In either of these cases, it is possible to determine the types of contaminants seen in various segments of the chicken breasts at accuracies above 90% for the contaminant categories considered in this work.

The performance of the KNN algorithm in predicting food contaminant presence on CZT detector-acquired X-rays, and classifying such contaminants was investigated. Chicken breast fillets containing either aluminum, copper, glass, rock, or a mix of these contaminants were scanned moving at speeds of 10, 20, and 30 mm/s between the CZT detector set to 1 ms sampling time at 5 s exposure and energy channels of 16–35, 35–50, 50–70, 70–90, and 90–120 keV and an X-ray tube with settings of 120 kVp, 2 mA. The resulting images were corrected for defective pixels, flat-field corrected, and summed using time-delay integration techniques at a batch size of 24 frames. Edge-detection techniques on the 35–50 keV channel images were used to segment out potential contaminants. The mean and variance of the entire frame and the supposed contaminant region's mean, contrast, and CNR were paired with manual contaminant labels (present or not and type) to create a labelled dataset. Among the best results, it was found that using 16–35 keV photons and whole-frame statistics, 100% of the frames not containing contaminants and 70% of frames containing contaminants were correctly identified. Alternatively, using segment statistics (ROI mean, contrast, CNR) for all segments in a given frame, 81% of frames not containing contaminants and 95% of frames containing contaminants were correctly identified. Furthermore, using the 50–70 keV channel, it was possible to correctly identify the type of contaminant present to an average accuracy of 95% in aluminum, 97% in copper, 91% in glass, and 93% in rock. The study results are summarized in Table 2.

Table 2 Summary of obtained results. Phantoms are easily recognized as expected with 100% accuracy using any energy bin (except Bin6)

	Accuracy	Best energy bin
Phantom	100%	Any (except bin 6)
Glass	85%	Bin 2
Steel	99%	Bin 2
PolyP	91%	Bin 4
PTFE	95%	Bin 2

Bin2 typically offers the best accuracy although for polypropylene Bin4 outperforms Bin2. Achieved accuracy is between 85% and 95%

In summary, two machine learning models of interest have been identified in this case study using real food contaminated data and their advantages are as follows. The first approach does not throw away any good chicken pieces; however, it captures only 70% contaminated ones. The second throws away 19% of the good chicken pieces but captures 95% of the contaminated ones. In either case, it is possible to identify detected contaminants among aluminum, copper, glass, and rock to an accuracy of 95%, 97%, 91%, and 93%, respectively. This requires PCD outputs in the 16–35, 35–50, and 50–70 keV energy bins only, which can speed up processing, as there is no need to process all six energy bins. While this study shows great potential of using AI for food contaminant detection, it is clear that further work is required to decrease false positives.

4.4 Case Study: Industrial Example

Early examples of commercial NDT equipment that uses artificial intelligence (AI) for X-ray food detection are already available. In this section, we will report on XSpectra Meat Inspection Solution (https://x-next.com/). XSpectra XIS4 Top-Down Inspection 600, the inspection machine specifically designed and developed to offer the best meat inspection, especially bulk meat, is shown in Fig. 13. Thanks to its multienergy X-ray inspection technology, it can detect high- and low-density foreign bodies such as bones and cartilage, plastics, and wood inside bulk meat directly in the production line.

The XIS4 Top-Down Inspection 600 incorporates proprietary XSpectra technology, which consists of a state-of-the-art detection unit capable of detecting up to 1024 energy levels of X-rays, while traditional inspection systems use a maximum of two. This way, it is possible to obtain much more information on the composition of products such as chicken breasts and pork by detecting small fragments of bone, cartilages, and plastics, among others, with extreme precision, minimizing the number of false rejects. XSpectra performs a multispectral analysis in milliseconds, detecting nonconformities and contaminants and immediately increasing compliance with the required standards.

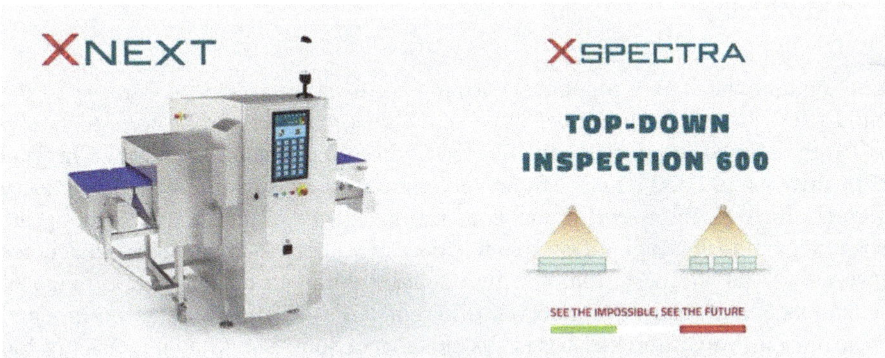

Fig. 13 Example of commercial X-ray food inspection system that utilizes spectral photon counting ((https://x-next.com/)

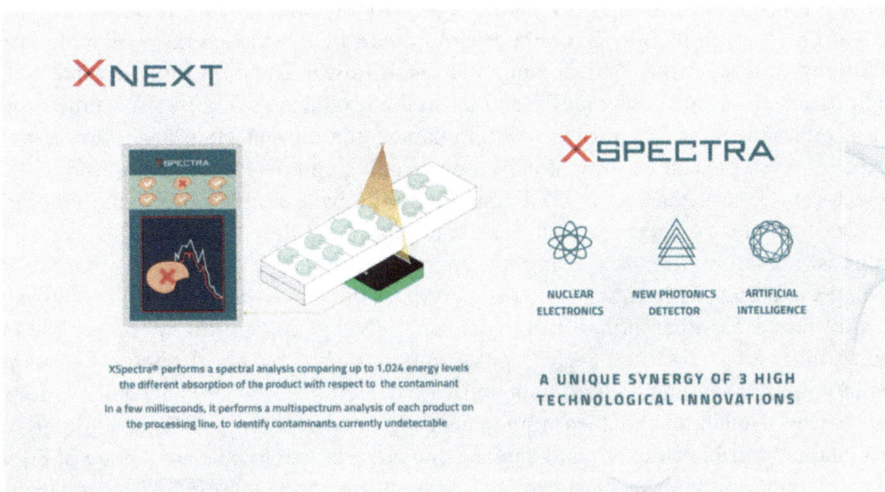

Fig. 14 Schematic illustration of the NDT food inspection system. (https://x-next.com/)

The machine's hardware features a large 60-cm-wide inspection area that allows for multirow scanning by scanning products arranged in multiple parallel lines at speeds of up to 50 m/min depending on the product's weight. The inspection geometry has a top-to-bottom orientation to ensure the best scan of the entire product is being scanned by the detection unit as shown in Fig. 14. On innovative hardware, the manufacturer has added artificial intelligence software (XInspector) based on deep learning that allows it to analyze a vast number of datasets by learning the composition of the product more and more accurately. The software is trained by acquiring data on various product/contaminant combinations and can be updated whenever the need arises to inspect new products or search for new contaminants. The system uses three energy bins using their photon-counting detector.

5 Conclusions

Food contamination is a regular occurrence all over the world. According to the Rapid Alert System for Food and Feed (RASFF) (https://ec.europa.eu/food/safety/rasff/), the number of alerts about foreign bodies and contaminants found in food keeps increasing. Food recalls can be very expensive; as a result, the food industry routinely inspects their food products at various stages of production. Among the broad range of technologies available for food inspection, X-ray detection machines represent by far the best solution, due to their ability to detect a wide range of contaminants and inspect products at different stages of their production process. These detectors are often line-based systems that acquire a continuous transmission image as the material moves via conveyor in front of the X-ray system.

Despite its widespread use, the existing X-ray technologies are currently not able to completely solve the issue of contaminated food as they can detect high-density foreign bodies, such as metal, lead, stones, bones, and glass, while they fail in detecting low-density contaminants made of a wide number of materials including plastic, wood, and organic and small unwanted foreign bodies made of light materials which accidentally end up in the products during their treatment on the production lines. As shown in this chapter, the current state-of-the-art X-ray detection systems can be dramatically improved using novel photon-counting CZT detectors. Recent advances in CZT manufacturing have enabled this technology to be used in industrial applications, thus enabling the ability to distinguish material composition from the processing of energy-dependent transmission images. It is expected that with time, spectral photon-counting X-ray detection is going to proliferate to other nondestructive testing (NDT) modalities in security and industrial testing. To fully explore possibilities offered by novel photon-counting technologies, a suitable contaminant software detection software is needed. Various approaches used in machine learning to aid in X-ray detection have been explored in this chapter. Further research and engineering are required to take advantage of both spectral photon-counting hardware technology and deep learning software techniques to achieve high-throughput, reliable, and cost-effective food inspection.

References

1. R. Haff, N. Toyofuku, X-ray detection of defects and contaminants in the food industry. Sens. Instrumen. Food Qual. **2**, 262 (2008) https://link.springer.com/article/10.1007/s11694-008-9059-8
2. J. Jiang et al., An adaptive image segmentation algorithm for X-ray quarantine inspection of selected fruits. Comput. Electron. Agric. **60**, 190 (2008) https://www.x-mol.net/paper/article/1414245899569401856
3. T. Schmidt, Future prospects of spectral CT: photon counting, in *Computed Tomography: Approaches, Applications, and Operations*, ed. by E. Samei, N.J. Pelc, (Springer, Cham, 2020), pp. 269–286. https://doi.org/10.1007/978-3-030-26957-9
4. L. Zhou et al., Application of deep learning in food: a review. Compr. Rev. Food Sci. Food Saf. **18**, 1793 (2019) https://ift.onlinelibrary.wiley.com/doi/abs/10.1111/1541-4337.12492

5. J. O'Connell, K. Iniewski, M. Bazalova-Carter, Optimal planar X-ray imaging soft tissue segmentation using a photon counting detector. J. Instrum. **14**, P01020. https://iopscience.iop.org/article/10.1088/1748-0221/14/01/P01020 (2019)
6. J. O'Connell, K. Murphy, S. Robinson, K. Iniewski, M. Bazalova-Carter, Unsupervised learning methods in X-ray spectral imaging material segmentation. J. Instrum. **14**, P06022. https://iopscience.iop.org/article/10.1088/1748-0221/14/06/P06022 & https://arxiv.org/abs/1904.03701 (2019)
7. D. Richtsmeier, E. Guliyev, K. Iniewski, M. Bazalova-Carter, Contaminant detection in non-destructive testing using a CZT photon counting detector. J. Instrum. **16**, P01011. https://iopscience.iop.org/article/10.1088/1748-0221/16/01/P01011/meta (2021)
8. J. Nguyen, P.-A. Rodesch, D. Richtsmeier, K. Iniewski, M. Bazalova-Carter, Optimization of a CZT photon counting detector for contaminant detection. J. Instrum. **16**, P11015. https://iopscience.iop.org/article/10.1088/1748-0221/16/11/P11015/meta (2021)
9. J. Nguyen, D. Richtsmeier, K. Iniewski, M. Bazalova-Carter, Contaminant detection using a CZT photon counting detector with TDI image reconstruction. J. Instrum. **17**, P05010. https://iopscience.iop.org/article/10.1088/1748-0221/17/05/P05010/pdf (2022)

Gamma-Ray and Cosmic Ray Muon Modalities for Cargo Inspection

JungHyun Bae, Stylianos Chatzidakis, and Reshma Ughade

1 Introduction

The increasing volume of international cargo transportation necessitates robust security measures to prevent illicit trafficking of unauthorized materials such as nuclear and radioactive materials, explosives, and drugs. Noninvasive inspection technologies play a critical role in thwarting the unauthorized transport of hazardous materials while ensuring the security and integrity of cargo containers. Traditional cargo inspection has predominantly relied on high-energy X-ray transmission systems, and it is one of core nondestructive testing (NDT) strategies at borders and secure facilities. The NDT approach has been exploited in examining the contents of shipments because it does not cause damage and ensures the efficiency of transportation. However, these methods come with significant challenges, such as high false-positive rates and beam attenuation issues. The need for more advanced inspection technologies has led to the exploration of alterative NDT modalities such as gamma-ray and cosmic ray muon inspection methods. Gamma-ray inspection offers several key benefits in the context of cargo inspection. (1) Gamma-rays, which are high-energy photons, can penetrate materials that would often absorb or deflect X-rays. This characteristic allows for the effective inspection of cargo containers, facilitating the detection of concealed items that might otherwise be unnoticed using traditional X-ray systems [1, 2]. (2) Gamma-ray inspection can provide a superior material discrimination capability by taking advantage of their unique interactions with nuclear materials. This advantage is particularly useful for identifying specific

J. Bae (✉)
Oak Ridge National Laboratory, Oak Ridge, TN, USA
e-mail: baejh@ornl.gov

S. Chatzidakis · R. Ughade
Purdue University, West Lafayette, IN, USA

K. Iniewski, H. Gadey (eds.), *Emerging Radiation Detection*,
https://doi.org/10.1007/978-3-031-63897-8_12

types of nuclear and radioactive substances within complex cargo loads [3]. (3) The better material identification ability of gamma-rays can lead to a reduction in false-positive rates. By more precisely distinguishing between harmless and potentially dangerous items, this technique improves the inspection process and reduces unnecessary delays in cargo scanning [4]. (4) Gamma-ray systems can scan cargo containers efficiently, maintaining a high throughput rate. This efficiency is crucial for busy ports and border crossings, where delays can lead to significant logistical challenges and economic costs [5, 6]. (5) Gamma-rays are particularly effective at detecting nuclear and radioactive materials, even when these materials are shielded. Gamma-ray inspection techniques, especially neutron-induced gamma-ray emission detection, are designed to identify characteristic gamma-ray emissions from target materials [7]. (6) Gamma-ray inspection can be used in conjunction with other NDT methods, such as X-ray transmission and cosmic ray muon imaging, to provide a more comprehensive security screening. This multimodal approach allows for the adaptation of inspection strategies based on the cargo's nature and the specific security requirements at hand [8, 9]. Meanwhile, cosmic ray muon scanning, a relatively novel technique in the field of noninvasive cargo inspection, offers several distinct advantages that complement traditional methods. The following advantages of using cosmic ray muon scanning significantly improve the effectiveness and operational performance of cargo security procedures:

1. Cosmic ray muons are highly penetrative, capable of passing through dense materials that might attenuate other forms of radiation such as X-rays or gamma-rays. Therefore, they are particularly effective in detecting shielded nuclear and radioactive materials. Their unique interaction with such materials can reveal the presence of illicit nuclear substances, even when attempts have been made to conceal them with radiation shielding [10, 11].
2. Muon scanning can differentiate between materials based on the scattering density, providing insights into the composition of the cargo without the need for direct visualization. This ability to identify various substances, including metals, plastics, and organic materials, enhances the detection of hazardous materials [12].
3. Unlike active interrogation methods that require external radiation sources, cosmic ray muons utilize naturally occurring radiation. Thus, external radiation sources are not required, which reduces the risk of exposure for both the inspection personnel and the public.
4. The operational and maintenance costs of muon scanning systems can be lower compared to systems that require the generation of high-energy particles.
5. Muon scanning systems can be adapted for various sizes and types of cargo, making them versatile tools for security and inspection applications. They can be scaled up to accommodate large shipping containers or scaled down for smaller packages [13].

In applications of nuclear material monitoring such as cargo inspection and spent nuclear fuel (SNF) cask monitoring, traditional X-ray systems encounter technical limitations because of their high attenuation and deflection within the high atomic number (i.e., high-Z) materials [14]. For decades, gamma-ray inspection techniques have been expected to address these limitations [15]. For example, active gamma-ray methods utilize the differential absorption or emission of gamma-rays by various materials to identify anomalies within cargo containers [16, 17]. Similarly, muon radiography exploits the natural occurrence and high penetration capability of cosmic ray muons to inspect cargo containers [18]. These techniques have gained attention to advance cargo inspection approaches by providing detailed three-dimensional images of container contents [19].

This chapter consists of gamma-ray radiography, cosmic ray radiography, and forthcoming technologies and advanced methods thereof. For gamma-ray radiography, operational principles of active and passive interrogation gamma-ray inspection techniques are discussed. For muon radiography, some essential features of cosmic ray muons are presented, and the technical details of two muon imaging techniques—muon transmission radiography and muon scattering tomography—are outlined. In the end, critical assessments and suggestions for directing future research and application of gamma-ray and muon imaging modalities are discussed. These technologies and methods will offer complementary benefits, address the limitations of traditional X-ray systems, and provide a more effective approach to cargo inspection. Future research should focus on further refining these technologies, improving their accuracy, and reducing operational costs to facilitate wider adoption. The continued development and implementation of advanced NDT methods are crucial for enhancing national security and facilitating safe and efficient international trade.

2 Gamma-Ray Radiography

Gamma-ray radiography is an NDT technique used across various industries to inspect the internal structure of materials without causing damage. Since the introduction of the first Anger camera in 1959 [20], significant advancements have been achieved in tomographic systems. These advancements have offered improved contrast, resolution, and reduced artifacts for three-dimensional image reconstruction. The applications of gamma-ray radiography include (1) examination to ensure structural integrity and prevent potential failures [21, 22]; (2) aircraft inspection to examine defects or damage of turbine blades, engine parts, and structural components [23]; (3) inspection of nuclear components such as reactor vessels, fuel rods, and piping systems to ensure the safety and reliability of nuclear power plants [24, 25]; and (4) cargo inspection at ports and borders to detect illicit materials such as unauthorized nuclear and radioactive materials, drugs, or weapons [26, 27]. In gamma-ray radiography, there are two main imaging techniques: active and passive interrogation methods. A general definition of active interrogation is the use of external radiation sources for material detection and/or imaging by analyzing the responses of the irradiated target materials [28]. Active interrogation gamma imaging techniques include gamma-ray transmission radiography and neutron-induced gamma-ray emission tomography. Although it is similar in principle to X-ray transmission radiography, the gamma-ray imaging exploits the high penetration capabilities of gamma-rays. Neutron-induced gamma-ray emission tomography is specifically designed for detecting nuclear and radioactive materials as it uses the characteristic gamma-ray signal emitted upon neutron absorption by the target material. In contrast, the passive interrogation technique involves the measurement and analysis of emitted gamma-ray signals from radioactive materials without external radiation sources. Therefore, this method depends on high-efficiency detectors to identify radiation emitted by concealed special nuclear or other radioactive materials. Within this context, two passive interrogation techniques are highlighted, gamma emission tomography and coded-aperture-based gamma imaging.

2.1 Active Interrogation Gamma-Ray Radiography

2.1.1 Gamma-Ray Transmission Tomography

In gamma-ray transmission tomography, or gamma computed tomography (γCT), a gamma-radiation source is positioned on one side of the object, emitting gamma-rays that pass through the target object. Detectors placed on the opposite side of the object measure the intensity of the gamma-rays that have traversed the object material. Areas of higher density attenuate more gamma-rays, and they result in lower-intensity readings on the detectors. By rotating the radiation source and detectors around the object and taking measurements from multiple angles, a series of two-dimensional projections is obtained. There are various methods of scanning in γCT, as shown in Fig. 1. For example, a parallel scanning method uses a collimated pencil gamma beam and an array of detectors that are positioned on the opposite side of the object. The gamma source moves linearly to scan the entire volume of the object and detectors measure the gamma-ray intensity. A fixed-detector, rotating-source configuration consists of a considerable number of detectors which are affixed to a stationary ring, surrounding the object under examination. The gamma-radiation source is positioned on a rotating arm or gantry to traverse the object during the scanning process. The detectors remain stationary throughout the entire scan.

In gamma transmission tomography, gamma-radiation beams are directed through the equipment to examine the object from multiple positions and various angles. Each beam is detected on the opposite side of the object from the beam source, and the measured intensity is analyzed to characterize the material and configuration of the object. If we define T as the transmittance of the object, the linear transformation of T is given by

$$T = \int_L \mu(x, y)\, dl, \tag{1}$$

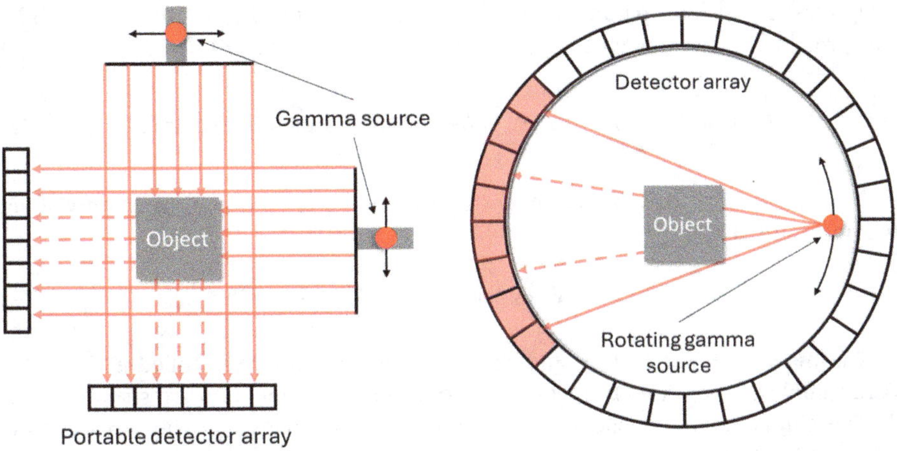

Fig. 1 A parallel scanning method (left) and a fixed detector-rotating source configuration (right) for gamma-ray transmission tomography

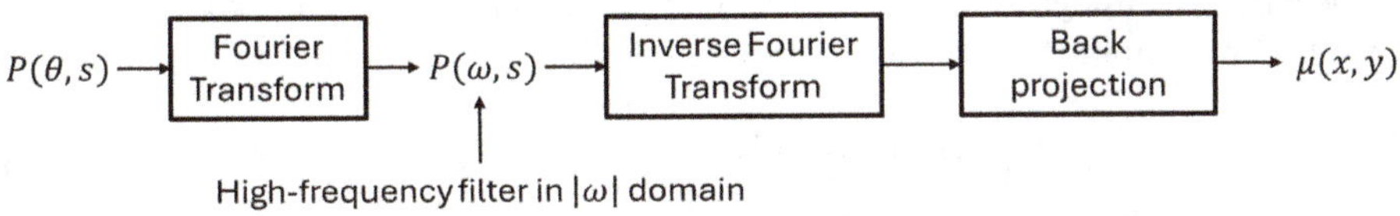

Fig. 2 A block diagram representing the signal flow and transforms of FBP algorithm. $P(\theta,s)$ is the projection at an angle, θ, and distance, s. $P(\omega,s)$ represents the Fourier transform of $P(\theta,s)$ within the frequency domain, ω

where $\mu(x,y)$ is the linear attenuation coefficient of the object at (x,y) and L is the beam travel length within the object. These projections are then reconstructed using mathematical algorithms to generate a three-dimensional image of the object. In general, the back-projection (BP) and filtered back-projection (FBP) algorithms, which utilize fast-Fourier and Radon transforms, are widely used. A block diagram depicting the signal flow and transforms of the FBP algorithm is shown in Fig. 2. The operational principle of the FBP algorithm is described in the following sections.

Back-Projection Algorithm

The BP algorithm is a computational method used for X-ray CT and γCT. It involves calculating the projections (Radon transform) of an image by summing the values along specific paths. Consider a two-dimensional image represented by a function, $\mu(x,y)$, where x and y are spatial coordinates in the plane of the image, and μ represents the attenuation coefficient of density at each point. The backward projection of the image along a line can be described by the Radon transform, which integrates μ over lines parameterized by the angle and the distance from the origin. The projection, $P(\theta,s)$ at an angle, θ, and distance, s, is given by

$$P(\theta, s) = \int_{-\infty}^{\infty} \int_{-\infty}^{\infty} \mu(x, y)\, \delta(x \cos\theta + y \sin\theta - s)\, dxdy, \tag{2}$$

where δ is the Dirac delta function, ensuring the integration is performed only along the line defined by

$$x \cos\theta + y \sin\theta = s. \tag{3}$$

An image of $P(\theta,s)$ with θ and s as rectilinear coordinate is called a *sinogram*. An example of the original image of the object and its sinogram are shown in Fig. 3. The sinogram image shows the accumulated BP images moving up from the bottom ($\theta = 0$) to top ($\theta = \pi$). In a discrete setting, where the image is represented as a grid of pixels, the backward projection can be approximated by summing the contributions of all pixels that lie along the line specified by θ and s. If μ_{ij} represents the value of the image at the pixel located at grid coordinates, (i,j), then the discrete forward projection, $P_d(\theta,s)$, can be approximated as

$$P_d(\theta, s) \approx \sum_{ij} \mu_{ij} \Delta_{ij}(\theta, s), \tag{4}$$

where $\Delta_{ij}(\theta,s)$ is a weighting factor that accounts for the portion of the path through pixel (i,j) for the line specified by θ and s. This factor can be determined based on the geometry of the line and the grid. An example of a reconstructed image of the object using the BP algorithm is shown in Fig. 3.

Filtered Back-Projection Algorithm

The FBP algorithm involves two main steps necessary to reconstruct the original image, applying a filter to the projection data and projecting filtered data back onto the image plane. In the first step, the projection data are acquired through the Radon transform. The acquired projection data are then filtered to enhance the high-frequency components, which correspond to edges and fine details in the image, and to mitigate the blurring effect inherent in the BP process. The filtering is performed in the frequency domain using a high-pass filter, typically the Ram-Lak filter [29], defined in the frequency domain by $|\omega|$, where ω is the spatial frequency. Given the Fourier transform, $\mathcal{F}$ of the projection $P(\theta,s)$, the filtered projection $P_f(\theta,s)$ is obtained as

$$P_f(\theta, s) = \mathcal{F}^{-1}\left[\mathcal{F}\{P(\theta, s)\} \bullet H(\omega)\right], \tag{5}$$

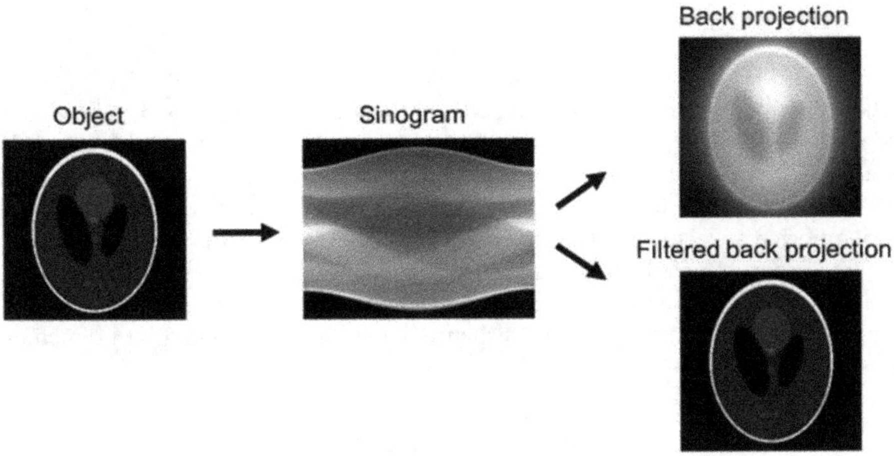

Fig. 3 Illustrations of (**a**) an example of targe object, (**b**) its sinogram image, (**c**) the reconstructed image using the BP algorithm, and (**d**) the reconstructed image using FBP algorithm. (Images are excerpted from Ref. [31])

where $H(\omega)$ is the filter function in the frequency domain and $\mathcal{F}^{-1}$ denotes the inverse Fourier transform. The filtered projections are then back projected onto the reconstruction plane. For each point (x,y) in the image plane, the reconstructed image $f_r(x,y)$ is obtained by integrating over all angles, θ, the FBP to the line that passes through (x,y):

$$f_r(x, y) = \int_0^{\pi} P_f(\theta, x\cos\theta + y\sin\theta)\, d\theta. \tag{6}$$

This step is effectively applied to the filtered projection data back across the image plane, with the superposition of all BPs. The further technical details of FBP can be found in [30]. The reconstructed images of our sample object, which were obtained through the BP and FBP algorithms, are shown in Fig. 3.

2.1.2 Neutron-Induced Gamma-Ray Emission Tomography

Neutron-induced gamma-ray emission tomography (NIGET) is an advanced imaging technique for identifying and characterizing materials, specifically for nuclear and radioactive materials. NIGET combines neutron activation analysis and gamma emission CT techniques. When a target material is irradiated by a thermal neutron with a flux, ϕ, for time, T_i, the anticipated detector response to gamma-ray D_γ is given by [16]

$$D_\gamma = \frac{Gmf\varepsilon\sigma_a\phi N_\upsilon\left(1-e^{-\lambda T_i}\right)}{A\lambda} \times e^{-\lambda T_o}\left(1-e^{-\lambda T_c}\right), \tag{7}$$

where G is the gamma-ray yield, m is the mass of the material, f is the abundance of the target isotope in the materials, ε is the detector efficiency, N_ν is Avogadro's number, T_o is the time elapsed between neutron irradiation and start of counting, T_c is the counting time, A is the atomic weight, σ_a is the microscopic neutron absorption cross section, and λ is the decay constant. To reconstruct images from the neutron-induced gamma-rays and detector responses, multiple neutron irradiation scanning passes with a well-collimated detector must be performed to exploit the CT techniques. The expected detector response due to the source in a given direction, L, is given by

$$P(s,\theta) = \int_L I(x,y)\, e^{-\int -k(x',y')dl}\, dl, \tag{8}$$

where $I(x,y)$ is the intensity distribution function of the source and $k(x',y')$ is the attenuation of the emitted radiation by the object. By multiplying Eqs. (7) and (8), the mass distribution of the object can be mapped [16]. Then, image reconstruction algorithms such as the FBP algorithm can be exploited to visualize the outcome. In many practical applications, however, the neutron-induced gamma-ray emission (NIGE) techniques can be used without the image reconstruction process. By detecting and analyzing the emission of gamma-rays with characteristic energies corresponding to the specific isotopes, a detailed material composition maps of the contents can be created. This method offers several advantages over traditional inspection approaches. For example, NIGE can be employed to scan dense or shielded nuclear materials more effectively, regardless of packaging, and it is less affected by the presence of other materials. These benefits make NIGE suitable for mixed cargo inspection [32].

2.2 *Passive Interrogation Gamma-Ray Radiography*

2.2.1 Passive Gamma Emission Tomography

Passive gamma emission tomography (PGET) is a nonintrusive imaging technique used for the detection and characterization of gamma-ray-emitting radioactive materials. Unlike active interrogation methods where an external radiation source is used to induce gamma-ray emissions from the material under inspection, PGET relies on detecting the naturally emitted gamma-radiation from the materials. The technical principle of PGET involves the deployment of gamma-ray detectors around the object or area being inspected. These detectors passively monitor the gamma-radiation emitted by radioactive sources present within the objects. By measuring the energy, intensity, and direction of the gamma-rays detected by each detector, a three-dimensional map of the gamma-ray emission distribution can be reconstructed using tomographic techniques. Similar to other radiography techniques, PGET typically utilizes the FBP algorithm and the linear inverse problem methods. The inverse problem method is a mathematical approach based on the conversion of the image reconstruction problem into a linear inverse problem [33]. Because the linear inverse problem approach follows the assumption that the vectorized simulated sinogram, ϕ, is a linear function of the vectorized image of the object, s, thus the vectorized simulated sinogram is given by

$$\boldsymbol{\phi} = \boldsymbol{A}\boldsymbol{s} + \boldsymbol{N}, \tag{9}$$

where $\boldsymbol{A}$ is the system response matrix and $\boldsymbol{N}$ is the random Gaussian observation noise. From the simulated sinogram and system response matrix, the reconstructed image matrix, $\hat{s}$, can be found by solving the following equation:

$$\hat{s} = \underset{s \geq 0}{\operatorname{argmin}} \left[\frac{1}{2} \|\boldsymbol{\phi} - \boldsymbol{A}\boldsymbol{s}\|_2^2 + \lambda \|\boldsymbol{s}\|_1 \right] = \underset{s \geq 0}{\operatorname{argmin}} \left[\frac{1}{2} (\boldsymbol{\phi} - \boldsymbol{A}\boldsymbol{s})^T (\boldsymbol{\phi} - \boldsymbol{A}\boldsymbol{s}) + \lambda \sum_{i=1}^{N} \boldsymbol{s}_i \right], \tag{10}$$

where the first term represents the feasibility of data assuming Gaussian observation noise and the second term denotes the regularization expression for the object image. The regularization parameter, λ, can be chosen based on the noise level [33]. The solution of Eq. (10) produces the reconstructed images.

PGET has been used in various applications, including nuclear security, environmental monitoring, and nuclear waste management [34, 35]. The characteristic passive emission gamma signals from nuclear and radioactive materials are still useful in safeguards and security applications without the image reconstruction process. In the context of cargo inspection, it can be deployed at ports, border crossings, or checkpoints to scan shipping containers, vehicles, or luggage for illicit radioactive materials. A common example is the radiation portal monitor (RPM), which typically consists of multiple gamma and neutron detectors [36]. For decades, the International Atomic Energy Agency (IAEA) has investigated the feasibility of gamma emission tomography for the inspection of spent nuclear fuel (SNF) to provide a more accurate verification method of fissile material contents in SNF than currently authorized instruments [34, 37]. Over the years, advancements in detector technology, data processing algorithms, and imaging techniques have improved the sensitivity, resolution, and efficiency of PGET systems. Today, PGET continues to be a valuable tool for nonintrusive inspection and monitoring of radioactive materials in various security and safety applications. A recent study demonstrates that the image quality is improved by simultaneously incorporating both emission and attenuation maps of gamma-rays in SNF assembly imaging [38]. For example, gamma-ray attenuation and emission maps obtained from a SNF canister are illustrated in Fig. 4. In this example, the highlighted areas show the presence of three 2×2 SNF assembly bundles. The gamma-ray attenuation map shows the reduction

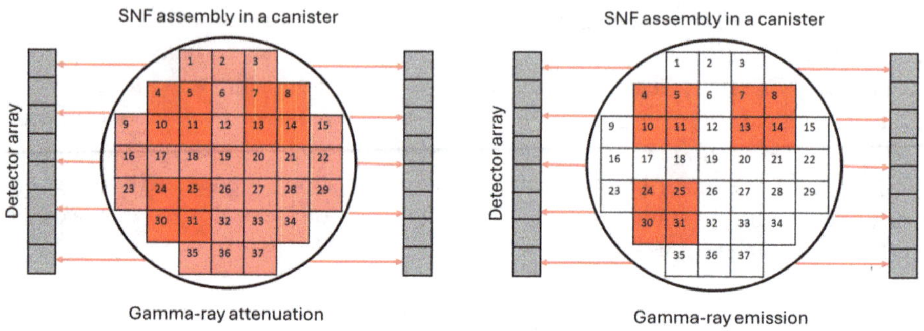

Fig. 4 Illustrations of gamma-ray attenuation (left) and emission (right) maps obtained from an SNF canister around the gamma-ray detector arrays. In this example, the highlighted areas reveal the presence of three 2 × 2 SNF assembly bundles

in gamma-ray intensity as it passes through the SNF canister, providing insights into the density and composition of the materials. Conversely, the gamma-ray emission map displays the distribution of gamma-ray emissions originating from the SNF assemblies, offering valuable information about the radioactive decay processes occurring within the canister. By analyzing both maps in conjunction, inspectors can accurately identify the location and configuration of SNF assemblies within the canister.

2.2.2 Coded-Aperture Gamma Imaging

Coded-aperture gamma imaging is a specialized technique used to locate radiation sources that is often employed in the field, including in cargo inspection and environment surveillance [39, 40]. In cargo inspection, the primary objective of coded-aperture gamma imaging is to identify and localize gamma-ray emitting radioactive sources concealed within cargo shipments while minimizing false positives and ensuring an efficient screening processes [41]. The operational principle of coded-aperture imaging involves the utilization of a patterned mask positioned in front of a gamma-ray detector. This mask features an array of openings or apertures arranged in a specific pattern. When gamma-rays emitted by a radioactive source pass through the apertures, they cast a shadow pattern onto the detector. A source localization method using coded-aperture gamma detection using an 8 × 8 patterned mask is illustrated in Fig. 5. The true position of a point gamma source and noise signals around the source are also included. By rotating the patterned mask with 90° intervals and employing a reconstruction algorithm, multiple shadow patterns are recorded. By analyzing the shadow pattern recorded by the detector, image reconstruction algorithms can be used to visualize the spatial distribution of gamma-ray emitting sources within the field of view. The unique design of the coded-aperture mask enables high-resolution imaging, allowing for improved sensitivity and resolution. Coded-aperture gamma imaging offers several advantages

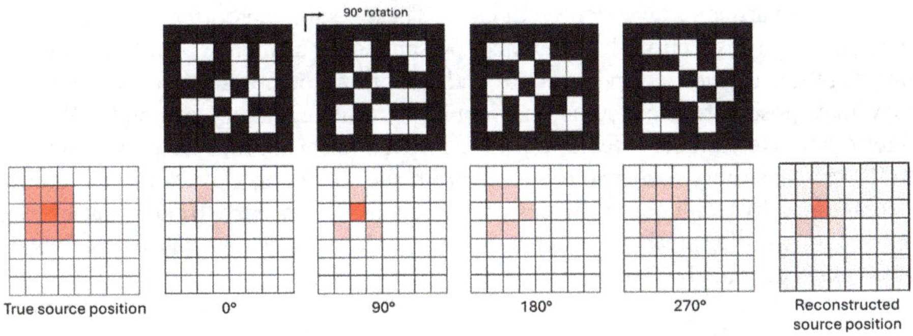

Fig. 5 Simplified operational principle of coded-aperture gamma imaging to find the gamma-ray emitting radiation source. In this example, an 8 × 8 patterned mask is used to localize a point gamma source. It successfully reconstructs the position of a point radiation source

over traditional gamma-ray imaging techniques. It provides a high spatial resolution and detection sensitivity without the need for complex focusing optics or bulky gamma detectors, enabling the localization of gamma-ray emitting sources with high precision. The applications of coded-aperture gamma imaging extend to various fields, including homeland security where it is used for monitoring cargo containers, vehicles, and luggage at ports and border crossings to detect nuclear and radioactive materials.

2.3 Applications

In gamma-ray, radiography gamma-rays emitted by a radioactive source such as Co-60 or Cs-137 are directed through the cargo. Detectors on the opposite side measure the transmitted radiation, allowing inspectors to visualize the contents of the cargo. For example, in a recent study, gamma-ray transmission radiography was used to detect illicit drugs hidden in cargo shipments [42]. Neutron-induced gamma-ray tomography utilizes a neutron source to induce gamma-ray emissions from the cargo material [43]. These emitted gamma-rays are then detected and analyzed to provide additional information for imaging purposes. Passive gamma emission tomography involves detecting gamma-rays naturally emitted by radioactive materials within the cargo. This technique enables inspectors to identify potential threats without the need for active radiation sources. An example of passive gamma emission tomography in action is the detection of radioactive materials in cargo shipments at border crossings [42, 44]. Coded-aperture gamma imaging utilizes a patterned mask in front of a gamma-ray detector to capture the shadow pattern cast by gamma-rays passing through the cargo. This technique allows for high-resolution imaging without the need for complex optics or bulky detectors. For example, in a study by Zhang [45], coded-aperture gamma imaging was utilized to detect hidden compartments in cargo containers. These techniques offer distinct advantages in cargo inspection, but they also present some challenges. Although active interrogation methods provide controlled radiation sources for precise imaging, they may pose safety concerns and require careful regulatory oversight. Passive interrogation techniques offer noninvasive inspection capabilities but may be limited by lower sensitivity and resolution. Overall, the integration of multiple gamma-ray radiography techniques provides a comprehensive approach to cargo inspection, enhancing security measures and ensuring the safe and efficient transport of goods across international borders.

3 Cosmic Ray Muon Radiography

Given the importance of national security and safeguarding of nuclear materials, innovative technologies that enhance detection capability to inspect and monitor sensitive materials are in constant demand. Cosmic ray muon radiography has emerged as a forefront technology in this domain, offering a noninvasive, high-penetration capability that can address the limitations of traditional inspection methods. This chapter explores the theoretical principles and applications of cosmic ray muon radiography, and its pivotal role in advancing nuclear security, nonproliferation, and safeguards is discussed. This section begins with a description of cosmic ray muons, focusing on the properties of muon particles that make them invaluable radiographic probes for imaging: their natural abundance, high-penetration capability, and the manner in which they interact with high-Z materials. Expanding on the basics, muon transmission radiography explores the applications in imaging the interiors of containers, vehicles, and large-scale structures without physical intrusion. By analyzing the attenuation of muons as they pass through objects, inspectors can better find hidden compartments and detect shielded materials. Muon scattering tomography represents an advanced approach that utilizes the scattering angles of muons to reconstruct detailed images of an object's internal structure. The enhanced image resolution and illicit material detection capabilities of this technique offer a reliable means for the detection of nuclear and radioactive materials as well as for high-resolution imaging of SNF dry casks, nuclear reactors, and large-scale structures. Several pioneering applications of muon radiography have demonstrated its potential in many fields, ranging from geology and archaeology to nuclear safety and homeland security.

3.1 Cosmic Ray Muons

3.1.1 Cosmic Ray Muons at Sea Level

Cosmic rays are high-energy particles originating from space that constantly bombard Earth. When these particles enter Earth's atmosphere, they interact with atomic nuclei at approximately 15 km above the Earth's surface. These interactions lead to the production of secondary particles, including pions and kaons. Subsequent decays of these particles result in the production of muons and neutrinos, contributing to a substantial flux of cosmic ray muons (CRMs) [46]. Muons are charged particles similar to electrons but possess an approximately 207 times greater mass. At sea level, muons are the most abundant charged particles, making them an invaluable particle for various scientific and practical applications. The predominant range of CRM momentum spans 0.1–10^3 GeV/c, as depicted in Fig. 6, with an average momentum of approximately 3–4 GeV/c. The muon vertical flux when their momentum is greater than 1 GeV/c is approximately 70 m^{-2} s^{-2} sr^{-1} at sea level [47]. The semiempirical CRM spectrum can be expressed as follows [48]:

$$\frac{dN_\mu}{dEd\Omega} \approx \frac{0.14E^{-2.7}}{\text{cm}^2\text{s sr GeV}}\left[\frac{1}{1+\frac{1.1E\cos\theta}{115\text{ GeV}}}+\frac{1}{1+\frac{1.1E\cos\theta}{850\text{ GeV}}}\right], \tag{11}$$

where N_μ is the expected CRM flux, E is the muon energy, Ω is the solid angle, and θ is the zenith angle. Each term in the right-hand side represents the contribution of the pions and kaons, respectively. After originating from the decay of pions and kaons, muons traverse tens to hundreds of kilometers across Earth's atmosphere before arriving at the Earth's surface. Although cosmic ray muons have high penetration capability, the flux of muons decreases as their travel distance increases. Therefore, the distribution of their flux at sea level follows the cosine power law [49]:

$$I(\theta) = I_0\cos^{n(h,E)}\theta, \tag{12}$$

where I_0 represents the vertical muon flux and the exponent, $n(h,E)$, varies as a function of muon energy, E, and its travel distance, h. Although the exponent n may exhibit minor variations depending on the azimuth angle, longitude, and solar activity, these effects are insignificant compared to zenith angle dependency. Therefore, for simplicity, n is often approximated as 2 at sea level. The measured CRM zenith angle distribution at sea level along with the cosine-squared model is shown in Fig. 6.

3.1.2 Muon Interaction with Matter

Muon Energy Loss Mechanism

Understanding the physical aspects of muon interaction with matter is fundamental for the muon radiography technique. As muons pass through matter, they lose energy via various mechanisms, including ionization, bremsstrahlung, pair production, and nuclear interactions. The most significant energy loss mechanism of muons is ionization, where muons interact with the electrons of atoms in the material, causing excitation or ejection of these electrons. For example, the mean rate of

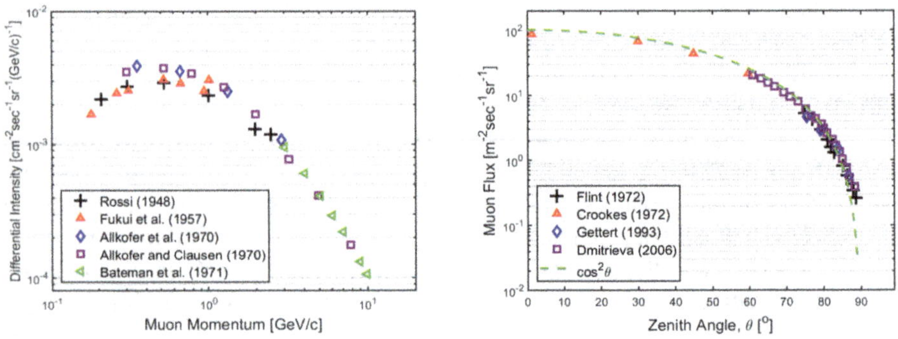

Fig. 6 Vertical differential cosmic ray muon momentum spectrum ranging from 0.2 to 10 GeV/c (left) and the variation of cosmic ray muon flux across zenith angles ranging from 0° to 89° at sea level (right) [50]. (Figures are reproduced from Ref. [51])

energy loss for a positive muon passing through copper, denoted as $<-dE/dx>$, as a function of muon momentum is plotted in Fig. 7 [46]. As shown in Fig. 7, the CRM momentum spectrum falls within the *Bethe–Bloch region,* where the Bethe–Bloch formula accurately describes the muon energy loss process. In this domain, the ionization energy loss mechanism dominates both nuclear and radiative energy loss mechanisms. Given the momentum range of CRMs, the mean rate of energy loss, or the mass stopping power of muons, can be estimated using the Bethe–Bloch formula [52].

Multiple Coulomb Scattering

Multiple Coulomb scattering (MCS) describes the phenomenon in which muons undergo multiple deflections as they electromagnetically interact with the nuclei of a material. The angular distribution of these scattering events is often modeled using a Gaussian distribution for simplicity and practical applicability. The foundational approximation for MCS of a singly charged particle was initially proposed by Rossi and Greisen [54]. This theoretical framework has been further refined and expanded by the work of Highland and Lynch–Dahl [55, 56]. The Lynch–Dahl model provides a more detailed and accurate description of the scattering of charged particles, like muons, as they pass through matter. The model is expressed through a formula that quantitatively predicts the scattering angle distribution for a charged particle after traversing a given thickness of material. The Lynch and Dahl formula for MCS is given by

$$\theta_0 = \frac{13.6\,\text{MeV}}{\beta cp} Z^{1/2} \sqrt{\frac{x}{X_0}} \left[1 + 0.038 \ln \frac{x}{X_0} \right], \tag{13}$$

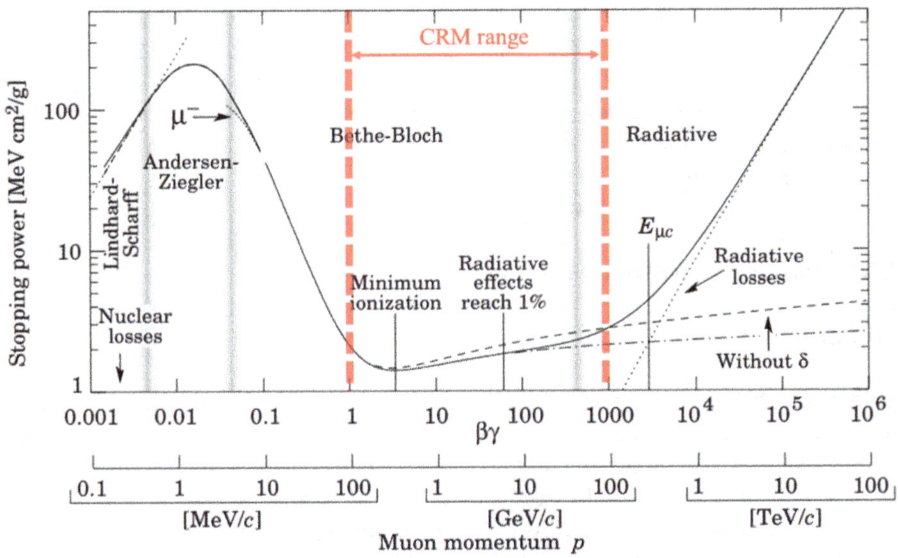

Fig. 7 The stopping power of muons in copper varies as a function of muon momentum, p. The solid curves represent the total stopping power. The approximate cosmic muon energy range (0.1–10^3 GeV/c) is indicated by vertical dashed lines. (Reproduced from Ref. [53])

where θ_0 is the root mean square scattering angle, βc is the velocity of the particle with respect to the speed of light, p is the particle momentum, Z is the atomic number of the material, x is the thickness of the material, X_0 is the radiation length of the material, and the factor 13.6 MeV comes from the constant term in the Bethe–Bloch formula for energy loss, adjusted for the context of scattering. This model enhances the predictive power of MCS calculations by incorporating corrections for the logarithmic increase in scattering with material thickness and for the dependence of scattering on the particle's speed and the atomic number of the material.

3.2 *Muon Detectors*

Muon detectors are sophisticated instruments that accurately identify and measure the properties of muons. The primary functions of muon detectors include determining the position, trajectory, and energy of muons as they pass through or interact with the detector material. In general, the muon position detector is utilized to reconstruct incoming and outgoing muon trajectories using typical radiation detectors such as drift chamber, silicon pixel, and scintillation detectors. Knowing the muon trajectories before and after interacting with objects provides important information for both muon attenuation and scattering in imaging applications. Enhancing the resolution of muon radiography images depends not only on precise muon position tracking but also on accurate measurement of muon energy. Various methodologies—calorimetry, Cherenkov radiation detector, time-of-flight analysis, and magnetic spectrometry—have been developed to measure muon energy. Each technique is designed for specific contexts and energy ranges. This section introduces examples of the instruments and methods employed for muon position and energy measurement.

3.2.1 Position Sensitive Detectors

Drift Chambers

A drift chamber operates on the principle of ionization. When a muon traverses the chamber, it ionizes the gas atoms along its path. The resultant free electrons drift toward an anode wire under the influence of an electric field. The time elapsed from the muon's passage to the electron's arrival at the wire is measured, and it offers a precise determination of the position at which the ionization occurred. A general drift chamber is composed of multiple parallel anode wires, each housed within a cathode structure. This assembly is typically filled with a gas mixture to aid in ionization and efficient electron drift. The choice of gas, electric field strength, and the geometrical arrangement are finely tuned to optimize the chamber's performance— such as spatial resolution, drift time, and gain. Drift chambers are widely used in particle physics experiments, such as at the Large Hadron Collider (LHC), where they provide precise measurements of charged particle trajectories [57, 58].

Scintillation Detectors

When a muon passes through the scintillator material, it deposits energy along its path. This energy excites the electrons in the scintillator to higher energy levels. As these electrons return to their ground state, they emit photons in the visible or near-visible spectrum. These photons are then detected by photodetectors such as photomultiplier tubes or silicon photomultipliers coupled with the scintillator. Then, they convert the light into an electrical signal that is proportional to the energy deposited by the muon. To estimate the muon position, scintillation detectors often use advanced configurations. By configuring scintillator materials into arrays composed of discrete segments or integrating them with wavelength-shifting fibers, it becomes possible to deduce the position of a muon based on which segments are activated. This strategic segmentation facilitates the detailed reconstruction of a muon's trajectory as it passes through the detector array and enables precise tracking of its path [59, 60].

Silicon Pixel Detectors

The silicon pixel detector typically consists of tens to hundreds of pixels, each pixel being micrometers in size. Each pixel functions as an individual detector, enabling the precise localization of muon interactions. The small size of the pixels enables the construction of a highly granular detector array, which is crucial for detailed tracking of particles across a wide range of incident angles. The advantages of silicon pixel detectors include high spatial resolution, rapid response capabilities, and high resistance to radiation damage [57, 58].

Resistive Plate Chambers

A resistive plate chamber (RPC) consists of two parallel resistive plates made of materials such as phenolic resin (Bakelite) or glass, separated by a thin, gas-filled gap. High-voltage electrodes are applied to the outer surfaces of these plates, creating a uniform electric field across the gap. When a muon passes through the detector, it ionizes the gas molecules in the gap, which results in the creation of electron ion pairs. The electric field causes these charges to drift toward the electrodes, inducing a signal on pickup strips or pads located on the outside of the resistive plates. The geometry of the pickup electrodes allows for both position determination and timing of the passing particles. RPCs have been successfully integrated into several high-profile particle physics experiments at the Compact Muon Solenoid (CMS) [61], A Toroidal LHC Apparatus (ATLAS) [62], and muon tomography systems for cargo containers [18].

3.2.2 Energy-Sensitive Detectors

Calorimetry

Calorimetry is a technique used to measure the energy of a particle by absorbing the entire energy of an incident particle in a material and converting it into heat or scintillation light. There are two primary types of calorimeters, electromagnetic and hadronic calorimeters. Muons typically pass through electromagnetic calorimeters (ECALs) with minimal energy loss and require hadronic calorimeters (HCALs) for energy measurement even though they often penetrate these as well. The operating principle of a calorimeter involves the complete absorption of the particle's kinetic energy within the detector material, leading to the production of a shower of secondary particles. The energy of these secondary particles is then converted into a measurable signal such as an electrical pulse or light. The signal amplitude is proportional to the energy of the incident particle. In the CMS detector, the ECAL utilizes lead tungstate crystals, whereas the HCAL is comprised of brass and scintillator plates. Similar to CMS, the ATLAS detector employs a liquid argon-based ECAL and a HCAL constructed from scintillating tiles [63, 64].

Cherenkov Radiation Detectors

Cherenkov radiation detectors are specialized instruments that measure the velocity of charged particles when they exceed the speed of light in a medium. The intensity and angle of Cherenkov radiation emission depend on the velocity of the charged particle and the refractive index of the medium [65]. The threshold velocity, v_{th}, for Cherenkov radiation in a medium with refractive index n is given by

$$v_{th} = c/n, \tag{14}$$

where c is the speed of light in a vacuum. The angle of emission, θ_C, relative to the particle's trajectory is described by the Cherenkov angle equation:

$$\cos\theta_C = c/nv, \tag{15}$$

where v is the particle's velocity. The velocity can be estimated by measuring the Cherenkov angle, θ_C. Cherenkov radiation detectors are widely used as a method for measuring particle energies, including muons. Examples of Cherenkov radiation detectors include a water Cherenkov detector at Super-Kamiokande experiment [66], the ring-imaging Cherenkov detectors (RICH) for the Large Hadron Collider beauty experiment (LHCb) [67], and the aerogel Cherenkov detector of the Belle II experiment [68].

Time-of-Flight Measurements

Time-of-flight (TOF) measurements offer a straightforward and powerful technique for determining the velocity and consequently the energy of muons. The basic principle of TOF measurement is to calculate a particle's velocity by dividing the traveling distance by the time. Given the muon mass, its kinetic energy can be determined using either nonrelativistic formula (or relativistic if it moves close to the speed of light). TOF systems are often employed in cosmic ray experiments to measure the velocity of particles such as muons. For example, the GRAPES-3 experiment uses a TOF system to study high-energy cosmic rays by measuring the arrival times of muons at different detector layers [69].

Magnetic Spectrometry

Magnetic spectrometry is a method for measuring the momentum of charged particles. This method relies on the principle that a charged particle's trajectory will bend when it moves through a magnetic field, and the degree of bending is inversely proportional to its momentum. In magnetic spectrometry, the trajectory of a muon is analyzed as it passes through a known magnetic field. The Lorentz force causes the particle to follow a curved path, and the radius of this curvature, r, is related to the particle momentum, p, by the equation

$$p = qBr, \tag{16}$$

where q is the charge of the particle and B is the magnetic field strength. Magnetic spectrometers are widely used for muon research such as the ATLAS and CMS detectors at the LHC and the Muon g-2 experiment [57, 58, 70].

3.3 Muon Transmission Radiography

Muon transmission radiography utilizes the continuous flux of CRMs to visualize the internal composition and density variations within geological formations, architectural structures, and nuclear materials. This technique mirrors the principles of X-ray CT but employs the naturally occurring, highly penetrative muons as probes. The foundational application of muon radiography traces back to the mid-twentieth century when it was devised by E. George to examine the depth of rock layers overlying underground tunnels in Australia [71]. A significant advancement occurred approximately a decade later when L. Alvarez applied this method to explore potential hidden chambers within the Great Pyramid of Giza [72].

3.3.1 Principle of Operation

Muon transmission radiography, also known as μCT, examines how muons are absorbed and transmitted through materials. To measure the transmission rate, muon detectors are deployed in either one or both sides of the object being scanned. One-sided μCT utilizes detectors placed on only one side, making it suitable for scenarios where access to both sides of the object is limited [73]. The operational principle of one-sided muon tomography involves measuring the flux of muons passing through the object from a single direction. As muons traverse through the object, their paths are affected by the density and composition of the materials they encounter, as shown in Fig. 8 [74]. Using the analyzed distribution of muons detected on the opposite side, image reconstruction algorithms can be applied to reconstruct a three-dimensional image of the interior structure of the objects. Typically, muon detectors are positioned on both sides of the object to gauge incoming and outgoing muon fluxes. By measuring the intensity of muons before and after traversing the object, inspectors can infer the material's density and composition. Areas of higher-density material will result in greater attenuation of the muon flux. The image reconstruction algorithm for muon transmission radiography involves processing the data collected by muon detectors to generate a two- or three-dimensional image of the interior of the inspected object. Several algorithms are commonly used for this purpose, including the FBP algorithm.

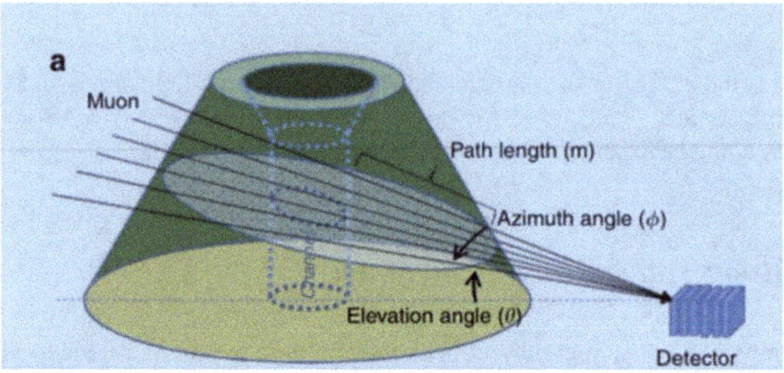

Fig. 8 An example of one-sided muon transmission radiography to scan a mountain. (Image is reproduced from Ref. [74])

3.4 *Muon Scattering Tomography*

Muon scattering tomography (MST) was initially devised by scientists at the Los Alamos National Laboratory in 2003 [12, 75]. Whereas muon transmission radiography utilizes the fraction of muon attenuation as it passes through objects, MST uses the scattering angle of muons to infer the internal structure and composition of objects. This method relies on the Coulomb scattering of muons, which occurs when muons interact with the atomic nuclei within a material. The degree of scattering is sensitive to the atomic number and density of the material, making MST particularly useful for identifying high-Z materials such as metals and nuclear materials.

3.4.1 Principle of Operation

As muons traverse an object, they interact with the atomic nuclei within the material, causing the muons to scatter and change trajectories. The degree of scattering angle depends on the Z number and density of the material; ρ—higher Z and ρ lead to increased muon scattering angle. An MST system typically consists of position-sensitive muon detectors placed above and below the object. These detectors record the incoming and outgoing muon trajectories. The upper detector records the initial trajectory as a muon enters, and the lower detector records the deflected trajectory as the muon exits. The scattering angle is the angular difference between the incoming and outgoing muon trajectories. By analyzing the scattering angles and positions of a statistically significant number of muons, the image of the object can be reconstructed using image reconstruction algorithms such as point-of-closest approach (PoCA).

Position and Density Mapping Algorithms

To effectively reconstruct images of objects utilizing MST, it is imperative to accurately locate muon scattering position and its angle. However, precise reconstruction of a muon's actual scattering trajectory within a medium is challenging because the observable scattering angle results from a series of consecutive and random deflections. Despite these challenges, advancements in computational analysis, the application of Bayesian analysis, and the maximum likelihood expectation maximization approaches have shown promise in reducing uncertainties associated with muon trajectory estimation [76, 77]. To address the intrinsic difficulties of precise muon trajectory reconstruction, the PoCA algorithm emerged as a straightforward and rapid solution for precisely locating a singular scattering event's position. The PoCA algorithm operates under the principal assumption that a muon is primarily deflected once within the medium. The process of determining the PoCA point involves measuring the muon's incoming and outgoing trajectories. The incoming trajectory—the initial path of the muon before it interacts with the target object—is reconstructed using two upper muon trackers. Each muon tracker records the x- and y-coordinates of a muon as illustrated in Fig. 9. Similarly, the outgoing trajectory—the muon's final path after interaction—is reconstructed using the same approach. Both trajectories are then extrapolated, and the shortest three-dimensional distance between these lines is calculated. The midpoint of this distance represents the PoCA point. Subsequently, a voxel containing the PoCA point is identified within a

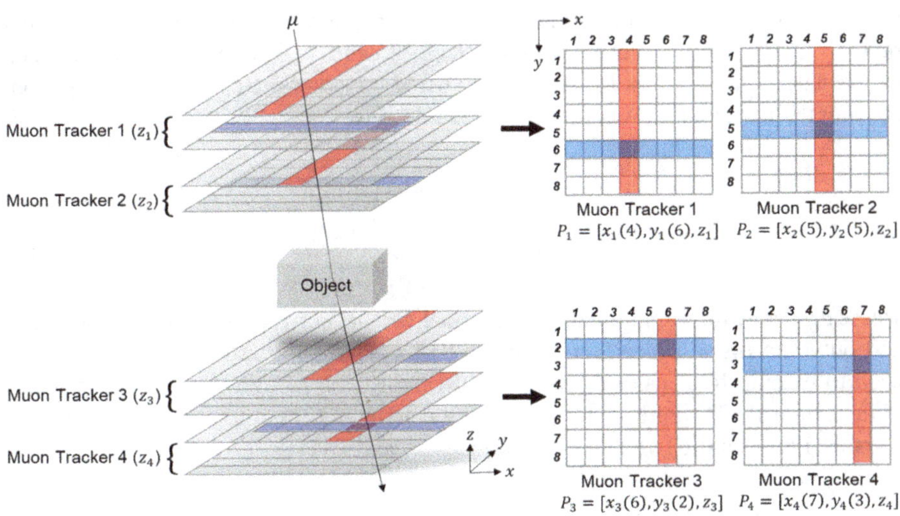

Fig. 9 The reconstruction of incoming and outgoing muon trajectories is facilitated through the deployment of both upper and lower muon trackers, arranged in a dual-layer configuration. An example shows four muon trackers, each with an 8 × 8 grid configuration. (Images are reproduced from Ref. [51])

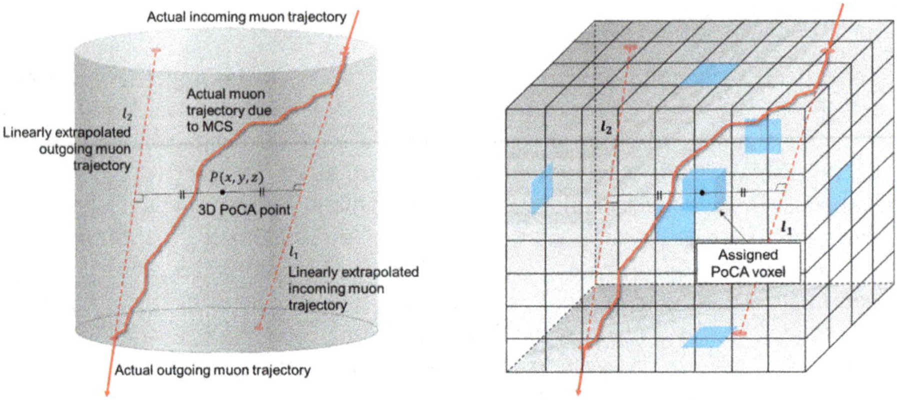

Fig. 10 The PoCA algorithm is visualized as it identifies a single scattering point by utilizing the incoming and outgoing muon trajectories (left) and the corresponding PoCA voxel within a three-dimensional voxelated volume of interest (right). (Images are reproduced from Ref. [51])

discretized volume of interest, as shown in Fig. 10. Based on the muon position data collected from the upper and lower muon tracking detectors, the three-dimensional coordinates of a muon can be determined by:

$$P = (x, y, z)_i, \tag{17}$$

where $(x, y, z)_i$ is the three-dimensional muon coordinate of the muon when i represents the muon tracker number, 1, 2, 3, and 4 from top to bottom. The extrapolated lines, L_1 and L_2, are derived using the coordinates of a muon position, P, from the upper and lower muon tracking detectors:

$$L_1 = \vec{k}_1 \bullet \vec{e}_1, \tag{18}$$

$$L_2 = \vec{k}_2 \bullet \vec{e}_2, \tag{19}$$

where $\boldsymbol{e_1}$ and $\boldsymbol{e_2}$ represent the unit directional vectors corresponding to the incoming and outgoing muon trajectories, respectively. $\boldsymbol{k_1}$ and $\boldsymbol{k_2}$ are the coefficient vectors used to extend the L_1 and L_2 lines toward the region of interest. Then, the three-dimensional PoCA point is determined by calculating the midpoint of the shortest distance between the two extrapolated lines L_1 and L_2:

$$P_{PoCA}(x, y, z) = \frac{1}{2} \min \|L_1, L_2\|. \tag{20}$$

This procedure is iteratively applied to each recorded muon event, resulting in PoCA voxels that may encapsulate multiple PoCA points. To finalize the image reconstruction process via MST, it is essential to visualize the assigned PoCA voxels. This is achieved by employing color scheme or density mapping, where the coloring of the PoCA voxel is determined based on the muon scattering angle. The scattering angle, θ_s, is an angular difference between L_1 and L_2 from Eqs. (18) and (19). The spatial scattering angle is given by

$$\theta_s = \sqrt{\Delta\theta_x^2 + \Delta\theta_y^2}, \tag{21}$$

where $\Delta\theta_x$ and $\Delta\theta_y$ represent the differences in angles between L_1 and L_2 with respect to the x-axis and y-axis, respectively. In instances where a single PoCA voxel contains multiple PoCA points, each exhibiting distinct scattering angles, the averaged scattering angle is utilized for that voxel:

$$\overline{\theta}_s = \frac{1}{N_V} \sum_{N_V} \theta_{s,n}, \tag{22}$$

where N_V represents the total number of recorded muon events within the PoCA voxel.

3.5 *Applications*

Muon radiography, including both transmission and scattering techniques, has found broad real-world application. It has been used to study the internal structure of volcanos, including searching magma chambers and understanding volcanic processes, and to predict and monitor a volcanic eruption [78, 79]. This technique has also been utilized to investigate underground features, including the detection of subterranean cavities and the characterization of soil composition [80, 81]. Muon radiography has enabled archaeologists to discover hidden chambers within ancient structures, such as the void discovered within the Great Pyramid of Giza [59, 72, 82]. One of most compelling applications of muon radiography is in the detection and investigation of nuclear materials, contributing to nuclear nonproliferation efforts and national security [12, 83]. For example, muon tomography offers innovative solutions for managing SNF and addressing challenges in verification, monitoring, and safeguarding [84–88]. Muon radiography systems have been deployed, importantly, at ports and border checkpoints for the inspection of cargo containers and vehicles, as shown in Fig. 11. These systems can effectively identify the presence of shielded nuclear materials without disrupting the flow of commerce, providing a nonintrusive means to enhance national security [19, 89, 90].

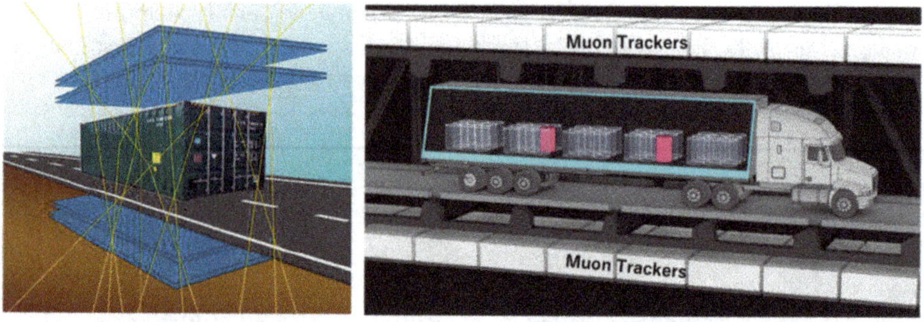

Fig. 11 Illustrations of the experimental setup for muon radiography applications of cargo container (left) and vehicle (right) inspections. (Images are reproduced from Ref. [19, 90])

4 Forthcoming Technologies and Advanced Methods

Although the applications of gamma-ray and cosmic ray muon tomography technologies offer valuable capabilities and benefits over traditional radiography methods in various fields, as described in the previous sections, they face several challenges that impact their effectiveness and widespread adoption. In gamma-ray and muon radiography applications, two primary challenges must be effectively addressed: (1) reducing radiation exposure and (2) achieving high image quality. To reduce unnecessary radiation exposure, optimization techniques can be implemented such as collimation, shielding, and exposure time control. By adjusting these parameters based on the specific inspection requirements, operators can minimize radiation doses while maintaining image quality. The challenge of unnecessary radiation exposure can also be effectively mitigated through the utilization of remote operation and automation technologies, which have seen rapid advancements across various industries [91, 92]. Implementing an unmanned monitoring system enables operators to safely conduct radiographic inspections from a considerable distance. Remote-controlled systems and robotic manipulators enable precise positioning of radiation sources and detectors without direct human intervention, reducing the risk of radiation exposure to personnel. For example, unmanned aerial vehicles (UAVs) or drones equipped with X-ray and gamma-ray imaging capabilities have been developed specifically to safely inspect areas with high levels of radiation exposure, as shown in Fig. 12.

Although radiation exposure from cosmic ray muons is generally not a concern due to their negligible radiation damage level at sea level, the challenge of muon tomography image quality remains significant. Achieving sufficient resolution and reducing noise are paramount in accurately visualizing the internal structures of the objects being inspected. Addressing this challenge requires advancements in detector technology, data processing algorithms, and experimental methodologies to enhance spatial resolution, reduce noise, and improve the reliability and accuracy of muon tomography inspections. For instance, in muon scattering tomography,

Fig. 12 A UAV equipped with a radiography camera, demonstrating an innovative solution for safe inspections in high-radiation exposure areas. (Photograph courtesy of the NDT Supply, Inc)

the scattering angle of muons passing through the object is a key parameter. Although it is commonly assumed that muon deflection angle is solely determined by density or Z number of the object, Eq. (13) demonstrates that the relationship is more complex. Indeed, muon scattering tomography should operate on the principle that the scattering angle of muons passing through an object depends on various factors, including material density, muon momentum, and the size of the object. This complexity underscores the challenges associated with accurately reconstructing images solely from muon scattering data. To enhance the quality of muon tomographic images, certain studies have proposed strategies such as (1) integrating muon momentum information and (2) developing algorithms for estimating muon trajectories and implementing them into existing muon image reconstruction algorithms. Recent efforts to estimate muon momentum within the muon tomography system have been made using various approaches based on multiple Coulomb scattering and energy loss [93, 94], Cherenkov detector [95, 96], and TOF [97]. Although the potential advantages of measuring muon momentum have been extensively demonstrated in simulations, as shown in Fig. 13 [98], this approach has not yet been widely deployed in the field due to significant uncertainties associated with the measurements. However, a strategic combination of various momentum estimation methods is anticipated to reduce measurement uncertainty because the systematic uncertainties of these methods are not correlated [89]. In addition, it is worthwhile to review the efforts to develop a muon trajectory estimation algorithm to overcome the limitation of the PoCA algorithm. Although the PoCA algorithm is widely used for online analysis because it is straightforward and fast [19], it has shown high false position estimation rates and ignores the

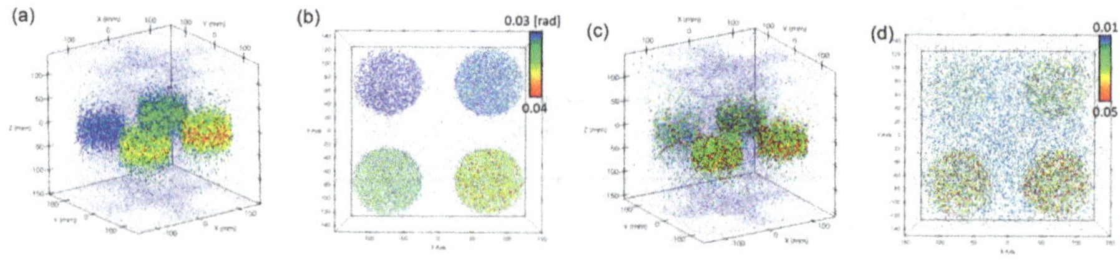

Fig. 13 Isometric and cross-sectional views of four cylindrical target materials—aluminum, steel, lead, and uranium—are depicted using a PoCA algorithm. The images are reconstructed with muon momentum information (left) and without muon momentum information (right). (Figures reproduced from Ref. [98])

second or further scattering points. To address these limitations, an advanced signal filtration algorithm [99] and muon trajectory reconstruction methods were developed based on maximum likelihood expectation maximization [100, 101] and Bayesian estimation [76, 77].

5 Summary

In this chapter, gamma-ray and cosmic ray muon modalities for various applications are explored, with a focus on monitoring nuclear and radioactive materials at ports or borders, such as in the case of cargo container inspection. It introduces two primary gamma-ray inspection and imaging techniques: active and passive interrogation methods. In the active interrogation method, external radiation sources are employed for detection and imaging to analyze the characteristic responses of target materials. Section 2.1 provides technical details and operational principles of two major active interrogation gamma imaging methods: a gamma-ray transmission radiography and neutron-induced gamma-ray emission tomography. Conversely, passive interrogation methods utilize emitted gamma-ray signals from concealed special nuclear or other radioactive materials. Section 2.2 presents overviews of two primary passive interrogation imaging methods: gamma emission tomography and coded-aperture-based gamma imaging.

Additionally, alongside gamma-ray radiography, this chapter introduces a newer nondestructive testing method, cosmic ray muon radiography. By utilizing the unique characteristics of cosmic ray muons, this technique represents a promising noninvasive examination method, one that can efficiently address the limitations of traditional inspection methods. Section 3.1 explores various fundamental properties of cosmic ray muons, including their flux at sea level, angular and energy distribution profiles, and mechanisms of energy loss and deflection within the matter. In Section 3.2, position-sensitive detectors such as drift chambers, scintillation detectors, silicon pixel detectors, and resistive plate chambers are discussed, along with energy-sensitive detectors including calorimetry, Cherenkov radiation detectors, TOF measurement detectors, and magnetic spectrometry. The principles of operation and technical details of image reconstruction algorithms for both muon transmission radiography and muon scattering tomography are discussed in Section 3.4.

Finally, forthcoming technologies and advanced methods of gamma-ray and cosmic ray muon imaging modalities are discussed. With considerable interest in unmanned aerial vehicles (UAVs) or drones, the potential for remote operation and automation of gamma-ray scanning and imaging is highlighted. This advancement promises safer inspection of inaccessible areas such as high-level radiation zones by reducing unnecessary radiation exposure to operators through remote UAV operation equipped with gamma-ray scanning or imaging equipment. Although radiation exposure from cosmic ray muons is typically negligible at sea level and poses little concern for radiation damage, the persistent challenge remains in the quality of muon tomography images due to the inherently low flux. To address this challenge, advanced imaging techniques have been studied, integrating muon scattering angle and momentum information and employing a sophisticated muon tracking estimation models to improve the current PoCA algorithm. The outcomes of these efforts are promising and demonstrate a clear path for the advancement of muon tomography methods. Nonetheless, the development of more advanced muon detectors and signal analysis methods remains a challenge.

References

1. M.R. Cleland, G.M. Pageau, Comparisons of X-ray and gamma-ray sources for industrial irradiation processes. Nucl. Instrum. Methods Phys. Res., Sect. B **24–25**, 967–972 (1987). https://doi.org/10.1016/S0168-583X(87)80290-2
2. G. Knoll, *Radiation Detection and Measurement*, 4th edn. (Wiley, Hoboken, 2010)
3. B.J. Micklich, D.L. Smith, Nuclear materials detection using high-energy γ-rays. Nucl. Instrum. Methods Phys. Res., Sect. B **241**(1–4), 782–786 (2005). https://doi.org/10.1016/j.nimb.2005.07.132
4. B.S. Henderson, H.Y. Lee, T.D. MacDonald, R.G. Nelson, A. Danagoulian, Experimental demonstration of multiple monoenergetic gamma radiography for effective atomic number identification in cargo inspection. J. Appl. Phys. **123**(16), 164901 (2018). https://doi.org/10.1063/1.5025805
5. V. V. Verbinski, J. Payne, M. Snell, Recent developments in the VACIS gamma radiography systems. Presented at the Enabling Technologies for Law Enforcement and Security, ed. by A.T. DePersia, J.J. Pennella, Boston, MA, Dec. 1998, pp. 368–374. https://doi.org/10.1117/12.335011
6. V.V. Verbinski, V.J. Orphan, Vehicle and cargo inspection system. Presented at the Fifth International Conference on Applications of Nuclear Techniques: Neutrons in Research and Industry, ed. by G. Vourvopoulos, Crete, Greece, Feb. 1997, p. 235. https://doi.org/10.1117/12.267908
7. J.L. Jones et al., Detection of shielded nuclear material in a cargo container. Nucl. Instrum. Methods Phys. Res., Sect. A **562**(2), 1085–1088 (2006). https://doi.org/10.1016/j.nima.2006.02.101

8. R.T. Bernardi, H.E. Martz, Nuclear waste drum characterization with 2 MeV x-ray and gamma-ray tomography. Presented at the SPIE's 1995 International Symposium on Optical Science, Engineering, and Instrumentation, ed. by R.B. Hoover, M.B. Williams, San Diego, CA, United States, Jun. 1995, p. 140. https://doi.org/10.1117/12.211897
9. W. Lee, D. Wehe, Hybrid gamma ray imaging—model and results. Nucl. Instrum. Methods Phys. Res., Sect. A **579**(1), 200–204 (2007). https://doi.org/10.1016/j.nima.2007.04.039
10. E. Aguayo Navarrete, R.T. Kouzes, A.S. Ankney, J.L. Orrell, T.J. Berguson, M.D. Troy, Cosmic Ray Interactions in Shielding Materials, PNNL-20693, 1025678, Sep. 2011. https://doi.org/10.2172/1025678
11. J. Bae, S. Chatzidakis, The Effect of Cosmic Ray Muon Momentum Measurement for Monitoring Shielded Special Nuclear Materials, INMM/ESARDA Joint Annual Meeting, 2021
12. L.J. Schultz et al., Image reconstruction and material Z discrimination via cosmic ray muon radiography. Nucl. Instrum. Methods Phys. Res., Sect. A **519**(3), 687–694 (2004). https://doi.org/10.1016/j.nima.2003.11.035
13. G. Bonomi, P. Checchia, M. D'Errico, D. Pagano, G. Saracino, Applications of cosmic-ray muons. Prog. Part. Nucl. Phys. **112**, 103768 (2020). https://doi.org/10.1016/j.ppnp.2020.103768
14. X. Liu, H.K. Lee, A simulation study of the spent nuclear fuel cask condition evaluation using high energy X-ray computed tomography. NDT&E Int. **80**, 58–64 (2016). https://doi.org/10.1016/j.ndteint.2016.02.008
15. K. Vetter, R. Barnowksi, A. Haefner, T.H.Y. Joshi, R. Pavlovsky, B.J. Quiter, Gamma-ray imaging for nuclear security and safety: towards 3-D gamma-ray vision. Nucl. Instrum. Methods Phys. Res., Sect. A **878**, 159–168 (2018). https://doi.org/10.1016/j.nima.2017.08.040
16. F.A. Balogun, N.M. Spyrou, C.A. Adesanmi, Neutron induced gamma-ray emission tomography. Nucl. Instrum. Methods Phys. Res., Sect. B **114**(3–4), 387–393 (1996). https://doi.org/10.1016/0168-583X(96)00150-4
17. K. Vetter, L. Mihailescu, K. Nelson, J. Valentine, D. Wright, Gamma-Ray Imaging Methods, UCRL-TR-225239, 1036848, Oct. 2006. https://doi.org/10.2172/1036848
18. P. Baesso, D. Cussans, C. Thomay, J. Velthuis, Toward a RPC-based muon tomography system for cargo containers. J. Instrum. **9**(10), C10041–C10041 (2014). https://doi.org/10.1088/1748-0221/9/10/C10041
19. H. Miyadera, C.L. Morris, Muon scattering tomography: review. Appl. Opt. **61**(6), C154 (2022). https://doi.org/10.1364/AO.445806
20. H. Anger, Scintillation camera. Rev. Sci. Instrum. **29**(1), 27–33 (1958)
21. H. Kasban, O. Zahran, H. Arafa, M. El-Kordy, S.M.S. Elaraby, F.E. Abd El-Samie, Welding defect detection from radiography images with a cepstral approach. NDT&E Int **44**(2), 226–231 (2011). https://doi.org/10.1016/j.ndteint.2010.10.005
22. S.K. Dwivedi, M. Vishwakarma, A. Soni, Advances and researches on non destructive testing: a review. Mater. Today Proc. **5**(2), 3690–3698 (2018). https://doi.org/10.1016/j.matpr.2017.11.620
23. C.G. Tui, B.S. Wong, X. Wang, Real-time Radiographic Non-destructive Inspection for Aircraft Maintenance, in *17th World Conference on Nondestructive Testing*, Shanghai, China, 25–28 (2008)
24. J.S. Beaumont, M.P. Mellor, M. Villa, M.J. Joyce, High-intensity power-resolved radiation imaging of an operational nuclear reactor. Nat. Commun. **6**(1), 8592 (2015). https://doi.org/10.1038/ncomms9592
25. L. Caballero *et al.*, "Gamma-ray imaging system for real-time measurements in nuclear waste characterisation," J. Instrum., 13, 03, P03016–P03016, Mar. 2018, https://doi.org/10.1088/1748-0221/13/03/P03016
26. J.Y. Park et al., Development of a dual-modality gamma-ray/fast neutron imaging system for air cargo inspection. Appl. Sci. **12**(19), 9775 (2022). https://doi.org/10.3390/app12199775
27. B.E. O'Day, Z.S. Hartwig, R.C. Lanza, A. Danagoulian, Initial results from a multiple monoenergetic gamma radiography system for nuclear security. Nucl. Instrum. Methods Phys. Res., Sect. A **832**, 68–76 (2016). https://doi.org/10.1016/j.nima.2016.05.117

28. R.T. Kouzes, E. Siciliano, J. Tanner, G. Warren, Testing requirements for active interrogation systems. Nucl. Instrum. Methods Phys. Res., Sect. A **922**, 222–229 (2019). https://doi.org/10.1016/j.nima.2018.12.088
29. G.N. Ramachandran, A.V. Lakshminarayanan, Three-dimensional reconstruction from radiographs and electron micrographs: application of convolutions instead of fourier transforms. Appl. Phys. Sci. **68**(9), 2236–2240. https://doi.org/10.1073/pnas.68.9.2236
30. J. Prince, J. Links, *Medical Imaging Signals and Systems*, 2nd edn. (Pearson, New York, 2014)
31. A. Vamvakeros, *Operando Chemical Tomography of Packed Bed and Membrane Reactors for Methane Processing* (University College London, London, 2017)
32. B. Perot et al., Measurement of 14MeV neutron-induced prompt gamma-ray spectra from 15 elements found in cargo containers. Appl. Radiat. Isot. **66**(4), 421–434 (2008). https://doi.org/10.1016/j.apradiso.2007.11.011
33. M. Fang, Y. Altmann, D. Della Latta, M. Salvatori, A. Di Fulvio, Quantitative imaging and automated fuel pin identification for passive gamma emission tomography. Sci. Rep. **11**(1), 2442 (2021). https://doi.org/10.1038/s41598-021-82031-8
34. M. Mayorov et al., Gamma emission tomography for the inspection of spent nuclear fuel, in *2017 IEEE Nuclear Science Symposium and Medical Imaging Conference (NSS/MIC)*, (IEEE, Atlanta, 2017), pp. 1–2. https://doi.org/10.1109/NSSMIC.2017.8533017
35. R. Virta, T.A. Bubba, M. Moring, S. Siltanen, T. Honkamaa, P. Dendooven, Improved passive gamma emission tomography image quality in the central region of spent nuclear fuel. Sci. Rep. **12**(1), 12473 (2022). https://doi.org/10.1038/s41598-022-16642-0
36. H. Al Hamrashdi, S.D. Monk, D. Cheneler, Passive gamma-ray and neutron imaging systems for national security and nuclear non-proliferation in controlled and uncontrolled detection areas: review of past and current status. Sensors **19**(11), 2638 (2019). https://doi.org/10.3390/s19112638
37. IAEA, Industrial Process Gamma Tomography, IAEA-TECDOC-1589, 2008
38. R. Backholm et al., Simultaneous reconstruction of emission and attenuation in passive gamma emission tomography of spent nuclear fuel. Inverse Probl. Imaging **14**(2), 317–337 (2020). https://doi.org/10.3934/ipi.2020014
39. M.J. Cieślak, K.A.A. Gamage, R. Glover, Coded-aperture imaging systems: past, present and future development—a review. Radiat. Meas. **92**, 59–71 (2016). https://doi.org/10.1016/j.radmeas.2016.08.002
40. K.-p. Ziock, M.A. Blackston, I. Garishvili, K.T. Schmitt, R. Venkataraman, Quantitative holdup determination using coded-aperture gamma-ray imaging, in *Proceedings of 2021 Joint ESARDA INMM Annual Meeting*, 2021
41. Y. Yue et al., Image reconstruction for the coded aperture system in nuclear safety and security using a Monte Carlo-based system matrix. Radiat. Detect. Technol. Methods **7**, 263–270 (2023)
42. M. Askari, A.M. Beigzadeh, A. Taheri, A new method for detecting the radioactive materials using X or γ -ray cargo inspection systems. Nucl. Instrum. Methods Phys. Res., Sect. A **1003**, 165325 (2021). https://doi.org/10.1016/j.nima.2021.165325
43. B.D. Sowerby et al., Recent developments in fast neutron radiography for the interrogation of air cargo containers, in *IAEA Conference*, 2009
44. L. Marques et al., Neutron and gamma-ray detection system coupled to a multirotor for screening of shipping container cargo. Sensors **23**(1), 329 (2022). https://doi.org/10.3390/s23010329
45. L. Zhang, *Near Field Three-Dimensional Coded Aperture Techniques: Theoretical and Experimental Exploration for Applications in Imaging and Detection Systems* (Massachusetts Institute of Technology, Cambridge, 1998)
46. R.L. Workman et al., The review of particle physics (2023). Prog. Theor. Exp. Phys., 083C01 (2022)
47. K. Hagiwara, K. Hikasa, K. Nakamura, M. Tanabashi, M. Aguilar-Benitez, C. Amsler, Review of particle properties. Phys. Rev. D **66**, 010001 (2002)

48. T. Gaisser, R. Engel, E. Resconi, *Cosmic rays and particle physics*, 2nd edn. (Cambridge University Press, Cambridge, 2016)
49. M. Bahmanabadi, A method for determining the angular distribution of atmospheric muons using a cosmic ray telescope. Nucl. Instrum. Methods Phys. Res., Sect. A **916**, 1–7 (2019). https://doi.org/10.1016/j.nima.2018.11.010
50. P.K.F. Grieder, *Cosmic rays at earth*, 1st edn. (Elsevier Science, Amsterdam, 2001)
51. J. Bae, *A Novel Muon Spectrometer Using Multi-Layer Pressurized Gas Cherenkov Radiators for Muon Tomography*, Purdue University, West Lafayette, 2022. [Online]. Available: https://doi.org/10.25394/PGS.19686633.v1
52. H.A. Bethe, Molière's theory of multiple scattering. Phys. Rev. **89**(6), 1256–1266 (1953)
53. D. Groom, N. Mokhov, S. Striganov, Muon stopping power and range tables 10 MeV—100 TeV. At. Data Nucl. Data Tables **76**(2), 183 (2001)
54. B. Rossi, K. Greisen, Cosmi-ray theory. Rev. Mod. Phys. **13**(240), 240 (1941)
55. V.L. Highland, Some practical remarks on multiple scattering. Nucl. Inst. Methods **129**(2), 497–499 (1975)
56. G. Lynch, O. Dahl, Approximations to multiple Coulomb scattering. Nucl. Inst. Methods Phys. Res. B **58**(1), 6–10 (1991)
57. ATLAS Collaboration, The ATLAS experiment at the CERN Large Hadron Collider. J. Instrum. **3**, S08003 (2008)
58. CMS Collaboration, The CMS Experiment at the CERN LHC. J. Instrum. **3**, S08004 (2008)
59. S. Aly et al., Simulation studies of a novel muography detector for the great pyramids. arXiv, Feb. 16, 2022. Accessed 1 Mar 2024. [Online]. Available: http://arxiv.org/abs/2202.07434
60. C. Liao, H. Yang, Z. Liu, J.P. Hayward, Design and characterization of a scintillator-based position-sensitive detector for muon imaging. Nucl. Technol. **205**(5), 736–747 (2019). https://doi.org/10.1080/00295450.2018.1522885
61. F. Thyssen, Performance of the resistive plate chambers in the CMS experiment. J. Instrum. **7**(01), C01104–C01104 (2012). https://doi.org/10.1088/1748-0221/7/01/C01104
62. G. Cattani, the RPC group, The resistive plate chambers of the ATLAS experiment: performance studies. J. Phys. Conf. Ser. **280**, 012001 (2011). https://doi.org/10.1088/1742-6596/280/1/012001
63. C. Biino, The CMS electromagnetic calorimeter: overview, lessons learned during Run 1 and future projections. J. Phys. Conf. Ser. **587**, 012001 (2015). https://doi.org/10.1088/1742-6596/587/1/012001
64. P. Puzo, ATLAS calorimetry. Nucl. Instrum. Methods Phys. Res., Sect. A **494**(1–3), 340–345 (2002). https://doi.org/10.1016/S0168-9002(02)01490-0
65. J.V. Jelley, Cerenkov radiation and its applications. Br. J. Appl. Phys. **6**, 227 (1958)
66. Y. Suzuki, Super-Kamiokande Results on Neutrino Oscillations. Physica Scripta T **T121**, 23 (2005)
67. F. Muheim, The ring imaging Cherenkov detectors of the LHCb experiment. Nucl. Instrum. Methods Phys. Res., Sect. A **639**(1), 11–14 (2011). https://doi.org/10.1016/j.nima.2010.09.158
68. R. Pestotnik et al., The aerogel Ring Imaging Cherenkov system at the Belle II spectrometer. Nucl. Instrum. Methods Phys. Res., Sect. A **876**, 265–268 (2017). https://doi.org/10.1016/j.nima.2017.04.043
69. B. Hariharan et al., Energy sensitivity of the GRAPES-3 EAS array for primary cosmic ray protons. Exp. Astron. **50**(2–3), 185–198 (2020). https://doi.org/10.1007/s10686-020-09671-y
70. A. Keshavarzi, K.S. Khaw, T. Yoshioka, Muon g−2: a review. Nucl. Phys. B **975**, 115675 (2022). https://doi.org/10.1016/j.nuclphysb.2022.115675
71. E.P. George, Cosmic rays measure overburden of tunnel, in *Commonwealth Engineer*, vol. 455, (1955)
72. L.W. Alvarez et al., Search for hidden chambers in the pyramids: the structure of the Second Pyramid of Giza is determined by cosmic-ray absorption. Science **167**(3919), 832–839 (1970). https://doi.org/10.1126/science.167.3919.832

73. A. Erlandson, K. Boniface, V.N.P. Anghel, G. Jonkmans, M. Thompson, S. Livingstone, One-sided muon tomography—a portable method for imaging critical infrastructure with a single muon detector. CNL Nucl. Rev., 1–9 (2016). https://doi.org/10.12943/CNR.2016.00014
74. D. Hatzifotiadou, Muon radiography–muon tomography, in *19th IPPOG Meeting*, 2020
75. K.N. Borozdin et al., Radiographic imaging with cosmic-ray muons. Nature **422**(6929), 277–277 (2003). https://doi.org/10.1038/422277a
76. S. Chatzidakis, Z. Liu, J.P. Hayward, J.M. Scaglione, A generalized muon trajectory estimation algorithm with energy loss for application to muon tomography. J. Appl. Phys. **123**(12), 124903 (2018). https://doi.org/10.1063/1.5024671
77. G. Wang, L.J. Schultz, J. Qi, Bayesian image reconstruction for improving detection performance of muon tomography. IEEE Trans. Image Process. **18**(5), 1080–1089 (2009). https://doi.org/10.1109/TIP.2009.2014423
78. H.K.M. Tanaka, T. Nakano, S. Takahashi, J. Yoshida, K. Niwa, Development of an emulsion imaging system for cosmic-ray muon radiography to explore the internal structure of a volcano, Mt. Asama. Nucl. Instrum. Methods Phys. Res., Sect. A **575**(3), 489–497 (2007). https://doi.org/10.1016/j.nima.2007.02.104
79. M. D'Errico et al., Muon radiography applied to volcanoes imaging: the MURAVES experiment at Mt. Vesuvius. J. Instrum. **15**(03), C03014–C03014 (2020). https://doi.org/10.1088/1748-0221/15/03/C03014
80. L. Cimmino et al., 3D muography for the search of hidden cavities. Sci. Rep. **9**(1), 2974 (2019). https://doi.org/10.1038/s41598-019-39682-5
81. G. Saracino et al., Imaging of underground cavities with cosmic-ray muons from observations at Mt. Echia (Naples). Sci. Rep. **7**(1), 1181 (2017). https://doi.org/10.1038/s41598-017-01277-3
82. K. Morishima et al., Discovery of a big void in Khufu's Pyramid by observation of cosmic-ray muons. Nature **552**(7685), 386–390 (2017). https://doi.org/10.1038/nature24647
83. V. Anghel et al., Cosmic ray muon tomography system using drift chambers for the detection of Special Nuclear Materials, in *IEEE Nuclear Science Symposium & Medical Imaging Conference*, (IEEE, Knoxville, TN, 2010), pp. 547–551. https://doi.org/10.1109/NSSMIC.2010.5873821
84. S. Chatzidakis, C.K. Choi, L.H. Tsoukalas, Interaction of cosmic ray muons with spent nuclear fuel dry casks and determination of lower detection limit. Nucl. Instrum. Methods Phys. Res. A **828**, 37–45 (2016)
85. J.M. Durham et al., Verification of spent nuclear fuel in sealed dry storage casks via measurements of cosmic-ray muon scattering. Phys. Rev. Appl. **9**(4), 044013 (2018). https://doi.org/10.1103/PhysRevApplied.9.044013
86. D. Poulson, J. Bacon, M. Durham, E. Guardincerri, C.L. Morris, H.R. Trellue, Application of muon tomography to fuel cask monitoring. Phil. Trans. R. Soc. A **377**(2137), 20180052 (2019). https://doi.org/10.1098/rsta.2018.0052
87. J. Bae, S. Chatzidakis, Monitoring spent nuclear fuel in a dry cask using momentum integrated muon scattering tomography. Trans. Am. Nucl. Soc. **127**, 828–832 (2022)
88. S. Chatzidakis, C.K. Choi, L.H. Tsoukalas, Analysis of spent nuclear fuel imaging using multiple coulomb scattering of cosmic muons. IEEE Trans. Nucl. Sci. **63**(6), 2866–2874 (2016). https://doi.org/10.1109/TNS.2016.2618009
89. S. Barnes et al., Cosmic-ray tomography for border security. Instruments **7**(1), 13 (2023). https://doi.org/10.3390/instruments7010013
90. P. Rocca, D. Presti, F. Riggi, Cosmic ray muons as penetrating probes to explore the world around us, in *Cosmic Rays*, (INTECH, 2018)
91. Y. Sato et al., Remote radiation imaging system using a compact gamma-ray imager mounted on a multicopter drone. J. Nucl. Sci. Technol. **55**(1), 90–96 (2018). https://doi.org/10.1080/00223131.2017.1383211
92. A. Stepanov, O. Ivanov, V. Stepanov, A. Volkovich, V. Potapov, V. Lukyanov, How useful are imaging and remote sensing technologies in decontamination & decommissioning?, *Waste Management 2017 Conference*, 2017

93. S. Vanini et al., Muography of different structures using muon scattering and absorption algorithms. Phil. Trans. R. Soc. A **377**(2137), 20180051 (2019). https://doi.org/10.1098/rsta.2018.0051
94. V. Anghel et al., A plastic scintillator-based muon tomography system with an integrated muon spectrometer. Nucl. Instrum. Methods Phys. Res., Sect. A **798**, 12–23 (2015). https://doi.org/10.1016/j.nima.2015.06.054
95. J. Bae, S. Chatzidakis, Fieldable muon spectrometer using multi-layer pressurized gas Cherenkov radiators and its applications. Sci. Rep. **12**(1), 2559 (2022). https://doi.org/10.1038/s41598-022-06510-2
96. J. Bae, S. Chatzidakis, Development of compact muon spectrometer using multiple pressurized gas Cherenkov radiators. Results Phys. **39**, 105771 (2022). https://doi.org/10.1016/j.rinp.2022.105771
97. J. Marteau et al., Implementation of sub-nanosecond time-to-digital convertor in field-programmable gate array: applications to time-of-flight analysis in muon radiography. Meas. Sci. Technol. **25**(3), 035101 (2014). https://doi.org/10.1088/0957-0233/25/3/035101
98. J. Bae, R. Montgomery, S. Chatzidakis, Nuclear material accountancy using momentum-informed muon scattering tomography. Ann. Nucl. Energy **197**, 110240 (2024). https://doi.org/10.1016/j.anucene.2023.110240
99. R. Sehgal, M.S. Mitra, T. Roy, S.T. Sehgal, L.M. Pant, B.K. Nayak, Voxelization based PoCA point cloud filtration algorithm for image reconstruction for Muon Tomography. J. Instrum. **15**(09), P09012–P09012 (2020). https://doi.org/10.1088/1748-0221/15/09/P09012
100. L.J. Schultz et al., Statistical reconstruction for cosmic ray muon tomography. IEEE Trans. Image Process. **16**(8), 1985–1993 (2007). https://doi.org/10.1109/TIP.2007.901239
101. M. Benettoni et al., Noise reduction in muon tomography for detecting high density objects. J. Instrum. **8**(12), P12007–P12007 (2013). https://doi.org/10.1088/1748-0221/8/12/P12007

Index

K. Iniewski, H. Gadey (eds.), *Emerging Radiation Detection*, https://doi.org/10.1007/978-3-031-63897-8

Printed in the USA
CPSIA information can be obtained
at www.ICGtesting.com
CBHW070315040924
13992CB00003B/6

9 783031 638961